TOWARD THE THEORY OF EVERYTHING: MRST'98

MRST '98

Toward the Theory of Everything

McGill University, Montréal, Québec, May 13-15, 1998

Editors:
James M. Cline, Marcia Knutt,
Gregory D. Mahlon, Guy D. Moore

E. Rutherford

TOWARD THE THEORY OF EVERYTHING: MRST'98

Montréal, Québec   May 1998

EDITORS
James M. Cline
Marcia E. Knutt
Gregory D. Mahlon
Guy D. Moore
McGill University, Montréal

American Institute of Physics

AIP CONFERENCE PROCEEDINGS 452

Woodbury, New York

Editors:

James M. Cline
Physics Department
McGill University
3600 University Street
Montréal, Québec H3A 2T8
Canada
Email: jcline@hep.physics.mcgill.ca

Marcia E. Knutt
Physics Department
McGill University
3600 University Street
Montréal, Québec H3A 2T8
Canada
Email: knutt@hep.physics.mcgill.ca

Gregory D. Mahlon
Physics Department
McGill University
3600 University Street
Montréal, Québec H3A 2T8
Canada
Email: mahlon@hep.physics.mcgill.ca

Guy D. Moore
Physics Department
McGill University
3600 University Street
Montréal, Québec H3A 2T8
Canada
Email: guymoore@hep.physics.mcgill.ca

Authorization to photocopy items for internal or personal use, beyond the free copying permitted under the 1978 U.S. Copyright Law (see statement below), is granted by the American Institute of Physics for users registered with the Copyright Clearance Center (CCC) Transactional Reporting Service, provided that the base fee of $15.00 per copy is paid directly to CCC, 222 Rosewood Drive, Danvers, MA 01923. For those organizations that have been granted a photocopy license by CCC, a separate system of payment has been arranged. The fee code for users of the Transactional Reporting Service is: 1-56396-845-2/ 98 /$15.00.

© 1998 American Institute of Physics

Individual readers of this volume and nonprofit libraries, acting for them, are permitted to make fair use of the material in it, such as copying an article for use in teaching or research. Permission is granted to quote from this volume in scientific work with the customary acknowledgment of the source. To reprint a figure, table, or other excerpt requires the consent of one of the original authors and notification to AIP. Republication or systematic or multiple reproduction of any material in this volume is permitted only under license from AIP. Address inquiries to Office of Rights and Permissions, 500 Sunnyside Boulevard, Woodbury, NY 11797-2999; phone: 516-576-2268; fax: 516-576-2499; e-mail: rights@aip.org.

L.C. Catalog Card No. 98-88416
ISBN 1-56396-845-2
ISSN 0094-243X
DOE CONF- 980590

Printed in the United States of America

Contents

Preface vii

Constant Curvature Effective Actions 1
    C. Burgess and M. Kamela

Deflection of Light in Conformal (Weyl) Gravity 10
    A. Edery and M. B. Paranjape

Leptoquark Production and Identification in High Energy Lepton Colliders ... 19
    M. A. Doncheski and S. Godfrey

Gluon Radiation in Top Quark Production and Decay at an e^+e^- Collider 28
    C. Macesanu and L. H. Orr

Single Pion Transitions of Charmed Baryons 37
    S. Tawfiq, P. J. O'Donnell, and J. G. Körner

Discrete Ambiguities in CP-violating Asymmetries in B Decays 46
    D. London

Large CP Violation in $B \to K^{(*)}X$ Decays 54
    T. E. Browder, A. Datta, X.-G. He, and S. Pakvasa

Vector Dilepton Production at Hadron Colliders in the 3—3—1 Model 63
    T. Grégoire

Two-Higgs-Doublet-Models and Radiative CP Violation 72
    O. C. W. Kong and F.-L. Lin

Minimal Ten-parameter Hermitian Texture Zeroes Mass Matrices
and the CKM Matrix 80
    M. Baillargeon, F. Boudjema, C. Hamzaoui, and J. Lindig

Power Counting in Non-Relativistic Effective Field Theories 91
    M. Luke

Photoproduction of h_c 101
    S. Fleming and T. Mehen

$\bar{\Lambda}$ and λ_1 from Inclusive B-decays 108
    C. Bauer

Hyperfine Interactions in Charm and Bottom 117
    P. J. O'Donnell

Electromagnetic Interactions in Quantum Hall Ferromagnets 125
    R. Ray

The Quantum Hall Effect, Skyrmions and Anomalies 134
    A. Travesset

Integrability in Classical and Quantum Field Theory and the Bukhvostov-
Lipatov Model 144
    B. Gerganov

Evidence for a Scalar $\kappa(900)$ Resonance in πK Scattering 153
    D. Black

The Influence of the Gauge Boson Mass on the Most Attractive
Channel Hypothesis 163
    F. S. Roux

The $\langle \bar{f}f \rangle$ Condensate Component of the Anomalous Magnetic Moment
of a Fundamental Dirac Fermion.................................... 169
    V. Elias and K. B. Sprague
Deriving $N=2$ S-dualities from Scaling................................ 178
    A. Buchel
Symmetry of Quantum Matrix Models.................................. 187
    C.-W. H. Lee and S. G. Rajeev
Perturbed Conformal Field Theory: A Tool for Investigating
Integrable Models.. 195
    M. Ameduri
Investigation by Monte Carlo Renormalization of 2-D Simplicial
Quantum Gravity Coupled to Gaussian Matter 202
    E. B. Gregory, S. M. Catterall, and G. Thorleifsson
Phase Structure of 3D Dynamical Triangulations with a Boundary........... 212
    S. Warner, S. Catterall, and R. Renken
The Experimental Status of the Deficit in Charmed Semileptonic B Decays ... 220
    R. Janicek
Nuclear Flow Inversion in the Boltzmann–Uehling–Uhlenbeck Model........ 227
    D. Persram and C. Gale
SUSY Threshold Corrections and the Up–Down Unification in the
Superpotential... 236
    C. Hamzaoui and M. Pospelov
From Super QCD to QCD 245
    F. Sannino
R-Parity "R" Us Rochester 254
    M. Bisset, O. C. W. Kong, C. Macesanu, and L. H. Orr
Low Energy Z_R^0 Based on an $SO(10)$ SUSY-GUT 263
    C. S. Kalman

Epilogue... 271
Schedule of Talks... 273
Conference Participants... 276
Author Index... 279

Preface

This year marked the 20th annual meeting of the Montréal-Rochester-Syracuse-Toronto (MRST) Conference on High Energy Physics. As promised, all the ice from the great storm of January had melted, and 65 particle physicists from 12 universities enjoyed perfect spring weather in Montréal, while bringing each other up to date about the latest developments in their research. The sessions covered a wide range of topics: gravitation, collider physics, heavy quark systems, new physics beyond the Standard Model, technicolor, supersymmetry and strings. In addition there was one session devoted to experiments of particular interest to the audience.

True to the spirit of this conference series, we had strong participation from graduate students and postdocs. We also benefitted from the financial support of the Institute for Particle Physics in Canada, a primarily experimental organization which nevertheless recognizes the importance of the interface between theory and experiment, as well as the Faculty of Graduate Studies and Research at McGill University. We wish to thank graduate students Tanvir Rahman and Abhijit Majumder for their hard work with the mundane tasks during the conference, and conference secretary Elizabeth Shearon for her invaluable organizational contributions.

Apologies are due Stephen Hawking for certain post-banquet remarks, made under the influence of Châteaux de MRST 1998. Interested readers will find the label of this fine vintage within these proceedings.

Missing was the late Professor Bernie Margolis of McGill, who was one of the strong supporters of MRST since its inception. We think he would be pleased to know that his efforts were well invested in that the conference continues to provide a friendly, vibrant atmosphere for the exchange of ideas in our field.

Jim Cline, Marcia Knutt, Greg Mahlon and Guy Moore
McGill University

Constant Curvature Effective Actions

Cliff Burgess[a] and Martin Kamela[ab]

[a] *Physics Department, McGill University*
3600 University St., Montréal, Québec, H3A 2T8, Canada,
[b] *Bishop's University, Lennoxville, Québec, J1M 1Z7, Canada*

Abstract. Closed forms are derived for the effective actions for free, massive spinless fields in maximally-symmetric spacetimes in arbitrary dimensions. The results have simple expressions in terms of elementary functions (for odd dimensions) or multiple Gamma functions (for even dimensions).

I INTRODUCTION

In this talk we give explicit expressions for the effective actions for free, massive scalar fields propagating within anti-de Sitter (AdS) spacetimes of arbitrary dimension. Besides their intrinsic interest as exact expressions for quantum systems interacting with nontrivial gravitational fields, or as the first terms in a derivative expansion for more complicated backgrounds, these actions may also have applications to the calculation of quantum effects within cosmologically-interesting spacetimes. Remarkably, their supersymmetric extensions in five-dimensions may prove useful for study of large-N corrections to nonabelian gauge theories, in view of the recently-proposed duality between these theories and AdS supergravity in five dimensions [1]. Our calculations extend a number of similar calculations which have been performed by others in the past. Much of the early interest was motivated by the questions of principle which arise when quantizing fields in these spacetimes [2], [3], and by vacuum-stability [3], [4] and divergence [5] issues associated with the appearance of AdS spacetimes as supersymmetric vacua in extended-supergravity models. Starting very early, the maximal symmetry of these spacetimes was harnessed to perform explicit effective-action calculations for scalar fields in both de Sitter [6], [7], and anti-de Sitter [7], [8], [9], [10], as well as calculations of the functional determinants which arise in higher-spin calculations [14], [15]. The main advantage of our expressions over those in the literature is their validity for general spacetime dimension. For odd dimensions the results may be expressed in closed form using elementary functions. For even dimensions we also obtain closed-form

TABLE 1. Maximally symmetric spaces.

Space	Signature	Curvature
Sphere	$(+,...,+)$	$R < 0$
Hyperbolic Space	$(+,...,+)$	$R > 0$
de-Sitter Space	$(-,+,...,+)$	$R < 0$
Anti-de-Sitter Space	$(-,+,...,+)$	$R > 0$

results in terms of a class of special functions — the multiple gamma functions, $\{G_n\}$ — whose properties have been extensively studied.

Our presentation is organized in the following way. In Section II we briefly review some properties of maximally symmetric spaces which are useful for obtaining the effective action. Section III contains our main result: the derivation of the scalar-field effective action in an AdS space-time of arbitrary dimension, n. Section IV specializes this result to various cases of particular interest. For even dimensions we display results for $n = 2$ and $n = 4$, where we reproduce previous calculations. We also present the odd-dimensional cases $n = 3$ and 5, which have not been previously calculated. Finally, we gather some useful definitions and properties of the multiple Gamma functions in an appendix.

II QUANTUM FIELDS ON MAXIMALLY SYMMETRIC SPACES

An n-dimensional spacetime which admits $\frac{1}{2}n(n+1)$ Killing vectors is said to be maximally symmetric [11], [12]. The Reimann curvature tensor for any such spacetime may be written in the following way (our conventions are those of ref. [12].)

$$R_{\lambda\rho\sigma\nu} = K(g_{\sigma\rho}g_{\lambda\nu} - g_{\nu\rho}g_{\lambda\sigma}) \qquad R = -n(n-1)K, \tag{1}$$

where K a real constant. The possible maximally-symmetric spaces which can be entertained may be characterized by the signatures of their metrics as well as the sign of their Ricci scalar R (or, K). For Riemannian and pseudo-Riemannian maximally symmetric spaces there are four possibilities, which are summarized in **Table 1**. The Euclideanization of de Sitter space is given by the sphere, while the hyperbolic space performs the same role for anti-de Sitter space.

Quantization of scalar field theory on both AdS and dS spacetimes involves additional complications over those which arise for flat Minkowski space. For de Sitter space the complication is associated with the existence of the apparent event horizon. For anti-de Sitter space — besides unrolling the compact time direction and working on the Universal Covering Space — the tricky feature is tied up with this spacetime not being globally hyperbolic [2], [3]. (That is, in order for the scalar-field equations to formulate a well-posed boundary-value problem, boundary information is required on a time-like surface at spatial infinity in addition to the usual initial conditions which would have been sufficient in Minkowski space.) For both spacetimes these complications lead to the existence of more than one Fock vacuum for the quantum field theory. As a consequence, different physical situations can lead to different boundary conditions, and so to different quantum field theories.

Given a scalar quantum field on dS or AdS spacetime, our goal is to compute the scalar-field contribution, Σ, to the effective action. This is given by the following path integral:

$$e^{i\Sigma(g,m^2)} = \int [DX]_{g_{\mu\nu}} \exp\left[-\frac{i}{2}\int d^n x \sqrt{g} X \left(-\Box_g + m^2\right) X\right]$$
$$= \mathrm{det}'\left[\Box_g - m^2\right]^{-1/2} \tag{2}$$

where $\Box_g := \frac{1}{\sqrt{g}}\partial_\mu g^{\mu\nu}\sqrt{g}\partial_\nu$ is the usual Laplacian operator acting on scalar fields, and the prime in the second equality indicates the omission of any zero modes. Rather than using eq. (2) directly in what follows, we instead use its derivative with respect to m^2, which implies:

$$\frac{d\Sigma}{dm^2} = \frac{i}{2}\mathrm{Tr}'\left(\frac{1}{-\Box_g + m^2}\right) = \frac{i}{2}\int d^n x \lim_{x'\to x} G(x,x'), \tag{3}$$

where $G(x,x')$ is the scalar Feynman propagator:

$$(-\Box_g + m^2)_x G(x,x') = \frac{\delta^n(x,x')}{\sqrt{g}}. \tag{4}$$

To obtain the effective action we integrate eq. (3) with respect to m^2:

$$\Sigma_1(g,m^2) - \Sigma_1(g,m_0^2) = \int_{m_0^2}^{m^2} dm^2 \left\{\frac{i}{2}\int d^n x \sqrt{g} G(x,x)\right\}. \tag{5}$$

The result will equal to the desired effective action up to terms independent of the mass m^2. The quantity m_0 is a reference mass, for which we imagine the functional determinant to have been explicitly evaluated using other means. Convenient choices for which this is often possible are $m_0 = 0$ or $m_0 \to \infty$.

In this way the problem reduces to the construction of the scalar-field Feynman propagator on a dS or AdS spacetime, whose form in n dimensions has long been known for both dS [6] and AdS [8] spacetimes.

III THE N-DIMENSIONAL EFFECTIVE ACTION

It only remains to evaluate the previous expression using the explicit expression for the Feynman propagator. To do so requires a choice of vacuum state. Although we quote results here for both signs of K, for definiteness we work with the propagator which satisfies the energy-conserving boundary conditions on anti-de Sitter space [8], which is given in terms of standard hypergeometric functions, $F(a,b;c;x)$ [13], by:

$$-\frac{i}{2} G_F(z) = \frac{C_{F,n}}{2\,z^\beta} F\left(\frac{\beta}{2}, \frac{\beta+1}{2}; \beta - \frac{n}{2} + \frac{3}{2}; z^{-2}\right), \qquad (6)$$

where $z = 1 + |K|\,\sigma(x,x')$ and $\sigma(x,x')$ is the square of the geodesic distance between the points x and x', and β denotes the expression

$$\beta = \frac{n-1}{2} \pm \sqrt{\frac{(n-1)^2}{4} + \frac{m^2}{|K|}}. \qquad (7)$$

Finally, the coefficient $C_{F,n}$ is a known constant, defined in equation (9) of ref. [8]:

$$C_{F,n} = \frac{|K|^{(n-2)/2}\,\Gamma(\beta)}{2^{\beta+1}\,\pi^{n/2-1/2}\,\Gamma\left(\beta - \frac{1}{2}(n-3)\right)} \qquad (8)$$

We require the coincidence limit ($\sigma \to 0$) of eq. (6), and so take $z \to 1$, from which the scalar Feynman propagator is found to be

$$-\frac{i}{2} G_F(1) = \frac{\Gamma\left(\frac{n}{2} - \frac{1}{2} + \sqrt{\frac{(n-1)^2}{4} + \frac{m^2}{|K|}}\right)\,\Gamma\left(1 - \frac{n}{2}\right)\,|K|^{(n-2)/2}}{2^{n+1}\,\pi^{n/2}\,\Gamma\left(-\frac{n}{2} + \frac{3}{2} + \sqrt{\frac{(n-1)^2}{4} + \frac{m^2}{|K|}}\right)}. \qquad (9)$$

To proceed, we now integrate eq. (9) with respect to m^2. The limit $n \to D$ of eq. (9), when D is an odd integer, is well-defined and so may be taken directly, and the result integrated with respect to m^2. When D is even, however, the pole from the Γ-function in the numerator gives a divergent result, which we may isolate by performing a Laurent series in powers of $(n - D)$. It is generally useful to perform this expansion first, and reserving until last the integration over m^2.

IV APPLICATIONS TO SPECIFIC DIMENSIONS

A D=2 Scalar Fields

Specializing eq. (9) to $D = 2$, the Laurent expansion of the scalar propagator becomes (neglecting terms which are $O(n-2)$):

$$\frac{i}{2} G_F(1) = \frac{1}{4\pi(n-2)} - \frac{1}{8\pi}\left[\ln\left(\frac{4\pi\Lambda^2}{|K|}\right) - \gamma - 2\Psi\left(\frac{1}{2} + \frac{1}{2}\sqrt{1 + \frac{4m^2}{|K|}}\right)\right], \qquad (10)$$

where $\Psi(x) := d\ln\Gamma(x)/dx$ and Λ is the usual arbitrary scale which enters when dimensions are continued to complex values.

Integrating eq. (10) with respect to mass, we obtain the effective action as the integral over an effective lagrangian density: $\Sigma = -\int d^2x \sqrt{-g}\, V_{eff}(g, m^2)$, with

$$V_{eff}(g, m^2) = V_{eff}(g, 0) - \left[-\frac{1}{4\pi(n-2)} + \frac{1}{8\pi}\left(-\gamma + \ln\left(\frac{4\pi\Lambda^2}{|K|}\right) - 2 \right) \right] m^2$$
$$+ \frac{|K|}{8\pi}\left[2\ln G_1\left(\frac{1}{2}\sqrt{1 + \frac{4m^2}{|K|}} + \frac{1}{2}\right) + 4\ln G_2\left(\frac{1}{2}\sqrt{1 + \frac{4m^2}{|K|}} + \frac{1}{2}\right) \right.$$
$$\left. + \left(1 - \sqrt{1 + \frac{4m^2}{|K|}}\right) \ln(2\pi) \right]. \qquad (11)$$

Here $G_n(x)$ denote the multiple Gamma functions [18], which are defined to satisfy the Gamma-function-like property: $G_n(z+1) = G_{n-1}(z)G_n(z)$, as explained in the Appendix.

Notice, in two dimensions, that the massless reference point is useful because the functional integral for massless scalars is known to give the Liouville action:

$$\Sigma(g, 0) = -\frac{1}{96\pi}\int d^2x \sqrt{-g}\, R\left(\frac{1}{\Box}\right) R, \qquad (12)$$

where $\Box^{-1} R$ denotes the convolution of R with the Feynman propagator of eq. (4): $\int d^2y \sqrt{-g}\, G(x,y)\, R(y)$.

Using the asymptotic expansions of the G_n [19], the small curvature limit ($|K| \ll m^2$) of eq. (11) is found to be:

$$V_{eff}(g, m^2) \sim V_{eff}(g, 0) - \frac{m^2}{8\pi}\left[\frac{2}{(n-2)} - \ln\left(\frac{4\pi\Lambda^2}{m^2}\right) + \gamma - 1 \right]$$
$$- \frac{|K|}{24\pi}\left[\ln\left(\frac{|K|}{8\pi^3 m^2}\right) + \frac{3}{2} - 12\zeta'(1) \right] + \frac{|K|^2}{120\pi m^2} + O\left(|K|^3\right), \qquad (13)$$

where $\zeta(x)$ denotes the usual Reimann zeta function.

B The Case $D = 4$

Evaluating eq. (9) for $n \to D = 4$ dimensions permits a comparison of this expression with previous work. The expansion of eq. (6) about $n = 4$ produces the following coincidence limit up to $O(n-4)$:

$$\frac{i}{2} G_F = -\frac{2|K| + m^2}{16\pi^2(n-4)} + \frac{m^2}{32\pi^2}$$
$$+ \left(\frac{2|K| + m^2}{32\pi^2}\right) \left[\ln\left(\frac{4\pi\Lambda^2}{|K|}\right) - \gamma - 2\Psi\left(\frac{1}{2} + \sqrt{\frac{9}{4} + \frac{m^2}{|K|}}\right) \right] \qquad (14)$$

Integrating with respect to mass then gives:

$$V_{eff}(g, m^2) = V_{eff}(g, 0)$$
$$- \frac{|K|^2}{64\pi^2} \left\{ \left(-\frac{2}{n-4} + \ln\left(\frac{4\pi \Lambda^2}{|K|}\right) - \gamma + \frac{1}{3} \right) \left(b^2 - \frac{9}{4} \right) \left(b^2 + \frac{7}{4} \right) \right.$$
$$+ \left[(6 + 8C_2)\left(\frac{1}{2} + b\right) - 9 + 24C_3 + 8C_2 \right] \left(b^2 - \frac{9}{4} \right)$$
$$+ (24C_2 + 11 + 48C_3 + 48C_4)\left(-\frac{3}{2} + b \right)$$
$$\left. - 72 \ln G_3\left(\frac{1}{2} + b\right) - 24 \ln G_2\left(\frac{1}{2} + b\right) - 48 \ln G_4\left(\frac{1}{2} + b\right) \right\} \quad (15)$$

where $b^2 := \frac{9}{4} + \frac{m^2}{|K|}$, and the C_n are as defined in the Appendix.

This expression can be compared with earlier calculations. In particular, we find agreement (up to a shift in the renormalization scale) with ref. [9], where ζ-function methods were used.

C Scalar Fields in Odd Dimensions

We now turn to the effective action for massive scalar fields in odd-dimensional anti-de Sitter spacetimes. As is usually the case for dimensionally-regularized one-loop quantities, the resulting expressions are easier to evaluate due to the absence in odd dimensions of logarithmic divergences at one loop.

For 3-dimensional AdS spacetimes the massive scalar effective lagrangian density becomes:

$$V_{eff}(K, m) - V_{eff}(K, 0) = -\frac{|K|^{3/2}}{12\pi}\left[\left(\frac{|K| + m^2}{|K|}\right)^{3/2} - 1 \right]. \quad (16)$$

The corresponding result for 5-dimensional AdS spacetimes is:

$$V_{eff}(K, m) - V_{eff}(K, 0) = \frac{|K|^{5/2}}{360\,\pi^2}\left[\left(\frac{4|K| + m^2}{|K|}\right)^{3/2} \left(\frac{7|K| + 3m^2}{|K|}\right) - 56 \right]. \quad (17)$$

V OUTLOOK

We have shown that effective actions for massive scalar fields in anti-de Sitter spacetime may be explicitly evaluated in terms of elementary functions for odd dimensions, and multiple Gamma functions for even dimensions. One advantage of our expressions stems from the availability of asymptotic expansions of the multiple Gamma functions, which may be used to derive the first few terms of the derivative expansion for arbitrary ($R < m^2$) curvature background.

Recently, Cruz [16] has proposed a canonical equivalence between massless and $m^2 = R$ massive scalar field in 2D AdS. Since our results are valid for arbitrary scalar masses and cosmological constants, they bear on the issue of the existence of (quantum) duality transformations relating different values of these parameters. For the discussion of Cruz' duality, as well as for the extension of the effective action calculations to scalar fields in de Sitter spacetime and to (constrained) arbitrary spin fields in 4D AdS, we refere the reader to our forthcoming paper [20].

APPENDIX

In this appendix we state some principal formulae pertaining to the multiple gamma function. We also derive an integral representation for these functions, and use it to obtain closed forms for the integral moments of the $\Psi = \partial_x \ln \Gamma(x)$ function.

In 1899, Barnes [17] introduced a generalization G of the Γ function with the property: $G(z+1) = \Gamma(z)G(z)$. This idea was further generalized by Vignéras [18] in 1979 who introduced a hierarchy of Multiple Gamma functions, $\{G_n\}$.

Theorem (Vigneras)

There exists a unique hierarchy of functions which satisfy

$$
\begin{aligned}
&(1) \quad G_n(z+1) = G_{n-1}(z)G_n(z), \\
&(2) \quad G_n(1) = 1, \\
&(3) \quad \frac{d^{n+1}}{dz^{n+1}} \log G_n(z+1) \geq 0 \quad for \quad z \geq 0, \\
&(4) \quad G_0(z) = z
\end{aligned}
\tag{18}
$$

In above notation, G_1 is the usual Γ-function, and G_2 is Barnes' G function.

For infinite product representation of the multiple Gammas, as well as for asymptotic expansions, we refer the reader to Ueno and Nishizawa [19]. Next, we prove the following:

Theorem

Line integral representation of the logarithm of the multiple Gammas

$$\ln G_n(z+1) = \int_0^\infty dt \frac{e^{-t}}{t}(-1)^n \left(\frac{1-e^{-zt}}{(1-e^{-t})^n} + \sum_{m=1}^n \frac{(-1)^m}{(1-e^{-t})^{n-m}} \binom{z}{m} \right) \tag{19}$$

Proof: We show explicitly that the defining conditions of theorem (18) are satisfied. The proof follows by induction on n and from the uniqueness of the hirarchy of the functions $\{G_n\}$.

i) $\ln G_n(z+2) = \ln G_{n-1}(z+1) + \ln G_n(z+1)$ follows from the binomial relation:

$$\binom{z+1}{m} = \binom{z}{m-1} + \binom{z}{m} \tag{20}$$

The integrand splits up as follows:
$$(-1)^n \left(\frac{1-e^{-zt}e^{-t}}{(1-e^{-t})^n} + \sum_{m=1}^{n} \frac{(-1)^m}{(1-e^{-t})^{n-m}} \binom{z+1}{m} \right) =$$
$$= (-1)^n \left(\frac{1-e^{-zt}}{(1-e^{-t})^n} + \sum_{m=1}^{n} \frac{(-1)^m}{(1-e^{-t})^{n-m}} \binom{z}{m} \right)$$
$$+ (-1)^{n-1} \left(\frac{1-e^{-zt}}{(1-e^{-t})^{n-1}} + \sum_{m=1}^{n-1} \frac{(-1)^m}{(1-e^{-t})^{n-m-1}} \binom{z}{m} \right) \quad (21)$$

where the index on the second sum has been shifted to bring it to the standard form.

ii) $\ln G_n(1) = 0$ follows from the vanishing integrand in the limit $z \to 0$;

iii) $\partial_z^{n+1} \ln G_n(z+1) \geq 0$ follows from the absolute positivity of the integrand:
$$\int_0^\infty dt \, \frac{e^{-t}}{t} \frac{-(-t)^n + 1}{(1-e^{-t})^n} \geq 0 \quad (22)$$

iv) Setting $n \to 0$ reduces to an integral representation of $\ln(z+1)$ and $n \to 1$ to a standard representation of the logarithm of the Γ function, thereby completing the proof by induction on n.

Corollary 1

Using the integral representation of G_n we derive the following tower of relations among the logarithmic derivatives $\psi_n(z+1) := \partial_z \ln G_n(z+1)$:

$$\psi_2(z+1) - z\psi_1(z+1) = C_2 - \frac{z}{2}$$

$$\psi_3(z+1) - z\psi_2(z+1) + \frac{z(z+1)}{2!}\psi_1(z+1) = C_3 + \frac{3}{4}z + \frac{1}{4}z^2$$

$$\psi_4(z+1) - z\psi_3(z+1) + \frac{z(z+1)}{2!}\psi_2(z+1) - \frac{z(z+1)(z+2)}{3!}\psi_1(z+1) =$$
$$= C_4 - \frac{11}{18}z - \frac{1}{3}z^2 - \frac{1}{18}z^3 \quad (23)$$

where $C_2 = -\zeta'(0) - \frac{1}{2} = \frac{1}{2}\ln(2\pi) - \frac{1}{2}$, $C_3 = -.3332237448...$, $C_4 = .2786248832...$, etc.

Corollary 2

Substituting lower order relations in the higher order ones, and integrating with respect to z, we find

$$\int_0^a z\psi_1(z+1)dz = \ln G_2(z+1) - aC_2 + \frac{1}{4}a^2$$

$$\int_0^a \frac{1}{2!} z(z-1)\psi_1(z+1)dz = \ln G_3(z+1) + \frac{1}{12}a^3 - \left(\frac{1}{2}C_2 + \frac{3}{8}\right)a^2 - aC_3$$

$$\int_0^a \frac{1}{3!} z(z-1)(z-2)\psi_1 dz = \ln G_4(z+1) + \frac{1}{72}a^4 - \left(\frac{1}{6}C_2 + \frac{2}{9}\right)a^3$$
$$- \left(\frac{1}{2}C_3 - \frac{11}{36} - \frac{1}{4}C_2\right)a^2 - C_4 a \quad (24)$$

The integrals (24) may be rewritten as follows:

$$\int^a z^n \psi(z+1) dz = \begin{cases} n=0: & \ln G_1(a+1) \\ n=1: & \ln G_2(a+1) - aC_2 + \frac{1}{4}a^2 \\ n=2: & \frac{1}{6}a^3 + \left(-\frac{1}{2} - C_2\right)a^2 + (-C_2 - 2C_3)a + \\ & 2\ln G_3(a+1) + \ln G_2(a+1) \\ n=3: & \frac{1}{12}a^4 + \left(-C_2 - \frac{5}{6}\right)a^3 + \left(-\frac{1}{6} - \frac{3}{2}C_2 - 3C_3\right)a^2 + \\ & (-6C_3 - C_2 - 6C_4)a + 6\ln G_4(a+1) + \\ & 6\ln G_3(a+1) + \ln G_2(a+1) \end{cases}$$

(25)

REFERENCES

1. Maldacena, J., hep-th/9712200
2. Avis, S., Isham, C., Storey, D., PRD Vol. 18, No. 10, p3565, (1978)
3. Breitenlohner, P., and Freedman, D., Ann. Phys. (NY) 144, 249 (1982)
4. Coleman, S., Nucl. Phys. B259, 170, 1985
5. Christensen, S., Duff, M., Nucl. Phys. B 170, 480, (1980)
6. Candelas, P., Raine, D., PRD Vol. 12, No. 4, p965, (1975), and PRD Vol. 15, No. 6, p1494, (1976)
7. Allen, B., Jacobson, T., Comm. Math. Phys., 103, p669 (1986)
8. Burgess, Lutken, Phys. Lett. 153B, No3, 137, (1984).
9. Camporesi, R., PRD Vol. 43, No. 12, p3958, (1991)
10. Buchbinder, S., Odintsov, S., Acta Phys. Polon., B18, 237, 1987.
11. See, for instance: Misner, Thorne and Wheeler, *Gravitation*
12. Weinberg, S., *Gravitation and Cosmology*, Chapter 13.
13. Whittaker, E.T., and Watson, G.N., *Modern Analysis*, 4th ed., Cambridge University Press, 1927, part II, 1934.
14. Allen, B., Nucl. Phys. B 226, p228, (1983)
15. Camporesi, R., Higuchi, A., PRD Vol. 47, No. 8, p3339, (1993)
16. Cruz, J. *Hidden conformal symmetry of a massive scalar field in AdS_2*, hep-th/9806145.
17. Barnes, E., Quaterly Journal of Mathematics, Vol. 31, p.264, (1899)
18. Vigneras, M., Asterisque 61, p. 235, (1979)
19. Ueno, K., Nishizawa, M., q-alg/9605002
20. Kamela, M., Burgess, C., *Massive-Scalar Effective Actions on Maximally-Symmetric Spaces*, in preparation.

Deflection of light in Conformal (Weyl) Gravity

A. Edery and M. B. Paranjape

Groupe de Physique des Particules, Département de Physique,
Université de Montréal, C.P. 6128,
succ. centreville, Montréal, Québec, Canada, H3C 3J7

Abstract. Conformal (Weyl) gravity has been advanced in the recent past as an alternative to General Relativity (GR). The theory has had some success in fitting galactic rotation curves without the need for copious amounts of dark matter. To check the viability of Weyl gravity, we calculate the deflection of light in the exterior of a static spherically symmetric source. The result for the deflection of light is remarkably simple: besides the usual positive (attractive) Einstein deflection of $4GM/r_0$ we obtain an extra deflection term of $-\gamma r_0$ where γ is a constant and r_0 is the radius of closest approach. With a negative γ, the extra term can increase the deflection on large distance scales (galactic or greater) and therefore imitate the effect of dark matter. Notably, the negative sign required for γ is opposite to the sign of γ used to fit galactic rotation curves by Mannheim and Kazanas. We explain why the signs in the two analyses are different.

INTRODUCTION

The higher-derivative conformally invariant Weyl action, the integral of the square of the Weyl tensor, has attracted much interest as a candidate action for quantum gravity . Unlike GR, the lack of scale in the theory probably implies that it is perturbatively renormalizable [1,2]. The theory is also asymptotically free [3,4].

Weyl gravity, as a classical theory, has attracted less attention because GR has been so remarkably successful at large distances i.e. on solar system scales, and therefore there seems no pressing need to study a higher-derivative alternative classical theory. However, GR may not be free of difficulties either theoretical or experimental. At present, it is faced with one long-standing problem: the notorious cosmological constant problem [5] whose solution is not yet in sight. There may however be an experimental problem with GR: the so-called dark matter problem. The clearest evidence for the existence of large amounts of dark matter comes from the flat rotation curves of galaxies, velocities of galaxies in clusters and the deflection of light from galaxies and clusters [6] (for short, we will call these observations "galactic phenomenology"). From this evidence, there is a consensus in the astrophysical community that most of the mass of galaxies (and of our universe) consists of non-luminous matter. However, the nature of this dark matter is still unknown and is one of the great unsolved problems in astrophysics. At first it was thought that it may be faint stars or other forms of baryonic matter i.e. the so-called massive compact halo objects (MACHOS). However, it is safe to say that observations have obtained much fewer events than required for an explanation of the galactic phenomenology with a dark halo dominated by MACHOS [7] (though there is still the possibility that future experiments might show otherwise). One is then left to consider non-baryonic forms of dark matter such as massive neutrinos, axions and WIMPS i.e. the weakly interacting massive particles as predicted for example by supersymmetric theories. The direct experimental observation of such non-baryonic candidates is of date singularly lacking (though many experiments are currently under development) [15]. Hence, to date, the nature of the dark matter that is thought to comprise most of the mass of our universe is still elusive. Is it possible that the copious amounts of dark matter we are searching for is simply not there? We believe it is reasonable at this juncture to consider such a possibility.

As far as we know, the deviation of galactic rotation curves from the Newtonian expectation occurs at distances way beyond the solar-system scale [14]. In other words, it is a galactic scale phenomena. Newton's gravity theory, which GR recovers in the non-relativistic weak gravity limit, was originally formulated to explain solar-system phenomenology and it may be incorrect to extrapolate this theory to galactic scales. It has therefore been suggested by a handful of authors [8,9,14] that there may not be large amounts of dark matter after all and that the "galactic phenomenology" may be signaling a breakdown of Newtonian gravity (and hence GR) on galactic scales.

Some authors have therefore proposed alternative classical theories of gravity. Most notably there is Milgrom's MOND program [8], Mannheim and Kazanas' Weyl (conformal) gravity program [9] and Bekenstein and Sander's scalar-tensor gravity theory [13]. In MOND, Newtonian dynamics are modified at low accelerations typical of orbits on galactic scales. It has had success in fitting galactic rotation curves without the need for dark matter [8,14]. MOND, however, is a non-relativistic theory and therefore cannot make any predictions on relativistic phenomena such as the deflection of light, cosmology, etc. In the scalar-tensor theory, it has been shown that the bending of light cannot exceed that which is predicted by GR [13], in conflict with the observations i.e. the observed bending is actually even greater than that predicted by GR. On esthetic grounds, conformal gravity is more appealing than other alternative theories because it is based on a local invariance principle i.e. conformal invariance of the metric. Weyl gravity encompasses the largest symmetry group which keep the light cones invariant i.e. the 15 parameter conformal group. It has already been stressed in the past that unlike Weyl gravity and gauge theories, GR is not based on an invariance principle. The Principle of General Covariance, which follows from the Principle of Equivalence, is not an invariance principle. It describes how physical systems behave in a given arbitrary gravitational field but it does not tell us much about the gravitational field itself beyond restricting the gravitational action to a scalar. The lack of an invariance principle is partly the reason why guesswork is inevitable in the derivation of Einstein's gravitational field equations (see [17] for details). In contrast, the Weyl action is unique due to its conformal invariance. Besides its esthetic appeal, Weyl gravity has many other attractive features not the least being that it is renormalizable owing to its lack of length scale. Since the early days of GR, it has been known that the vacuum GR equations $R_{\mu\nu} = 0$ are also vacuum solutions of the Weyl theory. One therefore expects the Schwarzschild metric to be one possible solution to the spherically symmetric Weyl vacuum equations. More recently, Weyl gravity has attracted some interest because it has had reasonable success in fitting galactic rotation curves without recourse to any dark matter [10].

The principal reason that Weyl gravity has not received general acceptance is because some solutions of the classical theory are expected to have no lower energy bound and therefore exhibit instabilities [20] i.e. runaway solutions common to higher-derivative theories. For example, there may exist some Weyl vacuum solutions other than $R_{\mu\nu} = 0$ which are not desirable. Though it has been shown that the Einstein-Hilbert action plus higher-derivative terms has a well posed initial value problem [16] this has yet to be shown for the pure fourth order Weyl gravity. Fortunately, however, the static spherically symmetric vacuum solutions [9], the analog to the Schwarzschild metric, has been found to be stable and to make important corrections to the Schwarzschild metric at large distances i.e. it contains a linear potential that plays a non-trivial role on galactic scales. It therefore becomes compelling and interesting to compare Weyl gravity to GR in their classical predictions.

GEODESIC EQUATIONS

Conformal (Weyl) gravity is a theory that is invariant under the conformal transformation $g_{\mu\nu}(x) \to \Omega^2(x) g_{\mu\nu}(x)$ where $\Omega^2(x)$ is a finite, non-vanishing, continuous real function. The metric exterior to a static spherically symmetric source (i.e. the analog of the Schwarzschild solution in GR) has already been obtained in Weyl gravity by Mannheim and Kazanas [9]. For a metric in the standard form

$$d\tau^2 = B(r)\,dt^2 - A(r)\,dr^2 - r^2\left(d\theta^2 + \sin^2\theta\,d\varphi^2\right) \tag{1}$$

they obtain the vacuum solutions

$$B(r) = A^{-1}(r) = 1 - \frac{2\beta}{r} + \gamma r - kr^2 \tag{2}$$

where β, γ and k are constants. The authors note that with $\beta = GM$, the Schwarzschild metric can be recovered on a certain distance scale (say the solar system) provided γ and k are small enough. The linear γ term would then be significant only on larger distance scales (say galactic or greater) and hence would deviate from Schwarzschild only on those scales. The constant k, which should be taken negative, can then be made even smaller so that the kr^2 term becomes significant only on cosmological scales (in fact, it has been shown [9] that k is proportional to the cosmological scalar curvature). It should be noted that the solution (2) is not unique. The Weyl gravitational field equations are conformally invariant so that any metric which is related to the standard metric (1) by a conformal factor $\Omega^2(r)$ is also a valid solution. This is in contrast to GR where the Schwarzschild solution is the unique vacuum solution for a spherically symmetric source. Two metrics that differ by a conformal factor of course have different curvatures. Remarkably, however, the geodesic equations for light are conformally invariant. Massive particles, on the other hand, have geodesics that depend on the conformal factor (though it is conceivable to envisage some spontaneous conformal symmetry breaking mechanism which gives rise to conformally covariant massive geodesics. e.g. see [11]. We do not entertain conformal symmetry breaking in this paper).

The geodesic equations along the equatorial plane ($\theta = \pi/2$) for a metric of the form (1) are [17]

$$r^2 \frac{d\varphi}{dt} = J\,B(r) \tag{3}$$

$$\frac{A(r)}{B^2(r)}\left(\frac{dr}{dt}\right)^2 + \frac{J^2}{r^2} - \frac{1}{B(r)} = -E \tag{4}$$

$$d\tau^2 = E\,B^2(r)\,dt^2 \tag{5}$$

where E and J are constants with $E = 0$ for null geodesics (photons) and $E > 0$ for massive particles. The above geodesic equations are only conformally invariant for photons and therefore one can calculate the deflection of light knowing the result is conformally invariant.

DEFLECTION OF LIGHT

The geodesic equations (3)-(5) enable one to express the angle φ as a function of r

$$\varphi(r) = \int \frac{A^{1/2}(r)}{r^2 \left(\frac{1}{J^2 B(r)} - \frac{E}{J^2} - \frac{1}{r^2}\right)^{1/2}}\,dr. \tag{6}$$

where the functions $A(r)$ and $B(r)$ are given by (2). To do a scattering experiment, the light is taken to approach the source from infinity. Unlike the Schwarzschild solution where the metric is Minkowskian at large distances from the source i.e. $B(r)$ and $A(r) \to 1$ as $r \to \infty$, $B(r)$ given by the solution (2) diverges as $r \to \infty$ and we do not recover Minkowski space at large distances. However, this is not a problem. At large r it has been shown that the metric is conformal to a Robertson Walker metric with three space curvature $K = -k - \gamma^2/4$ [9]. Hence, at large r the photon is simply moving in a "straight" line in this background geometry (i.e. with $B(r)$ given by (2) and $\varphi(r)$ given by (6), it is easy to see that $d\varphi/dr \to 0$ as $r \to \infty$). The photon then deviates from this "straight" line path as it approaches the source.

We now substitute the appropriate quantities in Eq. (6). For the photon we set $E = 0$. At the point of closest approach $r = r_0$, we have that $dr/d\varphi = 0$ and using equations (5) one obtains $(1/J^2) = B(r_0)/r_0^2$. ¿From the solutions (2) we know that $A^{1/2}(r) = B^{-1/2}(r)$. The deflection of the photon as it moves from infinity to r_0 and off to infinity can be expressed as

$$\Delta\varphi = 2\int_{r_0}^{\infty} \left(\frac{B(r_0)}{r_0^2} - \frac{B(r)}{r^2}\right)^{-1/2} \frac{dr}{r^2} - \pi \tag{7}$$

where π is the change in the angle φ for straight line motion and is therefore subtracted out. We now calculate the integral in (7) using $B(r) = 1 - \frac{2\beta}{r} + \gamma r - kr^2$. This yields

$$\int_{r_0}^{\infty} \left(\left(1 - \frac{2\beta}{r_0} + \gamma r_0\right)\frac{r^4}{r_0^2} - \gamma r^3 - r^2 + 2\beta r\right)^{-1/2} dr \tag{8}$$

The above integral, being the inverse of the square root of a fourth-degree polynomial, can be expressed in terms of elliptic integrals. However, this is not very illuminating. It will prove more instructive to evaluate the integral after expanding the integrand in some small parameters. Note that the constant k, important on cosmological scales, has canceled out and does not appear in the integral (8). The deflection of light is insensitive to the cosmology of the theory and in general would not be affected by a spherically symmetric Hubble flow. On the other hand, the motion of massive particles on galactic or greater scales is affected by the Hubble flow [10,19]. Hence, the bending of light is highly appropriate for testing Weyl gravity.

We now evaluate the integral (8). It can be rewritten in the form

$$\int_{r_0}^{\infty} \left(\frac{1}{r_0^2} - \frac{1}{r^2}\right)^{-1/2} \left\{1 - 2\beta\left(\frac{1}{r_0} + \frac{1}{r} - \frac{1}{r+r_0}\right) + \frac{\gamma r_0}{1 + r_0/r}\right\}^{-1/2} \frac{dr}{r^2}. \tag{9}$$

After making the substitution $\sin\theta = r_0/r$ the integral becomes

$$\int_0^{\pi/2} \left[1 - \frac{2\beta}{r_0}\left(1 + \sin\theta - \frac{\sin\theta}{1+\sin\theta}\right) + \frac{\gamma r_0}{(1+\sin\theta)}\right]^{-1/2} d\theta \tag{10}$$

For any realistic situation, such as the bending of light from the sun, galaxies or cluster of galaxies the deflection is of the order of arc seconds and therefore the parameters β/r_0 and γr_0, which measure the deviation from straight line motion in Eq. (10), must be much less than one. We will therefore expand the integrand to first order in the small parameters β/r_0 and γr_0. One obtains

$$\int_0^{\pi/2} \left[1 + \frac{\beta}{r_0}\left(1 + \sin\theta - \frac{\sin\theta}{1+\sin\theta}\right) - \frac{\gamma r_0}{2(1+\sin\theta)} \right] d\theta = \frac{\pi}{2} + \frac{2\beta}{r_0} - \frac{\gamma r_0}{2} \quad (11)$$

The deflection, given by (7), is therefore

$$\Delta\varphi = \frac{4\beta}{r_0} - \gamma r_0 \quad (12)$$

a simple modification of the standard "Einstein" result of $4GM/r_0$ (where $\beta = GM$). The constant γ must be small enough such that the extra term $-\gamma r_0$ is negligible compared to $4GM/r_0$ on solar distance scales. The linear γ term, however, can begin to make important contributions on larger distance scales where discrepancies between experiment and theory presently exist i.e. the "Einstein" deflection due to the luminous matter in galaxies or clusters of galaxies is less than the observed deflection. Of course, these discrepancies are usually taken as evidence for the existence of large amounts of dark matter in the halos of galaxies. If the extra term $-\gamma r_0$ is to ever replace or imitate this dark matter on large distance scales it would have to be positive (i.e. attractive), implying that γ must be negative. The sign of γ used to fit galactic rotation curves [10] however, is positive (the reason why the sign of γ is different for null and non-relativistic massive geodesics is discussed in the next section on potentials). Therefore there is a glaring incompatibility between these two analyses. This means that Weyl gravity does not seem to solve the dark matter problem, although this does not signal any inconsistency of Weyl gravity itself. In addition, the mechanism of conformal symmetry breaking is not well understood and it must be addressed in more detail before considering massive geodesics or just mass in general. The analysis of the deflection of light is more reliable since it is completely independent of any such conformal symmetry breaking mechanism.

THE POTENTIAL IN WEYL GRAVITY

In General Relativity, the Schwarzschild geodesic equations can be viewed as "Newtonian" equations of motion with a potential (see [18]). In Weyl gravity, a potential can also be extracted from the vacuum equations and for this purpose it is convenient to define a new "time" coordinate p such that $dp = B(r)dt$. The vacuum equations (3)-(5) in these new coordinates are

$$r^2 \frac{d\varphi}{dp} = J \quad (13)$$

$$\frac{1}{2}\left(\frac{dr}{dp}\right)^2 + \frac{J^2}{2r^2} B(r) - \frac{1}{2} = \frac{-E\, B(r)}{2} \quad (14)$$

$$d\tau^2 = E\, dp^2. \quad (15)$$

Let $B(r) \equiv 1+2\phi(r)$ where ϕ is not necessarily a weak field. Equation (14) becomes

$$\frac{1}{2}\left(\frac{dr}{dp}\right)^2 + \frac{J^2}{2r^2} + \phi\left(\frac{J^2}{r^2} + E\right) = \frac{1-E}{2}. \tag{16}$$

The above geodesic equation together with Eq. (13) can be viewed as a particle having energy per unit mass $(1-E)/2$ and angular momentum J moving in ordinary mechanics with a potential

$$V(r) = \phi\left(\frac{J^2}{r^2} + E\right). \tag{17}$$

The derivative of the potential is

$$V'(r) = \frac{\beta}{r^2}\left(\frac{3J^2}{r^2} + E\right) + \frac{\gamma}{2}\left(E - \frac{J^2}{r^2}\right) - krE. \tag{18}$$

where $\phi(r) = -\beta/r + \gamma r/2 - kr^2/2$ was used. There are three terms in Eq. (18): a β, γ and k term respectively. The k term vanishes for null geodesics in agreement with our results on the deflection of light. For massive geodesics the k term is non-zero but is negligible unless one is considering cosmological scales. Hence, this term will be ignored. The factor $3J^2/r^2 + E$ in front of the β term is always positive since $E \geq 0$. Therefore, the β term is attractive for both massive and null geodesics (which is the case in GR). On the other hand, the factor $E - J^2/r^2$ in front of the γ term, can be positive or negative depending on the physical situation. For a non-relativistic particle moving in a weak field, which is the case of galactic rotation curves, we obtain $E \approx 1$, $J^2/r^2 \ll 1$, and therefore the factor $E - J^2/r^2$ is positive. For light, E is zero and the factor is negative. The potential (17) is different for non-relativistic particles and light: the γr term in ϕ contributes a linear potential for non-relativistic particles but an inverse r potential for light. Their corresponding derivatives therefore have opposite sign and this explains why γ obtained through galactic rotation curves has the opposite sign to that obtained in the deflection of light.

Of course, a negative γ term is not reserved to null geodesics only. Any massive particle which is sufficiently relativistic will also have this property. For example consider a particle moving in a weak field ϕ with a negligible "radial velocity" dr/dp. One obtains from Eq. (16) that $J^2/r^2 \approx 1 - E - 2\phi$ and therefore $E - J^2/r^2 \approx 2E + 2\phi - 1$. It follows that if a particle is sufficiently relativistic such that $E < 1/2 - \phi \approx 1/2$ then we obtain a negative γ term.

We can actually reproduce the deflection of light result Eq. (12) in a most straightforward way using the potential Eq. (17). For null geodesics($E = 0$) the potential is given by

$$V_{null}(r) = \frac{-\beta J^2}{r^3} + \frac{\gamma J^2}{2r} + \frac{-kJ^2}{2}. \tag{19}$$

The deflection by a potential $V(r)$ is obtained by integrating along the straight line path the gradient of $V(r)$ (in the $\perp$ direction i.e. in the direction of r_0). As long as the deflection is very small, integrating along the straight line path instead of the curved path gives the same results. The deflection is given by

$$\Delta\varphi = \int_{-\infty}^{\infty} \nabla_\perp V(r)\, dZ. \tag{20}$$

where Z is the distance along the straight line path i.e. $r^2 = Z^2 + r_0^2$. In the potential V_{null}, the γ term is an inverse r potential. This is the reason why its contribution to the deflection of light Eq. (20) is finite and comes with a relative negative sign. If V_{null} had contained a linear potential, the integral for the deflection would diverge, implying that no scattering states could exist.

Using $J^2 = r_0^2/B(r_0)$ given in section III and V_{null} as the potential, the deflection Eq. (20) yields

$$\Delta\varphi = \frac{4\beta}{r_0} - \gamma\, r_0 \tag{21}$$

where only first order terms in β/r_0 and γr_0 were kept. The deflection of light result Eq. (12) is therefore reproduced in a straightforward fashion that allows one to trace clearly the origin of the negative sign in $-\gamma r_0$.

REFERENCES

1. K.S. Stelle, Phys. Rev. **D16**, 953 (1977).
2. E.S. Fradkin and A.A. Tseytlin, Nucl. Phys. **B201**, 469 (1982).
3. E.S. Fradkin and A.A. Tseytlin, Phys. Lett. **104B**, 377 (1981).
4. J. Julve and M. Tonin, Nuovo Cimento **46B**, 137 (1978).
5. For an excellent review see S. Weinberg, Rev. Mod. Phys. **61**(1), 1 (1989).
6. For a review of the evidence see V. Trimble, ARA& A **25**, 425 (1987).
7. C. Alcock et al., ApJ **445**, 133 (1995); A. Gould, astro-ph/9703050 and references therein.
8. M. Milgrom, ApJ **270**, 365 (1983); ApJ **270**, 371 (1983); ApJ **270**, 384 (1983).
9. D. Kazanas and P.D. Mannheim, ApJ **342**, 635 (1989).
10. P.D. Mannheim, ApJ **479**, 659 (1997) and references therein.
11. P.D. Mannheim, Gen. Rel. Grav. **25**, 697 (1993).
12. P.D. Mannheim and D. Kazanas, Gen. Rel. Grav. **26**, 337 (1994).
13. J.D. Bekenstein and R.H. Sanders, ApJ **429**, 480 (1994).
14. R.H. Sanders, ApJ **473**, 117 (1996).
15. For a nice review see D. Spergel in: *Unsolved Problems in Astrophysics*, eds. J.N. Bahcall and J.P. Ostriker (Princeton University Press, Princeton, 1997), 221 and references therein.
16. D.R. Noakes, J. Math. Phys. **24**(7), 1846 (1983).
17. S. Weinberg, *Gravitation and Cosmology* (Wiley, New York, 1972).

18. R. Wald, *General Relativity* (University of Chicago Press, Chicago, 1984).
19. M. Carmeli, astro-ph/9607142.
20. D.G. Boulware, G.T. Horowitz and A. Strominger, Phys. Rev. Lett. **50**, 1726 (1983) and references therein.

Leptoquark Production and Identification at High Energy Lepton Colliders[1]

Michael A. Doncheski[1] and Stephen Godfrey[2]

[1]*Department of Physics, Pennsylvania State University, Mont Alto, PA 17237 USA*
[2]*Ottawa-Carleton Institute for Physics
Department of Physics, Carleton University, Ottawa CANADA, K1S 5B6*

Abstract. Leptoquarks can be produced in substantial numbers for masses very close to the collider centre of mass energy in e^+e^-, $e\gamma$, and $\mu^+\mu^-$ collisions due to the quark content of the photon resulting in equivalently high discovery limits. Using polarization asymmetries in an $e\gamma$ collider the ten different types of leptoquarks listed by Buchmüller, Rückl and Wyler can be distinguished from one another for leptoquark masses essentially up to the kinematic limit. Thus, if a leptoquark were discovered, an $e\gamma$ collider could play a crucial role in determining its origins.

I INTRODUCTION

There is considerable interest in the study of leptoquarks (LQs) which are colour (anti-)triplet, spin 0 or 1 particles that carry both baryon and lepton quantum numbers [1]. Such objects appear in a large number of extensions of the standard model such as grand unified theories, technicolour, and composite models. Quite generally, the signature for leptoquarks is very striking: a high p_T lepton balanced by a jet (or missing p_T balanced by a jet, for the νq decay mode, if applicable). Present limits on leptoquarks have been obtained from direct searches at the HERA ep collider, the Tevatron $p\bar{p}$ collider, and at the LEP e^+e^- collider [2] and estimates of future discovery limits for the LHC also exist [3]. In this paper we consider single leptoquark production in $\mu^+\mu^-$, e^+e^-, and $e\gamma$ collisions which utilizes the quark content of either a backscattered laser photon for the $e\gamma$ case or a Weizacker-Williams photon radiating off of one of the initial leptons for the $\mu^+\mu^-$ or e^+e^- cases [4–6]. This process offers the advantage of a much higher kinematic limit

---

[1]) This research was supported in part by the Natural Sciences and Engineering Research Council of Canada.

than the LQ pair production process, is independent of the chirality of the LQ, and gives similar results for both scalar and vector leptoquarks.

Although the discovery of a leptoquark would be dramatic evidence for physics beyond the standard model it would lead to the question of which model the leptoquark originated from. Given the large number of leptoquark types it would be imperative to measure its properties to answer this question. There are 10 distinct leptoquark types which have been classified by Buchmüller, Rückl and Wyler (BRW) [7]: S_1, $\check{S}_1$ (scalar, iso-singlet); R_2, $\tilde{R}_2$ (scalar, iso-doublet); S_3 (scalar, iso-triplet); U_1, $\tilde{U}_1$ (vector, iso-singlet); V_2, $\tilde{V}_2$ (vector, iso-doublet); U_3 (vector, iso-triplet). The production and corresponding decay signatures are quite similar, though not identical, and have been studied separately by many authors. The question arises as to how to differentiate between the different types. We show how a polarized $e\gamma$ collider can be used to differentiate the LQs. (ie. a polarized e beam, like SLC, in conjunction with a polarized-laser backscattered photon beam.)

II LEPTOQUARK PRODUCTION

The process we are considering is shown if Fig. 1. The parton level cross section for scalar leptoquark production is trivial, given by:

$$\sigma(\hat{s}) = \frac{\pi^2 \kappa \alpha_e m}{M_s} \delta(M_s - \sqrt{\hat{s}}) \qquad (1)$$

where we have followed the convention adopted in the literature where the leptoquark couplings are replaced by a generic Yukawa coupling g which is scaled to electromagnetic strength $g^2/4\pi = \kappa \alpha_{em}$. We give results with κ chosen to be 1. [2] The cross section for vector leptoquark production is a factor of two larger. We only consider generation diagonal leptoquark couplings so that only leptoquarks which couple to electrons can be produced in $e\gamma$ (or e^+e^-) collisions while for the $\mu^+\mu^-$ collider only leptoquarks which couple to muons can be produced. Convoluting the parton level cross section with the quark distribution in the photon one obtains the expression

$$\sigma(s) = \int f_{q/\gamma}(z, M_s^2) \hat{\sigma}(\hat{s}) dz = f_{q/\gamma}(M_s^2/s, M_s^2) \frac{2\pi^2 \kappa \alpha_{em}}{s}. \qquad (2)$$

The cross section depends on the LQ charge since the photon has a larger u quark content than d quark content.

For $e\gamma$, e^+e^-, and $\mu^+\mu^-$ colliders the cross section is obtained by convoluting the expression for the resolved photon contribution to $e\gamma$ production of leptoquarks, Eqn. (2), with, as appropriate, the backscattered laser photon distribution [8] or the Weizsäcker-Williams effective photon distribution: [3]

---

[2] We note that the interaction Lagrangian used by Hewett and Pakvasa in Ref. [6] associates a factor $1/\sqrt{2}$ with the leptoquark-lepton-quark coupling.
[3] The effective photon distribution from muons is obtained by replacing m_e with m_μ.

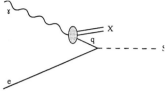

FIGURE 1. The resolved photon contribution for leptoquark production in $e\gamma$ collisions.

$$\sigma(\ell^+\ell^- \to XS) = \frac{2\pi^2 \alpha_{em} \kappa}{s} \int_{M_s^2/s}^{1} \frac{dx}{x} f_{\gamma/\ell}(x, \sqrt{s}/2) \, f_{q/\gamma}(M_s^2/(xs), M_s^2). \qquad (3)$$

Before proceeding to our results we consider possible backgrounds [9]. The leptoquark signal consists of a jet and electron with balanced transverse momentum and possibly activity from the hadronic remnant of the photon. The only serious background is a hard scattering of a quark inside the photon by the incident lepton via t-channel photon exchange; $eq \to eq$. By comparing the invariant mass distribution for this background to the LQ cross sections we found that it is typically smaller than the LQ signal by two orders of magnitude. Related to this process is the direct production of a quark pair via two photon fusion

$$e + \gamma \to e + q + \bar{q}. \qquad (4)$$

However, this process is dominated by the collinear divergence which is actually well described by the resolved photon process $eq \to eq$ given above. Once this contribution is subtracted away the remainder of the cross section is too small to be a concern [9]. Another possible background consists of τ's pair produced via various mechanisms with one τ decaying leptonically and the other decaying hadronically. Because of the neutrinos in the final state it is expected that the electron and jet's p_T do not in general balance which would distinguish these backgrounds from the signal. However, this background should be checked in a realistic detector Monte Carlo to be sure. The remaining backgrounds originate from heavy quark pair production with one quark decaying semileptonically and only the lepton being observed with the remaining heavy quark not being identified as such. All such backgrounds are significantly smaller than our signal in the kinematic region we are concerned with.

III LEPTOQUARK DISCOVERY LIMITS

In Fig. 2 we show the cross sections for a $\sqrt{s} = 1$ TeV e^+e^- operating in both the backscattered laser $e\gamma$ mode and in the e^+e^- mode. The cross section for leptoquarks coupling to the u quark is larger than those coupling to the d quark. This is due to the larger u quark content of the photon compared to the d quark content which can be traced to the larger Q_q^2 of the u-quark. There exist several different

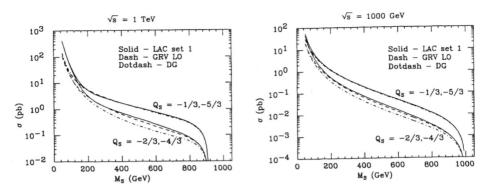

FIGURE 2. The cross sections for leptoquark production due to resolved photon contributions in $e\gamma$ collisions, for $\sqrt{s}_{e^+e^-} = 1$ TeV with κ chosen to be 1. In the left figure the photon beam is due to laser backscattering and in the right figure it is given by the Weizsäcker-Williams effective photon distribution. In both cases the solid, dashed, dot-dashed line is for resolved photon distribution functions of Abramowicz, Charchula and Levy [14], Glück, Reya and Vogt [13], Drees and Grassie [12], respectively.

quark distribution functions in the literature [10–14]. For the four different leptoquark charges we show curves for three different distributions functions: Drees and Grassie (DG) [12], Glück, Reya and Vogt (GRV) [13], and Abramowicz, Charchula and Levy (LAC) set 1 [14]. The different distributions give almost identical results for the $Q_{LQ} = -1/3, -5/3$ leptoquarks and for the $Q_{LQ} = -2/3, -4/3$ leptoquarks give LQ cross sections that vary by most a factor of two, depending on the kinematic region. In the remainder of our results we will use the GRV distribution functions [13] which we take to be representative of the quark distributions in the photon.

Comparing the cross sections for the two collider modes we see that the kinematic limit for the $e\gamma$ mode is slightly lower than the e^+e^- mode. This is because the backscattered laser mode has an inherent energy limit beyond which the laser photons pair produce electrons. On the other hand the backscattered laser cross sections is larger than the e^+e^- mode. This simply reflects that the backscattered laser photon spectrum is harder than the Weizacker-Williams photon spectrum. To determine the leptoquark discovery limits for a given collider we multiply the cross section by the integrated luminosity and use the criteria that a certain number of signature events would constitute a leptoquark discovery. When we do this we find that the discovery limits for the $e\gamma$ and e^+e^- modes are not very different. Although the $e\gamma$ mode has a harder photon spectrum, the e^+e^- mode has a higher kinematic limit.

In Fig. 3 we plot the number of events for various collider energies for e^+e^-, $e\gamma$, and $\mu^+\mu^-$ colliders. For the $\mu^+\mu^-$ collider we used the c and s quark distributions in the photon rather than the u and d quark distributions since we only consider

TABLE 1. Leptoquark discovery limits for e^+e^-, $e\gamma$, and $\mu^+\mu^-$ colliders. The discovery limits are based on the production of 100 LQ's for the energies and integrated luminosities given in columns one and two. The results were obtained using the GRV distribution functions [13].

		e^+e^- Colliders			
$\sqrt{s}$ (TeV)	L (fb^{-1})	Scalar		Vector	
		-1/3, -5/3	-4/3, -2/3	-1/3, -5/3	-4/3, -2/3
0.5	50	490	470	490	480
1.0	200	980	940	980	970
1.5	200	1440	1340	1470	1410
5.0	1000	4700	4200	4800	4500

		$e\gamma$ Colliders			
$\sqrt{s}$ (TeV)	L (fb^{-1})	Scalar		Vector	
		-1/3, -5/3	-4/3, -2/3	-1/3, -5/3	-4/3, -2/3
0.5	50	450	450	450	440
1.0	200	900	900	910	910
1.5	200	1360	1360	1360	1360
5.0	1000	4500	4400	4500	4500

		$\mu^+\mu^-$ Colliders			
$\sqrt{s}$ (TeV)	L (fb^{-1})	Scalar		Vector	
		-1/3, -5/3	-4/3, -2/3	-1/3, -5/3	-4/3, -2/3
0.5	0.7	250	170	310	220
0.5	50	400	310	440	360
4.0	1000	3600	3000	3700	3400

generation diagonal leptoquark couplings. For $\sqrt{s} = 500$ GeV a e^+e^- collider will have about a 25% higher reach than a $\mu^+\mu^-$ collider due to the larger u and d distributions arising from the smaller quark masses. For the highest energy lepton colliders considered the differences become relatively small. For the high luminosities being envisaged, the limiting factor in producing enough leptoquarks to meet our discovery criteria is the kinematic limit. Because, for a given e^+e^- centre of mass energy, an e^+e^- collider will have a higher energy than an $e\gamma$ collider using a backscattered laser, the e^+e^- collider will have a higher discovery limit. Finally, note that the discovery limit for vector leptoquarks is slightly higher than the discovery limit for scalar leptoquarks. This simply reflects the fact that the cross section for vector leptoquarks is a factor of two larger than the cross section for scalar leptoquarks. We summarize the discovery limits for the various colliders in Table 1. The OPAL [17] and DELPHI [18] collaborations at LEP have obtained leptoquark limits using the process we have described.

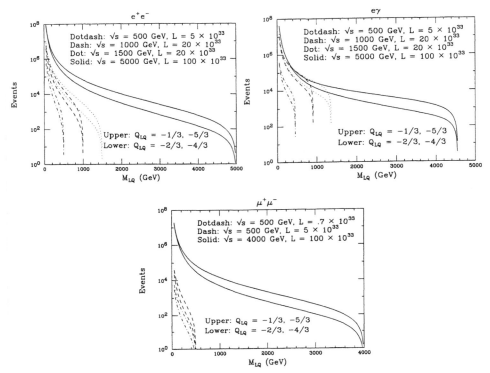

FIGURE 3. Event rates for single leptoquark production in e^+e^-, $e\gamma$, and $\mu^+\mu^-$ collisions. The centre of mass energies and integrated luminosities are given by the line labelling in the figures. The results were obtained using the GRV distribution functions [13]. Note that only 2nd generation LQ's would be produced in $\mu^+\mu^-$ collisions with our assumption of generation diagonality.

IV   LEPTOQUARK IDENTIFICATION

If a leptoquark were actually discovered the next step would be to determine its properties so that we could determine which model it originated from. We will assume that a peak in the $e + jet$ invariant mass is observed in some collider (*i.e.*, the existance of a LQ has been established), and so we need simply to identify the particular type of LQ. We assume that the leptoquark charge has not been determined and assume no intergenerational couplings. Furthermore, we will assume that only one of the ten possible types of LQs is present. Table 2 of BRW [7] gives information on the couplings to various quark and lepton combinations; the missing (and necessary) bit of information in BRW is that the quark and lepton have the same helicity (RR or LL) for scalar LQ production while they have opposite helicity (RL or LR) for vector LQ production. It is then possible to construct the cross sections for the various helicity combinations and consequently the double spin asymmetry [5], for the different types of LQs.

Thus, a first step in identifying leptoquarks would be to determine the coupling chirality, ie. whether it couples to e_L, e_R, or e_U. This could be accomplished by using electron polarization either directly or by using a left-right asymmetry measurement:

$$A^{+-} = \frac{\sigma^+ - \sigma^-}{\sigma^+ + \sigma^-} = \frac{C_L^2 - C_R^2}{C_L^2 + C_R^2}$$

This divides the 10 BRW leptoquark classifications into three groups:

e_L^-: $\tilde{R}_2$, S_3, U_3, $\tilde{V}_2$

e_R^-: $\tilde{S}_1$, $\tilde{U}_1$

e_U^-: U_1, V_2, R_2, S_1

We can further distinguish whether the leptoquarks are scalar or vector. This could be accomplished in two ways. In the first one can study the angular distributions of the leptoquark decay products. In the second we can use the double asymmetry:

$$A_{LL} = \frac{(\sigma^{++} + \sigma^{--}) - (\sigma^{+-} + \sigma^{-+})}{(\sigma^{++} + \sigma^{--}) + (\sigma^{+-} + \sigma^{-+})}$$

where the first index refers to the electron helicity and the second to the quark helicity. Because scalars only have a non-zero cross section for σ^{++} and σ^{--} for scalar LQ's the parton level asymmetry for eq collisions is $\hat{a}_{LL} = +1$. Similarly, since vectors only have a non-zero cross section for σ^{+-} and σ^{-+} for vector LQ's $\hat{a}_{LL} = -1$.

To obtain observable asymmetries one must convolute the parton level cross sections with polarized distribution functions. Doing so will reduce the asymmetries from their parton level values of ±1 so one must determine whether the observable asymmetries can distinguish between the leptoquark types. The expressions for the double longitudinal spin asymmetry A_{LL} are given in Ref. [5]. In Figure 4 we plot A_{LL} for the $e\gamma$ collider which started with $\sqrt{s_{e^+e^-}}=1$ TeV. To obtain these curves we used parameterizations of the *asymptotic* polarized photon distribution functions [15,16], where it is assumed that Q^2 and x are large enough that the Vector Meson Dominance part of the photon structure is not important, but rather the behavior is dominated by the point-like $\gamma q\bar{q}$ coupling. In order to be consistent, we used a similar asymptotic parameterization for the unpolarized photon distribution functions as well [10], even though various sets of more correct photon distribution functions exist (*e.g.*, [11–14]). We only used this asymptotic approximation in the unpolarized case for the calculation of the asymmetry, where it is hoped that in taking a ratio of the asymptotic polarized to the asymptotic unpolarized photon distribution functions, the error introduced will be minimized. Still, we suggest that our results be considered cautiously, at least in the relatively small LQ mass

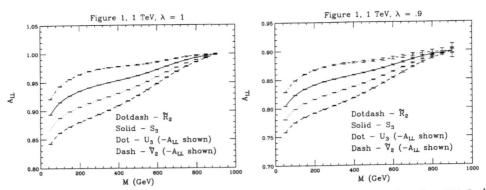

FIGURE 4. A_{LL} vs M_{LQ} for a 1 TeV $e\gamma$ collider. The statistical errors are based on 200 fb^{-1}. The left figure is for 100% polarization and the right for 90% polarization.

region. We note that in the asymptotic approximation, the unpolarized photon distribution functions have (not unexpectedly) a similar form to the polarized photon distribution functions.

In Fig. 4 we show asymmetries for 100% polarization and for 90% polarization which is considered to be achievable given the SLC experience. The error bars are based on an total integrated luminosity of 200 fb^{-1} for $\sqrt{s}_{e^+e^-} = 1$ TeV. In these figures note that we are showing $-A_{LL}$ for the vector cases so that we can use a larger scale. Quite clearly, polarization would enable us to distinguish between vector and scalar. For the cases where there are two types of leptoquarks of the same chiral couplings, for example the scalar isodoublet $\tilde{R}_2$ and scalar isotriplet S_3, we could distinguish between them up to about 3/4 the kinematic limit.

Finally, one additional bit of information to further differentiate among the various possible leptoquarks is to search for the $\nu q'$ decay mode. This signature is quite similar to a supersymmetric particle decay (high p_T jet plus missing p_T) so that although it cannot be used to unambiguously determine the existence on leptoquarks, it can be used, in conjunction with the observation of an approximately equal number of eq events (as expected in some models) to provide further information on leptoquark couplings. Taken together, the leptoquark type can be uniquely determined. If more than one leptoquark were discovered, determining their properties would tell us their origin and therefore, the underlying theory.

V SUMMARY

To summarize, we have presented results for single leptoquark production in $e\gamma$, e^+e^-, and $\mu^+\mu^-$ collisions. The discovery limits for leptoquarks is very close to the centre of mass energy of the colliding particles. It also appears that a polarized $e\gamma$ collider can be used to differentiate between the different models of LQs that can exist, essentially up to the kinematic limit. Furthermore, it is quite easy to distinguish scalar LQs from vector LQs for all LQ mass (given that the LQ is kinematically allowed). Thus e^+e^-, $e\gamma$, and $\mu^+\mu^-$ colliders have much to offer in the searches for leptoquarks. If leptoquarks were discovered, $e\gamma$ colliders could play a crucial role in unravelling their properties, and therefore the underlying physics.

REFERENCES

1. See, for example, J.L. Hewett and T.G. Rizzo, hep-ph/9708419 and references therein.
2. Particle Data Group, C. Caso et al, The European Physical Journal C3, 1 (1998).
3. See for example the ATLAS Technical Proposal, W.W. Armstrong et al., CERN Report CERN/LHCC/94-43; J.L. Hewett, T.G. Rizzo, S. Pakvasa, H.E. Haber and A. Pomarol, *Proceedings of the Workshop on Physics at Current Accelerators and Supercolliders*, ed. J.L. Hewett, A.R. White, and D. Zeppenfeld, Argonne report ANL-HEP-CP-93-92, p. 539.
4. M. A. Doncheski and S. Godfrey, Phys. Rev. **D49**, 6220 (1994) [hep-ph/9311288]; Phys. Lett. **B393**, 355(1997), [hep-ph/9608368]; Mod. Phys. Lett. **A12**, 1859(1997), [hep-ph/9704380].
5. M. A. Doncheski and S. Godfrey, Phys. Rev. **D51**, 1040 (1995) [hep-ph/9407317].
6. O.J. Éboli, E.M. Gregores, M.B. Magro, P.G. Mercadante, and S.F. Novaes, Phys. Lett. **B311**, 147 (1993); H. Nadeau and D. London, Phys. Rev. **D47**, 3742 (1993); G. Bélanger, D. London and H. Nadeau, Phys. Rev. **D49**, 3140 (1994); J.L. Hewett and S. Pakvasa, Phys. Lett. **B227**, 178 (1989); F. Cuypers, [hep-ph/9602355]. Other related papers are: J. E. Cieza Montalvo and O.J.P. Éboli, Phys. Rev. **D47**, 837 (1993); T.M. Aliev and Kh.A. Mustafaev, Yad. Fiz. **58**, 771 (1991); V. Ilyin et al., Phys. Lett. **B351**, 504 (1995); erratum **B352**, 500 (1995); Phys. Lett. **B356**, 531 (1995).
7. W. Buchmüller, R. Rückl, and D. Wyler, Phys. Lett. **B191**, 442 (1987).
8. I.F. Ginzburg et al., Nucl. Instrum. Methods, **205**, 47 (1983); **219**, 5 (1984); C. Akerlof, Ann Arbor report UM HE 81-59 (1981; unpublished).
9. Hadronic backgrounds in e^+e^- and $e\gamma$ collisions and associated references are given in M.A. Doncheski, S. Godfrey, and K.A. Peterson, Phys.Rev. **D55**, 183 (1997) [hep-ph/9407348].
10. A. Nicolaidis, Nucl. Phys. **B163**, 156 (1980).
11. D.W. Duke and J.F. Owens, Phys. Rev. **D26**, 1600 (1982).
12. M. Drees and K. Grassie, Z. Phys. **C28**, 451 (1985); M. Drees and R. Godbole, Nucl. Phys. **B339**, 355 (1990).
13. M. Glück, E. Reya and A. Vogt, Phys. Lett. **B222**, 149 (1989); Phys. Rev. **D45**, 3986 (1992); Phys. Rev. **D46**, 1973 (1992).
14. H. Abramowicz, K. Charchula, and A. Levy, Phys. Lett. **B269**, 458 (1991).
15. J.A. Hassan and D.J. Pilling, Nucl. Phys. **B187**, 563 (1981).
16. Z. Xu, Phys. Rev. **D30**, 1440 (1984).
17. OPAL Collaboration, OPAL Physics Note 288 (1997).
18. DELPHI Collaboration, DELPHI 97-112 Conf 94 (1997).

Gluon Radiation in Top Quark Production and Decay at an e^+e^- Collider[1]

Cosmin Macesanu and Lynne H. Orr

*Department of Physics and Astronomy, University of Rochester
Rochester, New York 14627-0171*

Abstract. We study the effects of gluon radiation on top production and decay processes at an e^+e^- collider. The matrix elements are computed without any approximations, using spinor techniques. We use a Monte Carlo event generator which takes into account the infrared singularity due to soft gluons and differences in kinematics associated with radiation in the production versus decay process. The calculation is illustrated for several strategies of top mass reconstruction.

INTRODUCTION

The study of the top quark is one of the most important goals for future high energy experiments. Top has several characteristics which set it apart from the other quarks, the most obvious being its large mass: at 175 GeV top is 35 times heavier than its partner the b. The large mass of the top quark has a number of interesting consequences. It makes $t\bar{t}$ loop corrections to electroweak processes important, so that a good measurement of m_t coupled with extremely precise values of the W mass will give limits on the Higgs mass. A heavy top may decay into supersymmetric (SUSY) particles, providing a way of testing predictions of SUSY models. Perhaps most important, large mass means that the top Yukawa coupling to the Higgs is large; actually it is close to unity. This means that top studies can offer insight into electroweak symmetry breaking, as well as into the fermion mass generation process.

Another consequence of the top quark's large mass is that its decay width is about 1.5 GeV, which means its lifetime is of order 10^{-24} seconds. In this short a time, the top quark doesn't have time to hadronize. As a consequence, the top can be completely described by perturbative QCD. Also, the spin information is transmitted to the decay products (b and a W^+). This allows us to get informa-

---

[1] Presented by C. Macesanu.

tion about the top couplings from angular distributions of the resulting particles; differences between measured and SM values would indicate new physics.

The top was discovered in 1995 at the Fermilab Tevatron proton-antiproton collider. This machine is now being upgraded, and in 2000 the study of the top here will start again. Also, at the Large Hadron Collider precise measurements of the top parameters (mass, couplings, production cross-section) will be one of the priorities.

Besides studies at these machines, it would be interesting to study top at a high-energy electron-positron collider. The principal advantage of a lepton collider versus a hadronic collider is a much cleaner environment; the task of performing certain precision measurements will be much easier in the absence of large QCD backgrounds. In particular, studies at the $t\bar{t}$ threshold can be done at a lepton collider, but such studies are impossible at a hadron collider. Measurements of top quark couplings are especially difficult at a hadron collider. At an e^+e^- collider, however, besides smaller background, the possibility of using polarized electrons in the initial state will be an advantage; also, the coupling of the top to neutral currents (γ and Z) will be accessible. In general, the capabilities of a high energy lepton collider are complementary to those of a hadron collider.

To be able to perform these studies, though, a good theoretical understanding of top quark's production and decay processes is essential. An important issue is QCD corrections due to real or virtual gluon emission because of the large value of the strong coupling constant (≈ 0.1). In the following, we will be concerned with corrections due to the radiation of a real gluon.

CALCULATIONAL PROCEDURE

We are interested in top production and decay with emission of a gluon:

$$e^+e^- \to \gamma^*, Z^* \to t\bar{t}(g) \to bW^+\bar{b}W^-(g)$$

(the top decays into a $b - W$ pair with a branching ratio close to unity). The gluon radiation can take place during the $t\bar{t}$ production process or during the decay of either the t or $\bar{t}$. Studies of this subject have focused on a single portion of the entire process (such as corrections to production [2,3] or decay [4] only), or imposed approximations such as soft gluons [5] or intermediate on-shell top quarks and massless b quarks [6]. The results presented here are obtained using an exact computation of the entire matrix element for real gluon emission, including all spin correlations, top width, and b mass effects [1].

Let us briefly describe the procedure for obtaining these results. We start by evaluating the amplitude for the process; there are four diagrams which contribute, (see Fig. 1), so we will have:

$$M = \frac{B}{P_{tg} * P_{\bar{t}}} + \frac{BB}{P_t * P_{\bar{t}g}} + \frac{T}{P_t * P_{\bar{t}g} * P_{\bar{t}}} + \frac{TB}{P_t * P_{\bar{t}} * P_{\bar{t}g}} \quad (1)$$

with

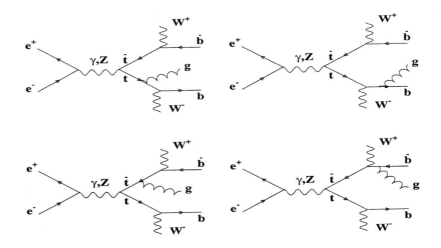

FIGURE 1. Feynman diagrams for top quark production with gluon radiation

$$P_t = p_t^2 - m_t^2 + im_t\Gamma_t \; ; \; p_t = p_{W^+} + p_b \tag{2}$$

$$P_{tg} = p_{tg}^2 - m_t^2 + im_t\Gamma_t \; ; \; p_{tg} = p_{W^+} + p_b + p_g$$

and similar definitions for the denominators of $\bar{t}$ propagators.

The B, BB terms correspond to the gluon being radiated by the b, $\bar{b}$ quarks. The last two terms correspond to a gluon radiated by the t or $\bar{t}$, either in the production or decay stage. To separate these cases, we rewrite the propagator products as follows:

$$\frac{1}{P_t * P_{tg}} = \frac{1}{2p_g p_t}[\frac{1}{P_t} - \frac{1}{P_{tg}}] \; , \; \frac{1}{P_{\bar{t}} * P_{\bar{t}g}} = \frac{1}{2p_g p_{\bar{t}}}[\frac{1}{P_{\bar{t}}} - \frac{1}{P_{\bar{t}g}}]$$

and the amplitude can be written

$$M = A1 + A2 + A3 \tag{3}$$

with

$$A1 = \frac{1}{P_{tg} * P_{\bar{t}}}[B - \frac{T}{2p_g p_t}] \; , \; A2 = \frac{1}{P_t * P_{\bar{t}g}}[BB - \frac{TB}{2p_g p_{\bar{t}}}]$$

$$A3 = \frac{1}{P_t * P_{\bar{t}}}[\frac{T}{2p_g p_t} + \frac{TB}{2p_g p_{\bar{t}}}]$$

We identify these three terms as corresponding to gluon radiation in the t decay ($A1$), $\bar{t}$ decay ($A2$) or production stage ($A3$).

The three separate parts of the amplitude are calculated using helicity amplitudes and can be evaluated numerically. We can then compute each of the six resulting terms separately:

$$\sum_{helicities} |M|^2 = \sum_{helicities} \left\{ |A1|^2 + |A2|^2 + |A3|^2 + 2Re[A1A2^* + A1A3^* + A2A3^*] \right\}$$

To integrate this formula over the phase space we use a three-channel Monte Carlo with one channel for each of the diagonal terms $|A1|^2, |A2|^2, |A3|^2$ and a combination of channels for the interference terms $2Re[A1A2^* + A1A3^* + A2A3^*]$. The phase space region where the gluon energy E_g goes to 0 presents some problems, though, because the amplitude has a singularity there. Even with cuts on E_g, the rapid variation of the integrand can spoil the integration procedure. To eliminate this problem, we tailor the momentum generator to the production of a gluon in association with two massive particles ($\gamma^*, Z^* \to t\bar{t}g$ or $t \to bWg$).

NUMERICAL RESULTS

In this section, we present some preliminary results obtained using the procedure described above. As input, we use the following values: $m_t = 175$ GeV, $\Gamma_t = 1.5$ GeV, $m_b = 5$ GeV, $\alpha_s = 0.1$, and a center of mass energy $W = 600$ GeV. Also, unless otherwise specified, we impose a cut on gluon energy $E_g > 10$ GeV.

First, we note that the radiation of a gluon plays an important role in $t\bar{t}$ production. The lowest-order, tree-level cross section is $\sigma_0 = 0.43$ pb, while the cross section for the process with emission of a gluon with energy greater than 10 GeV is $\sigma_0 = 0.39$ pb.[2] We see that the two quantities have the same order of magnitude.

An interesting (and important) issue is top mass reconstruction. This is closely related to the problem of assigning the gluon to the correct process in which it was radiated. For example, if the gluon was radiated in the production stage, the mass of the top is given by

$$m_t^2 = p_t^2 = (p_b + p_{W^+})^2 \tag{4}$$

On the other hand, if the gluon was radiated in t decay stage, the above formula will give us a low estimate of m_t; the correct formula will be

$$m_t^2 = p_{tg}^2 = (p_b + p_{W^+} + p_g)^2 \tag{5}$$

In Fig. 2 we present the mass distributions obtained using (4) and (5) respectively. While there are clear peaks at the input values of the top mass, we see significant tails due to wrong gluon assignments. The existence of such tails can increase measurement uncertainties on the top mass or confound attempts to identify top

---

[2] Note that this number includes contributions from decay-stage radiation which are not corrections to the total top production cross section.

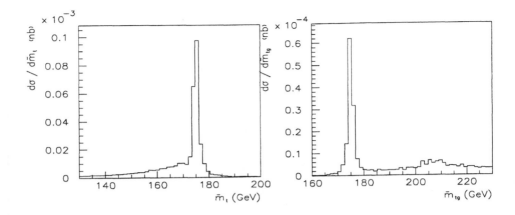

FIGURE 2. Distributions in $\tilde{m}_t$ and $\tilde{m}_{tg}$ as defined in the text.

events by mass reconstruction. It is apparent that, for good mass reconstruction, we need to find a method to assign the gluon to the correct process in which it was radiated.

Using the variables $\tilde{m}_t = \sqrt{p_t^2}$, $\tilde{m}_{tg} = \sqrt{p_{tg}^2}$, $\tilde{m}_{\bar{t}} = \sqrt{p_{\bar{t}}^2}$ and $\tilde{m}_{\bar{t}g} = \sqrt{p_{\bar{t}g}^2}$ we define four types of events:

-type 1 : $172\ GeV < \tilde{m}_{tg}, \tilde{m}_{\bar{t}} < 178\ GeV$

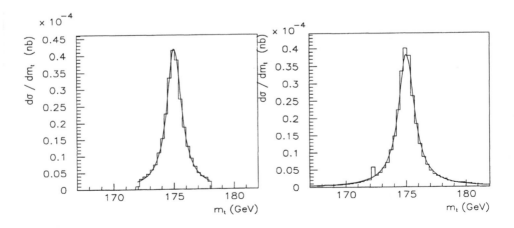

FIGURE 3. Mass distributions using mass cuts (left) and gluon angle cuts (right).

-type 2 : $172\ GeV < \tilde{m}_t, \tilde{m}_{\bar{t}g} < 178\ GeV$

-type 3 : $172\ GeV < \tilde{m}_t, \tilde{m}_{\bar{t}} < 178\ GeV$

-type 4 : any other event

We identify type 1 events as corresponding to gluons radiated in t decay stage. In this case we expect $p_{\bar{t}}$, p_{tg} to be on shell; however, as we use a Breit-Wigner distribution, we accept deviations from the exact value $m_t = 175$ GeV by about $2\Gamma_t = 3$ GeV. Furthermore, we identify type 2 events as corresponding to gluons radiated in $\bar{t}$ decay stage and type 3 as corresponding to gluons radiated in $t\bar{t}$ production stage. As the gluon energy is quite big (> 10 GeV) compared with Γ_t, we expect this interpretation to work; meaning that there will be very few events which satisfy one of the conditions 1, 2, 3 but which are open to more than one interpretation. Finally, type 4 would correspond to events for which there is no compelling evidence for either the production- or decay-stage radiation case; it will actually be a mixing of both.

In the first plot Fig. 3 we present the top mass distribution using events of types 1, 2, and 3 defined above. We have generated 250,000 events for each case. The smooth line is a Breit-Wigner function fitted over the mass distribution. The fit reproduces the input parameters with remarkable accuracy; it actually gives us 175.04 GeV for m_t and 1.46 GeV for Γ_t.

Another method for reconstructing the process makes use of the gluon-b and gluon-$\bar{b}$ angle distributions. It is known that a gluon radiated by a quark tends

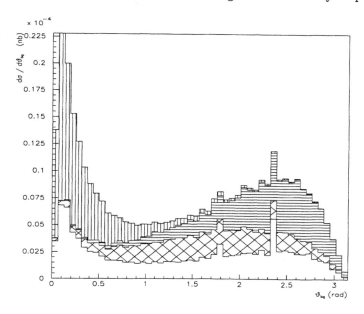

FIGURE 4. gluon-b angle distribution

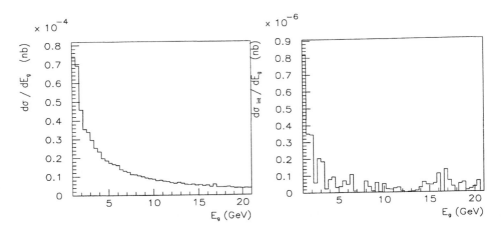

FIGURE 5. Gluon energy distribution; total cross-section and interference contribution, respectively.

to go in the same direction as that quark.[3] Then, the gluons close to the b quark probably came from the top decay process; similarly, the gluons close to the $\bar{b}$ are likely to have come from $\bar{t}$ decay. We might guess that the rest of the gluons may be assigned to the production process.

In Fig. 4 we present the distribution of the angle θ_{bg} between the gluon and the b quark. The vertical-hatched part corresponds to type 1 events; and they are close to the b direction. The horizontal-hatched part corresponds to type 2 events; as they will cluster near the $\bar{b}$ quark, and b and $\bar{b}$ are mostly in opposite hemispheres, these events tend to gather at large angles. The cross-hatched part corresponds to type 3 events, and they are distributed uniformly. Finally, the non-hatched part corresponds to type 4 events, and, as they are an admixture of the first three, it looks like the whole distribution on a smaller scale.

Using this figure we can make the following conventions:
 -if $\theta_{bg} < 0.7$ *rad* assign gluon to t decay
 -if $\theta_{\bar{b}g} < 0.7$ *rad* assign gluon to $\bar{t}$ decay
 -if $\theta_{bg}, \theta_{\bar{b}g} > 1$ *rad* assign gluon to $t\bar{t}$ production.

With these definitions, we construct the top mass distribution presented in the second plot in Fig. 3. Again, a fit with a Breit-Wigner function gives us values very close to our input parameters: $m_t = 175.05$ GeV , $\Gamma_t = 1.6$ GeV.

Fig. 5 presents the gluon energy distribution. The soft gluon singularity is visible (we have used a $E_g > 1$ GeV cut). This raises an interesting problem: soft gluons are not visible in the detector; therefore, the process in which a soft gluon

---

[3] Since we keep the b mass, there is strictly speaking no collinear singularity; however gluons tend to be radiated from the b at small angles nonetheless.

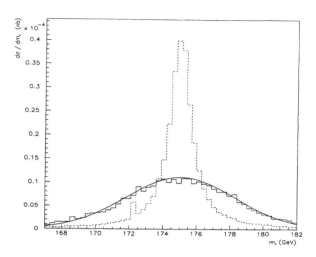

FIGURE 6. Mass reconstruction distribution, taking into account detector resolution (the dashed line corresponds to exact energy values).

is radiated will give the same signal as the process in which there is no gluon. The study of the experimental cross section for this last type of process will have to take into account radiation of gluons with energies up to 5 GeV, maybe. This also requires including virtual corrections, as in [6].

At this point, it is interesting to see if the interference terms contribute. We can integrate separately the contribution of diagonal terms and the contribution of interference terms; it turns out that this last contribution amounts only to around 1% of the total distribution, as shown in Fig. 5.

Up to here, we have worked at the parton level and presumed exact information about the momenta of particles in the final state. In a real experiment, though, there are a lot of complications: the electrons in the initial state can lose energy through radiation before interacting, thus having a lower CM energy; we don't see neutrinos, don't know which jet corresponds to which quark (or gluon), and the energies measured are not exact. One can expect that the detector resolution will have one of the biggest effects on the top mass (and width) reconstruction.

In Fig. 6 we present the distribution for the top mass (using gluon angle cuts) obtained after taking into account this effect. The spread in the measured energies is parameterized by Gaussians with widths $\sigma = 0.4\sqrt{E}$ for quarks and gluon, and $\sigma = 0.15\sqrt{E}$ for the W's. The m_t value obtained in this case is still very good (175.03 GeV) but the width of the distribution in this case reflects rather the detector resolution effects than the top width.

CONCLUSIONS

We have presented preliminary results of a calculation of real gluon radiation in top production and decay. [1] Mass reconstruction using various mass cuts and gluon-bottom quark angle cuts was performed. In an ideal situation, (no initial state radiation, perfect particle identification, exact energy and angle measurements) the results were found to be very good, *i.e.*, the distributions faithfully reproduced the input mass and width. In a further study, we will have to take into account the effects of energy smearing due to detector resolution, initial state radiation, undetected neutrinos and jet identification problems. Effects due to interference between gluon radiation in the production and decay stages do not appear to be experimentally visible, at least for hard gluons. Soft gluon radiation (the experimental signal in this case will mimic that of a process without any gluon) and virtual gluon corrections have to be studied in more detail.

We thank C.R. Schmidt and W.J. Stirling for helpful correspondence and discussions. This work was supported in part by the U.S. Department of Energy and the National Science Foundation.

REFERENCES

1. C. Macesanu and L.H. Orr, in preparation.
2. J. Jersak, E. Laerman, and P. Zerwas, Phys. Rev. **D25**, 1218 (1982).
3. Yu. L. Dokshitzer, V.A. Khoze, and W.J. Stirling, Nucl. Phys. **B428**, 3 (1994).
4. M. Jeżabek and J. Kühn, Phys. Lett. **B329**, 317 (1994).
5. V. Khoze, W. Stirling and L. Orr, Nucl. Phys. **B378**, 413 (1992).
6. C.R. Schmidt, Phys. Rev. **D54**, 3250 (1996).

Single Pion Transitions of Charmed Baryons

Salam Tawfiq,* Patrick J. O'Donnell* and J.G. Körner[†]

Department of Physics, University of Toronto, 60 St. George Street, Toronto, Ontario, M5S 1A7, Canada
[†] *Institut für Physik, Johannes Gutenberg-Universität, Staudinger Weg 7, D-55099 Mainz, Germany*

Abstract. The $SU(2N_f) \otimes O(3)$ constituent quark model symmetry of the light diquark system are used to analyze single pion transitions of S-wave to S-wave and P-wave to S-wave heavy baryons. We show that the Heavy Quark Symmetry (HQS) coupling factors are given in terms of the three independent couplings $g_{\Sigma_Q \Lambda_Q \pi}$, $f_{\Lambda_{Q1} \Sigma_Q \pi}$ and $f_{\Lambda^*_{Q1} \Sigma_Q \pi}$. Light-Front quark model spin wave functions are, then, employed to calculate these couplings and to predict decay rates of single pion transitions between charm baryon states.

Heavy Quark Symmetry (HQS) and $SU(2N_f) \times O(3)$ light diquark symmetry can be used to construct heavy baryon spin wave functions a la Bethe-Salpeter [1,2]. These covariant wave functions were employed [3] to analyze current-induced bottom baryon to charm baryon transitions. Similar procedure will be followed here to study heavy baryon S-wave to S-wave and P-wave to S-wave single pion transitions. In single–pion transitions between heavy baryons, the pion is emitted by each of the light quarks while the heavy quark is unaffected. In fact, the heavy baryon velocity will not be changed when emitting the pion since it is infinitely massive and will not recoil.

In a heavy baryon, a light diquark system with quantum numbers j^P couples with a heavy quark with $J_Q^P = 1/2^+$ to form a doublet with $J^P = (j \pm 1/2)$. Heavy quark symmetry allows us to write down a general form for the heavy baryon spin wave functions [3,4]. Ignoring isospin indices, one has

$$\chi_{\alpha\beta\gamma} = (\phi_{\mu_1 \cdots \mu_j})_{\alpha\beta} \psi_\gamma^{\mu_1 \cdots \mu_j}(v) \ . \tag{1}$$

Here, $v = \frac{P}{M}$ is the baryon four velocity, $\mu_1 \cdots \mu_j$ are Lorentz indices, the spinor indices α and β refer to the light quark system and the index γ refers to the heavy quark. In the heavy quark limit, the $\chi_{\alpha\beta\gamma}$ satisfy the Bargmann-Wigner equation on the heavy quark index

$$(\not{v})_\gamma^{\gamma'} \chi_{\alpha\beta\gamma'} = \chi_{\alpha\beta\gamma} \ . \qquad (2)$$

In general the light degrees of freedom spin wave functions $(\phi_{\mu_1\cdots\mu_j})_{\alpha\beta}$ are written in terms of the two bispinors $[\chi^0]_{\alpha\beta}$ and $[\chi^1_\mu]_{\alpha\beta}$. The matrix $[\chi^0]_{\alpha\beta} = [(\not{v}+1)\gamma_5 C]_{\alpha\beta}$, projects out a spin-0 object, is symmetric when interchanging α and β. However, $[\chi^1_\mu]_{\alpha\beta} = [(\not{v}+1)\gamma_{\perp\mu} C]_{\alpha\beta}$ which projects out a spin-1 object is antisymmetric. Here, C is the charge conjugation operator and $\gamma_\mu^\perp = \gamma_\mu - \not{v} v_\mu$. On the other hand the "superfield" $\psi_\gamma^{\mu_1\cdots\mu_j}(v)$ stands for the two spin wave functions corresponding to the two heavy quark symmetry degenerate states with spins $(j \pm 1/2)$. They are generally written in terms of the Dirac spinor u and the Rarita-Schwinger spinor u_μ.

The S-wave heavy-baryon spin wave functions are given by

$$\chi^{\Lambda_Q}_{\alpha\beta\gamma} = (\chi^0)_{\alpha\beta} u_\gamma \qquad (3)$$

and

$$\chi^{\Sigma_Q}_{\alpha\beta\gamma} = (\chi^{1,\mu})_{\alpha\beta} \left\{ \begin{array}{c} \frac{1}{\sqrt{3}} \gamma_\mu^\perp \gamma_5 u \\ u_\mu \end{array} \right\}_\gamma . \qquad (4)$$

To represent the orbital excitation for P-wave heavy baryon states, one can use the relative momenta $K = \frac{1}{\sqrt{6}}(p_1 + p_2 - 2p_3)$ and $k = \frac{1}{\sqrt{2}}(p_1 - p_2)$, symmetric and antisymmetric respectively under interchange of the constituent light quark momenta p_1 and p_2. The Λ_{Q1} degenerate state spin wave functions can be written as

$$\chi^{\Lambda_{Q1}}_{\alpha\beta\gamma} = (\chi^0 K^\mu)_{\alpha\beta} \left\{ \begin{array}{c} \frac{1}{\sqrt{3}} \gamma_\mu^\perp \gamma_5 u \\ u_\mu \end{array} \right\}_\gamma . \qquad (5)$$

S-wave and P-wave heavy baryon spin wave functions are summarized in Table (1). The heavy baryon total wave functions are constructed ensuring overall symmetry with respect to $flavour \otimes spin \otimes orbital$.

The one-pion transition amplitudes between heavy baryons can then be written as

$$M^\pi = \langle \pi(\vec{p}), B_Q(v) \mid T \mid B_Q(v) \rangle$$
$$= \bar{\psi}_2^{\nu_1\cdots\nu_{j_2}}(v) \psi_1^{\mu_1\cdots\mu_{j_1}}(v) \mathcal{M}_{\mu_1\cdots\mu_{j_1};\nu_1\cdots\nu_{j_2}} . \qquad (6)$$

The light diquark tensors $\mathcal{M}_{\mu_1\cdots\mu_{j_1};\nu_1\cdots\nu_{j_2}}$ of rank $(j_1 + j_2)$, describe $j_1 \to j_2 + \pi$ transitions, should have the correct parity and project out the appropriate partial wave amplitude.

HQS predicts that S-wave to S-wave transitions involve two p-wave coupling constants. However, each of the single pion transitions from the K-multiplet and from the k-multiplet down to the ground state are determined in terms of seven

TABLE 1. spin wave functions of S-wave and P-wave heavy baryons. The symbol $\{AB\}^0$ refers to a traceless symmetric tensor.

	j^P	J^P	$\chi_{\alpha\beta\gamma}$
S-wave states			
Λ_Q	0^+	$\frac{1}{2}^+$	$(\chi^0)_{\alpha\beta} u_\gamma$
Σ_Q	1^+	$\frac{1}{2}^+$, $\frac{3}{2}^+$	$(\chi^{1,\mu})_{\alpha\beta} \left\{ \begin{array}{c} \frac{1}{\sqrt{3}}\gamma_\mu^\perp \gamma_5 u \\ u_\mu \end{array} \right\}_\gamma$
Symmetric P-wave states			
Λ_{QK1}	1^-	$\frac{1}{2}^-$, $\frac{3}{2}^-$	$(\chi^0 K_\perp^\mu)_{\alpha\beta} \left\{ \begin{array}{c} \frac{1}{\sqrt{3}}\gamma_\mu^\perp \gamma_5 u \\ u_\mu \end{array} \right\}_\gamma$
Σ_{QK0}	0^-	$\frac{1}{2}^-$	$\frac{1}{\sqrt{3}}(\chi^{1,\mu} K_{\perp\mu})_{\alpha\beta} u_\gamma$
Σ_{QK1}	1^-	$\frac{1}{2}^-$, $\frac{3}{2}^-$	$\frac{i}{\sqrt{2}}(\varepsilon_{\mu\nu\rho\delta}\chi^{1,\nu} K_\perp^\rho v^\delta)_{\alpha\beta} \left\{ \begin{array}{c} \frac{1}{\sqrt{3}}\gamma_\mu^\perp \gamma_5 u \\ u_\mu \end{array} \right\}_\gamma$
Σ_{QK2}	2^-	$\frac{3}{2}^-$, $\frac{5}{2}^-$	$\frac{1}{2}(\{\chi^{1,\mu_1} K_\perp^{\mu_2}\}^0)_{\alpha\beta} \left\{ \begin{array}{c} \frac{1}{\sqrt{10}}\gamma_5\{\gamma_{\mu_1}^\perp u_{\mu_2}\}^0 \\ u_{\mu_1\mu_2} \end{array} \right\}_\gamma$
Antisymmetric P-wave states			
Σ_{Qk1}	1^-	$\frac{1}{2}^-$, $\frac{3}{2}^-$	$(\chi^0 k_\perp^\mu)_{\alpha\beta} \left\{ \begin{array}{c} \frac{1}{\sqrt{3}}\gamma_\mu^\perp \gamma_5 u \\ u_\mu \end{array} \right\}_\gamma$
Λ_{Qk0}	0^-	$\frac{1}{2}^-$	$\frac{1}{\sqrt{3}}(\chi^{1,\mu} k_{\perp\mu})_{\alpha\beta} u_\gamma$
Λ_{Qk1}	1^-	$\frac{1}{2}^-$, $\frac{3}{2}^-$	$\frac{i}{\sqrt{2}}(\varepsilon_{\mu\nu\rho\delta}\chi^{1,\nu} k_\perp^\rho v^\delta)_{\alpha\beta} \left\{ \begin{array}{c} \frac{1}{\sqrt{3}}\gamma_\mu^\perp \gamma_5 u \\ u_\mu \end{array} \right\}_\gamma$
Λ_{Qk2}	2^-	$\frac{3}{2}^-$, $\frac{5}{2}^-$	$\frac{1}{2}(\{\chi^{1,\mu_1} k_\perp^{\mu_2}\}^0)_{\alpha\beta} \left\{ \begin{array}{c} \frac{1}{\sqrt{10}}\gamma_5\{\gamma_{\mu_1}^\perp u_{\mu_2}\}^0 \\ u_{\mu_1\mu_2} \end{array} \right\}_\gamma$

coupling constants. In fact, there are three s−wave and four d−wave couplings for each. Matrix elements of these transitions are explicitly given in [4].

To go beyond HQS predictions, we invoke the $SU(6) \otimes O(3)$ symmetry of the light degrees of freedom to calculate the light-side transition matrix elements $\mathcal{M}$. In S-wave to S-wave single-pion decays, the transitions involved are $1^+ \to 0^+$ and $1^+ \to 1^+$ with light diquark transition tensors given by

$$\mathcal{M}_{\mu_1;\nu_{j_2}} = \left(\bar{\phi}_{\nu_{j_2}}\right)^{\alpha\beta} (\mathcal{O})_{\alpha\beta}^{\alpha'\beta'} (\phi_{\mu_1})_{\alpha'\beta'}, \tag{7}$$

where the operator $\mathcal{O}$ is given in terms of an overlap integral which is unknown. Constraints on the operator $\mathcal{O}$ come from parity conservation and from the partial wave involved in the emission process. Since $l_\pi = 1$, therefore, it is easy to show that $\mathcal{O}$ must be a pseudoscalar operator involving one power of the pion momentum p. In the constituent quark model the pion is emitted by one of the light quarks, hence, the transition operator $\mathcal{O}$ must be a one-body operator. Possible two-body emission operators are non leading in large-N_C [9] and are thus neglected in the constituent quark model approach [10]. One then has the unique operator

$$(\mathcal{O}(p))_{\alpha'\beta'}^{\alpha\beta} = \frac{1}{2}\left((\gamma_\sigma\gamma_5)_{\alpha'}^{\alpha} \otimes (1)_{\beta'}^{\beta} + (1)_{\alpha'}^{\alpha} \otimes (\gamma_\sigma\gamma_5)_{\beta'}^{\beta}\right) f_p\, p_\perp^\sigma \tag{8}$$

The relevant transition tensors for P-wave to S-wave single-pion decays, which involve $1^- \to \{0^+, 1^+\}$, $0^- \to 0^+$ and $2^- \to \{0^+, 1^+\}$, are given by

$$\mathcal{M}_{\mu_1\cdots\mu_{j_1};\nu_{j_2}} = \sum_{l_\pi=0,2} \left(\bar{\phi}_{\nu_{j_2}}\right)^{\alpha\beta} \left(\mathcal{O}_\lambda^{(l_\pi)}\right)_{\alpha\beta}^{\alpha'\beta'} \left(\phi_{\mu_1\cdots\mu_{j_1}}^\lambda\right)_{\alpha'\beta'}, \tag{9}$$

the appropriate operators for these transitions are given by

$$(\mathcal{O}_\lambda(p))_{\alpha'\beta'}^{\alpha\beta} = \frac{1}{2}\left((\gamma^\sigma\gamma_5)_{\alpha'}^{\alpha} \otimes (1)_{\beta'}^{\beta} \pm (1)_{\alpha'}^{\alpha} \otimes (\gamma^\sigma\gamma_5)_{\beta'}^{\beta}\right)(f_s\, g_{\sigma\lambda} + f_d\, P_{\sigma\lambda}), \tag{10}$$

with, $P_{\sigma\lambda}(p) = p_{\perp\sigma}p_{\perp\lambda} - \frac{1}{3}p_\perp^2 g_{\perp\sigma\lambda}$. The plus sign has to be used for transitions from the Symmetric (K-multiplet) and the minus one for transitions from the Antisymmetric (k-multiplet). P-wave to P-wave transitions were analyzed in [4] and the generalization to transitions involving higher orbital excitations is straightforward.

The matrix elements, Eq. (7) and Eq. (9), of the operators Eq. (8) and Eq. (10), can be readily evaluated using the light diquark spin wave functions in Table(1). The two couplings of the ground state transitions are not independent. They are, actually, related to the single p-wave coupling f_p by

$$f_p^1 = -f_p^2 = f_p. \tag{11}$$

Using PCAC the coupling constant f_p can be related to the axial vector current coupling strength g_A, one obtains $f_p = g_A/f_\pi$.

For P-wave (K-multiplet) to S-wave transitions, the evaluation of the matrix elements leads to the following relations

$$f_s^{1(K)} = f_s \ ; \ f_s^{2(K)} = -\sqrt{3}f_s \ ; \ f_s^{3(K)} = \sqrt{2}f_s \qquad (12)$$

$$f_d^{1(K)} = f_d \ ; \ f_d^{3(K)} = -\frac{1}{\sqrt{2}}f_d \ ; \ f_d^{4(K)} = -f_d \ ; \ f_d^{5(K)} = f_d \qquad (13)$$

The number of independent coupling constants, therefore, has been reduced from seven to the two constituent quark model s-wave and d-wave coupling factors f_s and f_d. Similar relations, with two different couplings, hold for transitions from the P-wave (k-multiplet) to S-wave.

The first important conclusion we have reached so far is that, S-wave to S-wave and P-wave (K-multiplet) to S-wave single pion transitions are given in terms of the three independent couplings f_p, f_s and f_d. For charmed baryons, they can be identified by the three strong couplings $g_{\Sigma_c \Lambda_c \pi}$, $f_{\Lambda_{c1} \Sigma_c \pi}$ and $f_{\Lambda_{c1}^* \Sigma_c \pi}$ respectively. We would like to mention that, after taking into account the different normalizations, the results Eqs. (11) and (12-13) are in agreement with corresponding results using HHCPT [7].

The three independent couplings can be written in terms of Light-Front (LF) [12] matrix elements of the strong transition current $\hat{j}_\pi(q)$ between LF heavy baryon helicity states. Working in the Drell-Yan frame, we get [13]

$$g_{\Sigma_c \Lambda_c \pi} = -\frac{2\sqrt{3 M_{\Lambda_c} M_{\Sigma_c}}}{(M_{\Sigma_c}^2 - M_{\Lambda_c}^2)} \langle \Lambda(P', \uparrow) | \hat{j}_\pi(0) | \Sigma(P, \uparrow) \rangle \qquad (14)$$

$$f_{\Lambda_{c1} \Sigma_c \pi} = \langle \Sigma(P', \uparrow) | \hat{j}_\pi(0) | \Lambda_{c1}(P, \uparrow) \rangle , \qquad (15)$$

and

$$f_{\Lambda_{c1}^* \Sigma_c \pi} = \frac{3\sqrt{2}}{(M_{\Lambda_{c1}^*} - M_\Sigma)^2} \frac{M_{\Lambda_{c1}}^2}{(M_{\Lambda_{c1}^*}^2 - M_{\Sigma_c}^2)} \langle \Sigma(P', \uparrow) | \hat{j}_\pi(0) | \Lambda_{c1}^*(P, \tfrac{1}{2}) \rangle \qquad (16)$$

In the LF formalism the total baryon spin-momentum distribution function can be written in the following general form

$$\Psi(x_i, \mathbf{p}_{\perp i}, \lambda_i; \lambda) = \chi(x_i, \mathbf{p}_{\perp i}, \lambda_i; \lambda) \psi(x_i, \mathbf{p}_{\perp i}). \qquad (17)$$

Here, $\chi(x_i, \mathbf{p}_{\perp i}, \lambda_i; \lambda)$ and $\psi(x_i, \mathbf{p}_{\perp i})$ represent the spin and momentum distribution functions respectively. Assuming factorization of longitudinal and transverse momentum distribution functions, one can write

$$\psi(x_i, \mathbf{p}_{\perp i}) = \prod_{i=1}^{3} \delta(x_i - \bar{x}_i) \exp\left[-\frac{\vec{k}^2}{2\alpha_\rho^2} - \frac{\vec{K}^2}{2\alpha_\lambda^2} \right]. \qquad (18)$$

The longitudinal momentum distribution functions are approximated by Dirac-delta functions which are peaked at the constituent quark longitudinal momenta mean values $\bar{x}_i = \frac{m_i}{M}$. This assumption is justified since in the weak binding [14] and the valence [15] approximations, the constituent quarks are moving with the same velocity inside the baryon. The heavy baryons spin wave functions, which are the LF generalization of the conventional constituent quark model spin-isospin functions, are explicitly given by [13]

$$\chi^{\Lambda_Q}(x_i, \mathbf{p}_{\perp i}, \lambda_i; \lambda) = \bar{u}(p_1, \lambda_1)[(\not{P} + M_\Lambda)\gamma_5]v(p_2, \lambda_2)\bar{u}(p_3, \lambda_3)u(P, \lambda). \qquad (19)$$

For the Σ_Q-like baryons, one has

$$\chi^{\Sigma_Q}(x_i, \mathbf{p}_{\perp i}, \lambda_i; \lambda) = \bar{u}(p_1, \lambda_1)[(\not{P} + M_\Lambda)\gamma_\perp^\mu]v(p_2, \lambda_2)\bar{u}(p_3, \lambda_3)\gamma_{\perp\mu}\gamma_5 u(P, \lambda), \qquad (20)$$

The excited states Λ_{Q1}, with $J^P = \frac{1}{2}^-$, and Λ_{Q1}^*, with $J^P = \frac{3}{2}^-$, have spin functions of the forms

$$\chi^{\Lambda_{QK1}}(x_i, \mathbf{p}_{\perp i}, \lambda_i; \lambda) = \bar{u}(p_1, \lambda_1)[(\not{P} + M_{\Lambda_{c1}})\gamma_5]v(p_2, \lambda_2)\bar{u}(p_3, \lambda_3)\not{K}\gamma_5 u(P, \lambda), \qquad (21)$$

and

$$\chi^{\Lambda^*_{QK1}}(x_i, \mathbf{p}_{\perp i}, \lambda_i; \lambda) = \bar{u}(p_1, \lambda_1)[(\not{P} + M_{\Lambda^*_{c1}})\gamma_5]v(p_2, \lambda_2)\bar{u}(p_3, \lambda_3)K_\mu u^\mu(P, \lambda) \ . \qquad (22)$$

The three charmed baryons strong couplings $g_{\Sigma_c \Lambda_c \pi}$, $f_{\Lambda_{c1}\Sigma_c\pi}$ and $f_{\Lambda^*_{c1}\Sigma_c\pi}$ are calculated[1] to be

$$g_{\Sigma_c\Lambda_c\pi} = 6.81 \text{ GeV}^{-1} \ , \ f_{\Lambda_{c1}\Sigma_c\pi} = 1.16 \ , \ f_{\Lambda^*_{c1}\Sigma_c\pi} = 0.96 \times 10^{-4} \text{ MeV}^{-2} \ . \qquad (23)$$

These values can be used to determine the corresponding HHCPT couplings, one gets

$$g_2 = 0.52 \ , \ h_2 = 0.54 \ , \ h_8 = 3.33 \times 10^{-3} \text{MeV}^{-1} \ . \qquad (24)$$

Assuming that the width of Σ_c, Λ_{c1} and Λ^*_{c1} are saturated by strong decay channels one can estimate the values of the three couplings using the experimental decay rates. CLEO [16] results for $\Gamma_{\Sigma_c^{*++} \to \Lambda_c^+ \pi^+} = 17.9^{+3.8}_{-3.2}$ MeV and $\Gamma_{\Sigma_c^{*0} \to \Lambda_c^+ \pi^-} = 13.0^{+3.7}_{-3.0}$ MeV can be used to determine the coupling $g_{\Sigma_c\Lambda_c\pi}$. One, therefore, respectively gets

$$g_{\Sigma_c\Lambda_c\pi} = 8.03^{+1.97}_{-1.92} \text{ GeV}^{-1} \qquad (25)$$

and

---

[1] The numerical values for the constituent quark masses and the oscillator couplings are taken to be $m_u = m_d = 0.33$ GeV, $m_c = 1.51$ GeV, $\alpha_\rho = 0.40$ GeV/c and $\alpha_\lambda = 0.52$ GeV/c. The charmed baryon masses will be taken from Table 1 of [8].

$$g_{\Sigma_c \Lambda_c \pi} = 6.97^{+1.84}_{-1.74} \text{ GeV}^{-1} \tag{26}$$

To estimate $f_{\Lambda_{c1} \Sigma \pi}$ we use the Particle Data Group [17] average value for $\Lambda_{c1}(2593)$ width ($\Gamma_{\Lambda_{c1}(2593)} = 3.6^{+2.0}_{-1.3}$ MeV) to obtain

$$f_{\Lambda_{c1} \Sigma \pi} = 1.11^{+0.31}_{-0.20}. \tag{27}$$

Finally, taking the upper bound on the $\Lambda_{c1}^+(2625)$ width obtained by CLEO [16] ($\Gamma_{\Lambda_{c1}^+(2625)} < 1.9$ MeV) one gets

$$f_{\Lambda_{c1}^* \Sigma \pi} = 1.66 \times 10^{-4} \text{ MeV}^{-2}. \tag{28}$$

The LF quark model predictions for the numerical values of the single-pion couplings Eq. (23) are in good agreement with estimates obtained using the available experimental data Eqs. (25-28).

We are now in a position to predict charmed baryons strong decay rates using the general formula

$$\Gamma = \frac{1}{2J_1 + 1} \frac{|\vec{q}|}{8\pi M_{B_Q}^2} \sum_{spins} |M^\pi|^2, \tag{29}$$

with $|\vec{q}|$ being the pion momentum in the rest frame of the decaying baryon. The numerical values for S-wave to S-wave and P-wave (K-multiplet) to S-wave single pion decay rates and the updated experimental values of the Review of Particle Physics [17] are summarized in Table 2. To predict the total decay width of these states, one has to include decay rates for the two-pion transitions reported by [7]. Table 2 shows that most of the predicted decay widths agree quite well or they are within the range of the corresponding experimental data. We, also, notice that the $\Sigma_{c0}(2760)$, $\Sigma_{c1}(2770)$ and $\Sigma_{c2}(2800)$ widths are relatively broad and it might be difficult to measure them experimentally.

To summarize, we have used the $SU(2N_f) \times O(3)$ symmetry, of the light diquark system, to reduce the number of HQS coupling factors of heavy baryon single-pion decays. These result, which are obtained using covariant spin wave functions for the light diquark system, agree with the HHCPT [7]. We also calculated the three independent couplings $g_{\Sigma_c \Lambda_c \pi}$, $f_{\Lambda_{c1} \Sigma_c \pi}$ and $f_{\Lambda_{c1}^* \Sigma_c \pi}$ using a Light-Front (LF) quark model functions. Most of the predicted decay rates agree with the available experimental data. Like other models, our numerical result will depend on the values of the the constituent quark masses and the harmonic oscillator constants α_ρ and α_λ which are free parameters.

ACKNOWLEDGMENTS

One of us (S.T.) would like to thank Patrick J. O'Donnell and the Department of Physics, University of Toronto for hospitality. This research was supported in part by the National Sciences and Engineering Research Council of Canada.

TABLE 2. Decay rates for charmed baryon states.

$B_Q \to B'_Q \pi$	Γ (MeV)	$\Gamma_{expt.}$ (MeV)
Ground state transitions		
$\Sigma_c^+ \to \Lambda_c \pi^0$	1.70	
$\Sigma_c^0 \to \Lambda_c \pi^-$	1.57	
$\Sigma_c^{++} \to \Lambda_c \pi^+$	1.64	
$\Sigma_c^{*0} \to \Lambda_c \pi^-$	12.40	$13.0^{+3.7}_{-3.0}$
$\Sigma_c^{*++} \to \Lambda_c \pi^+$	12.84	$17.9^{+3.8}_{-3.2}$
$\Xi_c^{*0} \to \Xi_c^0 \pi^0$	0.72	< 5.5
$\Xi_c^{*0} \to \Xi_c^+ \pi^-$	1.16	
$\Xi_c^{*+} \to \Xi_c^0 \pi^+$	1.12	< 3.1
$\Xi_c^{*+} \to \Xi_c^+ \pi^0$	0.69	
P-wave to S-wave transitions		
$\Lambda_{c1}(2593) \to \Sigma_c^0 \pi^+$	2.61	
$\Lambda_{c1}(2593) \to \Sigma_c^+ \pi^0$	1.73	$3.6^{+2.0}_{-1.3}$
$\Lambda_{c1}(2593) \to \Sigma_c^{++} \pi^-$	2.15	
$\Lambda_{c1}^*(2625) \to \Sigma_c^0 \pi^+$	0.77	
$\Lambda_{c1}^*(2625) \to \Sigma_c^+ \pi^0$	0.69	$\Gamma_{\Lambda_{c1}^*} < 1.9$
$\Lambda_{c1}^*(2625) \to \Sigma_c^{++} \pi^-$	0.73	
$\Xi_{c1}^*(2815) \to \Xi_c^{*0} \pi^+$	4.84	$\Gamma_{\Xi_{c1}^*} < 2.4$
$\Xi_{c1}^*(2815) \to \Xi_c^{*+} \pi^0$	2.38	
$\Xi_{c1}^*(2815) \to \Xi_c^0 \pi^+$	0.30	
$\Xi_{c1}^*(2815) \to \Xi_c^+ \pi^0$	0.15	
$\Sigma_{c0}(2760) \to \Lambda_c \pi$	110.36	
$\Sigma_{c1}(2770) \to \Sigma_c \pi$	50.92	
$\Sigma_{c2}(2800) \to \Sigma_c^* \pi$	50.21	

REFERENCES

1. F. Hussain, G. Thompson and J.G. Körner, preprint IC/93/314, MZ-TH/93-23 and hep-ph/9311309, to appear in the proceedings of the 6th Regional Conference in Mathematical Physics, Islamabad, February 1994.
2. J.G. Körner, M. Krämer and D. Pirjol, Progr. Part. Nucl. Phys. **33**, 787 (1994).
3. F. Hussain, J.G. Körner, J. Landgraf and Salam Tawfiq, Z. Phys. **C69**, 655 (1996).
4. F. Hussain, J.G. Körner and Salam Tawfiq, ICTP preprint, IC/96/35 and Mainz preprint MZ-TH/96-10, 1996.
5. P. Cho, Nucl Phys. **B396**, 183 (1993); Phys. Rev. **D50**, 3295 (1994).
6. M.-Q. Huang, Y.-B. Dai and C.-S. Huang, Phys. Rev. **D52**, 3986 (1995).
7. D. Pirjol and T.-M. Yan, Phys. Rev. **D56**, 5483 (1997).
8. G. Chiladze and A. Falk, Phys. Rev. **D56**, 6738 (1997).
9. E. Witten, Nucl. Phys. **B223**, 483 (1983); C. Carone, H. Georgi and S. Osofski, Phys. Lett. **B322** 493, (1994); M. Luty and J. March-Russel, Nucl. Phys. **B246**, 71 (1994); R.F. Dashen, E. Jenkins and A.V. Manohar, Phys. Rev. **D49**, 4713 (1994); R.F. Dashen, E. Jenkins and A.V. Manohar, Phys. Rev. **D51**, 3697 (1995).
10. C. Carone, H. Georgi, L. Kaplan and D. Morin, Phys. Rev. **D50**, 5793 (1994).
11. T.-M. Yan et. al., Phys. Rev. **D46**, 1148 (1992); Erratum, to be published.
12. For a recent review and references see S.J. Brodsky, H.-S. Pauli and S.S. Pinsky, hep-ph/9705477, 1997.
13. Salam Tawfiq, Patrick J. O'Donnell and J.G. Körner, hep-ph/9803246, University of Toronto preprint UTPT-98-03 and Mainz preprint MZ-TH/98-08, 1998, to appear in Phys. Rev. D.
14. F. Hussain, J.G. Körner and G. Thompson, Ann. Phys. **C59**, 334 (1993).
15. Z. Dziembowski, Phys. Rev. **D37**, 778 (1988); H.J. Weber, Phys. Lett. **B209**, 425 (1988); Z. Dziembowski and H.J. Weber, Phys. Rev. **D37**, 1289 (1988); W. Konen and H.J. Weber, Phys. Rev. **D41**, 2201 (1990).
16. G. Brandenburg et. al., CLEO Collaboration, Phys. Rev. Lett. **78**, 2304 (1997); L. Gibbons et. al., CLEO Collaboration, Phys. Rev. Lett. **77**, 810 (1996); P.L. Frabetti et. al., E687 Collaboration, Phys. Lett. **B365**, 461 (1996); Phys. Rev. Lett. **72**, 961 (1994); K.W. Edwards et.al., CLEO Collaboration, Phys. Rev. Lett. **74**, 3331 (1995); P. Avery, et. al., CLEO Collaboration, Phys. Rev. Lett. **75**, 4364 (1995); H. Albrecht et. al., ARGUS Collaboration, Phys. Lett. **B317**, 227 (1993).
17. R.M. Barnett et. al., Phys. Rev. **D54**, 1 (1996) and 1997 off-year partial update for the 1998 edition available on the PDG WWW pages (URL: http://pdg.lbl.gov).

Discrete Ambiguities in CP-violating Asymmetries in B Decays[1]

David London

Laboratoire de René J.-A. Lévesque, Université de Montréal
C.P. 6128, succ. centre-ville, Montréal, QC, Canada

Abstract. The CP-angles α, β and γ can be extracted from CP-violating asymmetries in the B system, but only up to discrete ambiguities. These discrete ambiguities make it difficult to determine with certainty whether or not new physics is present. I show that, if the condition $\alpha + \beta + \gamma = \pi$ is imposed, there remains a twofold ambiguity in the CP-angle set (α, β, γ), and I discuss ways to cleanly resolve this final discrete ambiguity.

Within the Standard Model (SM), CP violation is due to a complex phase in the Cabibbo-Kobayashi-Maskawa (CKM) mixing matrix. In the Wolfenstein parametrization of the CKM matrix [1], only the elements V_{ub} and V_{td} have non-negligible phases:

$$V_{CKM} = \begin{pmatrix} 1 - \frac{1}{2}\lambda^2 & \lambda & A\lambda^3(\rho - i\eta) \\ -\lambda & 1 - \frac{1}{2}\lambda^2 & A\lambda^2 \\ A\lambda^3(1 - \rho - i\eta) & -A\lambda^2 & 1 \end{pmatrix}. \quad (1)$$

It is convenient to parametrize V_{ub} and V_{td} as follows:

$$V_{ub} = |V_{ub}|e^{-i\gamma} \ , \quad V_{td} = |V_{td}|e^{-i\beta} \ . \quad (2)$$

Even though these elements are written in terms of two complex phases β and γ, it must be remembered that in fact there is only a single phase η in the CKM matrix; if η were to vanish, both β and γ would vanish as well.

The phase information in the CKM matrix can be displayed using the unitarity triangle [2]. The orthogonality of the first and third columns gives

$$V_{ud}V_{ub}^* + V_{cd}V_{cb}^* + V_{td}V_{tb}^* = 0 \ , \quad (3)$$

which is a triangle relation in the complex ρ-η plane, shown in Fig. 1. The angles β and γ are two of the interior angles of the unitarity triangle, with the third angle α satisfying $\alpha + \beta + \gamma = \pi$.

---

[1] Talk based on work done in collaboration with Boris Kayser.

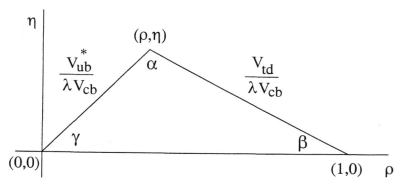

FIGURE 1. The unitarity triangle. The angles α, β and γ can be measured via CP violation in the B system.

There are constraints on the unitarity triangle due to (i) the extraction of $|V_{cb}|$ and $|V_{ub}|$ from semileptonic B decays, (ii) the measurements of $|V_{td}|$ and $|V_{ts}|$ in B mixing, and (iii) CP violation in the kaon system (ϵ). Unfortunately, there are substantial theoretical uncertainties in all of the constraints. For example, $|V_{ub}/V_{cb}|$ is measured to be 0.08, but with a 20% uncertainty, due principally to theoretical model dependence. In addition, the theoretical expressions for ϵ and B_d^0-$\overline{B_d^0}$ mixing depend respectively on the bag parameter $B_K = 0.90 \pm 0.15$ and $f_{B_d}\sqrt{B_{B_d}} = 200 \pm 40$ MeV. Combining the experimental errors and theoretical uncertainties in quadrature [3], the constraints on the unitarity triangle are shown in Fig. 2 (note: this figure does not include the increasingly-stringent constraints from $\Delta M_s/\Delta M_d$). It is clear that we really know rather little about the unitarity triangle at present, due to the theoretical uncertainties.

The reason that there has been so much interest in measurements of CP violation in the B system is that the angles α, β and γ can be extracted from CP-violating asymmetries in B decays with *no* theoretical uncertainties [4]. Due to B^0-$\overline{B^0}$ mixing, for any state f to which both B^0 and $\overline{B^0}$ can decay, there are (at least) two amplitudes which can interfere: $B^0 \to f$ and $B^0 \to \overline{B^0} \to f$. This gives rise to the possibility of CP violation. If the rates for $B_d^0(t) \to f$ and $\overline{B_d^0}(t) \to f$ are not equal, this is a signal for CP violation.

For certain final states f, one can cleanly extract the CP angles. In particular, $\sin 2\alpha$ and $\sin 2\beta$ can be obtained from time-dependent measurements of the decays $B_d^0(t) \to \pi^+\pi^-$ and $B_d^0(t) \to \Psi K_s$, respectively, and $\sin^2\gamma$ can be measured in $B^\pm \to DK^\pm$ [5] or $B_s^0(t) \to D_s^\pm K^\mp$ [6]. The idea is therefore to cleanly measure α, β and γ in CP-violating asymmetries in B decays, thus overconstraining the CKM matrix. If we are lucky, this will reveal the presence of new physics.

When talking about new physics, there are immediately two questions which

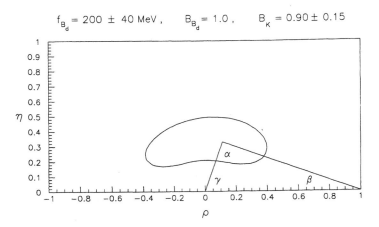

FIGURE 2. Allowed region (95% C.L.) in the ρ-η plane, from a simultaneous fit to all experimental and theoretical data. The theoretical errors are treated as Gaussian for this fit. The triangle shows the best fit.

spring to mind. First, how will we know if there is physics beyond the SM? There are several ways [7]. For example, if we find that $\alpha+\beta+\gamma \neq \pi$, it is clear that there must be new physics. And even if $\alpha+\beta+\gamma = \pi$, new physics will be revealed if the triangle so constructed is in disagreement with independent measurements of the sides. Also, lattice calculations show that $B_K > 0$ [8], meaning that the unitarity triangle points up in the SM. Thus, a measurement of $\sin 2\beta < 0$ would imply the presence of new physics.

The second question is: how can physics beyond the SM enter the CP-violating asymmetries? There are two possible ways: the new physics can affect B decays or B mixing. Most B decays are dominated by a W-mediated tree-level diagram. In most models of new physics, there are no contributions to B decays which can compete with the SM. Thus, in general, the new physics cannot significantly affect the decays[2]. However, the CP asymmetries *can* be affected if there are new contributions to B^0-$\overline{B^0}$ mixing [10]. The presence of such new-physics contributions will affect the extraction of V_{td} and V_{ts}. And if there are new phases, the measurements of α, β and γ will also be affected. Thus, new physics enters principally through new contributions to B^0-$\overline{B^0}$ mixing.

So far, so good. However, there are two problems with the plan. First, B-factories such as BaBar and Belle will measure α, β and γ via $B_d^0(t) \to \pi^+\pi^-$, $B_d^0(t) \to \Psi K_S$, and $B^\pm \to DK^\pm$, respectively. Note that only the first two decays involve B^0-$\overline{B^0}$ mixing. Thus, if there is new physics, only the measurements of α and β will be affected. However, they will be affected in opposite directions [11].

---

[2]) There is an exception: if the decay process is dominated by a penguin diagram, rather than a tree-level diagram, then new physics *can* significantly affect the decay, see Ref. [9].

That is, in the presence of a new-physics phase ϕ_{NP}, the CP angles are changed as follows: $\alpha \to \alpha + \phi_{NP}$ and $\beta \to \beta - \phi_{NP}$. The key point is that ϕ_{NP} cancels in the sum $\alpha + \beta + \gamma$, so that this sum is *insensitive* to the new physics, i.e. B-factories will always find $\alpha + \beta + \gamma = \pi$. (Note that hadron colliders do not suffer from the same problem – if γ is measured in $B_s^0(t) \to D_s^\pm K^\mp$, then $\alpha + \beta + \gamma \neq \pi$ can be found if there is new physics in B_s^0-$\overline{B_s^0}$ mixing.)

Thus, B-factories cannot discover new physics via $\alpha + \beta + \gamma \neq \pi$. Still, new physics can be found if the measurements of the angles are inconsistent with the measurements of the sides.

However, this brings us to the second problem. The CP angles α, β and γ are not measured directly – in fact it is the functions $\sin 2\alpha$, $\sin 2\beta$ and $\sin^2 \gamma$ which are obtained. But $\sin 2\alpha$ determines the angle α only up to a 4-fold ambiguity, and similarly for $\sin 2\beta$ and $\sin^2 \gamma$. Thus, there are in fact 64 possibilities for the CP-angle set (α, β, γ). Of these, some sets may be consistent with the SM, while others may not. In other words, due to the discrete ambiguities in the extraction of the angles α, β and γ, it is difficult to know whether or not new physics really is present [12].

However, there is a possible resolution of this problem. From the discussion above, we know that B-factories *must* find $\alpha + \beta + \gamma = \pi$, even if there is new physics. So we can impose this condition on the angles. Most of the 64 CP-angle sets will not satisfy this requirement. The question is: does this define the set (α, β, γ) uniquely?

Unfortunately, the answer is no. Here I present a partial proof (for the full proof, see Ref. [13]). Assume that α, β, γ are interior angles of the unitarity triangle, and take the allowed values to range between $-\pi$ and π. Thus, either

$$0 < \alpha, \beta, \gamma < \pi \quad \text{and} \quad \alpha + \beta + \gamma = \pi \,, \tag{4}$$

or

$$-\pi < \alpha, \beta, \gamma < 0 \quad \text{and} \quad \alpha + \beta + \gamma = -\pi \,. \tag{5}$$

Eqs. 4 and 5 correspond to the unitarity triangle pointing up and down, respectively. Note that a triangle pointing down is inconsistent with the assumption that the kaon bag parameter B_K is positive [8]. If it does turn out that measurements yield a unitarity triangle which points downward, this implies either that new physics is present, or that the assumption regarding the sign of the bag parameter is wrong. The point of measuring CP asymmetries in the B system is that the SM explanation for CP violation can be tested without theoretical input. In this spirit, I make no apriori assumption about the sign of B_K.

The quantities $\sin 2\alpha$ and $\sin 2\beta$ will be measured, and they can each be either positive or negative. Consider the case where both $\sin 2\alpha$ and $\sin 2\beta$ are positive. Then 2α and 2β both take values in the domain $(-2\pi, -\pi)$ or $(0, \pi)$, giving 16 possibilities for the pair (α, β). However, the number of possibilities can be reduced. As noted above, α and β must be of like sign, so both are in the domain $(-2\pi, -\pi)$, or both are in $(0, \pi)$. But the $(-2\pi, -\pi)$ domain can be immediately excluded: since $|2\alpha| > \pi$ and $|2\beta| > \pi$, this implies that $|\alpha| + |\beta| > \pi$, in violation of Eqs. 4 and 5. This leaves 4 possibilities for (α, β), which can be written as follows:

$$\alpha = \frac{\pi}{4} + \delta_\alpha ,$$
$$\beta = \frac{\pi}{4} + \delta_\beta ,$$
$$\gamma = \frac{\pi}{2} - (\delta_\alpha + \delta_\beta) . \tag{6}$$

The quantities $\delta_{\alpha,\beta}$ can each be positive or negative, but must satisfy $|\delta_{\alpha,\beta}| < \pi/4$.

Now consider $\sin^2 \gamma = \sin^2\left[\frac{\pi}{2} - (\delta_\alpha + \delta_\beta)\right] = \cos^2(\delta_\alpha + \delta_\beta)$. The measurement of this quantity gives us the relative sign of δ_α and δ_β, but not the overall sign. Thus, a twofold discrete ambiguity remains, corresponding to $\delta_{\alpha,\beta} \leftrightarrow -\delta_{\alpha,\beta}$, or, equivalently,

$$(\alpha, \beta, \gamma) \leftrightarrow \left(\frac{\pi}{2} - \alpha, \frac{\pi}{2} - \beta, \pi - \gamma\right). \tag{7}$$

In this case, this is a discrete ambiguity between two upward-pointing unitarity triangles.

For other choices of the signs of $\sin 2\alpha$ and $\sin 2\beta$, one is still left with a twofold discrete ambiguity, albeit a different one than in Eq. 7. I won't present the proofs for these other cases, but rather will just give the results.

When $\sin 2\alpha$ and $\sin 2\beta$ are both negative, the discrete ambiguity is between two downward-pointing unitarity triangles:

$$(\alpha, \beta, \gamma) \to \left(-\frac{\pi}{2} - \alpha, -\frac{\pi}{2} - \beta, -\pi - \gamma\right). \tag{8}$$

Of course, as noted above, the assumption that $B_K > 0$ requires the unitarity triangle to point up. Thus, in this case, one has to decide whether new physics is present, or whether the assumption is faulty.

For the case of $\sin 2\alpha > 0$ and $\sin 2\beta < 0$, there are again two possible solutions for (α, β, γ), one with positive values for the angles, the other with negative values. Denoting (α, β, γ) as the positive-angle solution, the discrete ambiguity is

$$(\alpha, \beta, \gamma) \to \left(-\frac{\pi}{2} - \alpha, \frac{\pi}{2} - \beta, -\gamma\right). \tag{9}$$

In this case, the first solution corresponds to a unitarity triangle pointing up, while the second corresponds to a downward-pointing triangle.

Finally, when $\sin 2\alpha < 0$ and $\sin 2\beta > 0$, the roles of α and β are reversed compared to the previous paragraph, and we have the following discrete ambiguity:

$$(\alpha, \beta, \gamma) \to \left(\frac{\pi}{2} - \alpha, -\frac{\pi}{2} - \beta, -\gamma\right). \tag{10}$$

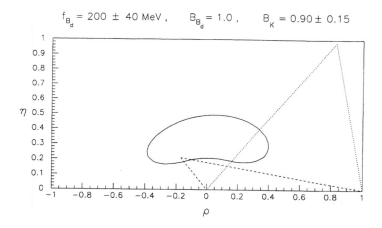

FIGURE 3. An example of the twofold discrete ambiguity. The true solution $[(\alpha,\beta,\gamma) = (40°, 10°, 130°)$, dashed line] is in agreement with the SM constraints, but the second solution $[(\alpha,\beta,\gamma) = (50°, 80°, 50°)$, dotted line] contradicts the SM. Is there new physics or not?

There is one complication which I haven't mentioned yet. The decay $B_d^0(t) \to \pi^+\pi^-$ may receive important contributions from penguin diagrams [14]. In the presence of penguins, the extraction of $\sin 2\alpha$ is no longer clean – there are hadronic uncertainties. Fortunately, this penguin "pollution" can be removed by the use of an isospin analysis [15]. But the net result is that $\sin 2\alpha$ is itself obtained with a fourfold discrete ambiguity, which depends on the relative magnitude and phase of the penguin and tree amplitudes. However, this ambiguity can be ignored (in principle), since in general only one of the four values of $\sin 2\alpha$ yields values of α which can satisfy $\alpha+\beta+\gamma = \pi$. Furthermore, $\sin 2\alpha$ can be extracted independently with no discrete ambiguity from a study of $B \to \rho\pi$ decays [16], as I will discuss below.

We have therefore seen that, even when one imposes the condition $\alpha+\beta+\gamma = \pi$, there is a twofold discrete ambiguity in the CP-angle set (α,β,γ). The presence of this discrete ambiguity causes problems – how can we be sure whether or not there is new physics? For example, suppose that the true values of (α,β,γ) are $(40°, 10°, 130°)$. From Eq. 7, there is another solution, also satisfying $\alpha+\beta+\gamma = \pi$: $(\alpha,\beta,\gamma) = (50°, 80°, 50°)$. These two solutions are shown in Fig. 3. In this case the spurious solution lies well outside the allowed region of the SM. If this were the true solution, it would be an obvious signal of new physics. Clearly if we want to confirm or invalidate the SM, it will be necessary to remove the discrete ambiguity.

In fact, there are a variety of ways to do this. There are other methods of measuring the CP angles which probe different trigonometric functions of these angles, and which can be used to remove the twofold discrete ambiguity.

One example is the measurement of α from an examination of the time-dependent

Dalitz plots of $B_d^0(t) \to \pi^+\pi^-\pi^0$ [16]. Ignoring nonresonant contributions, this final state can be fed by the intermediate states $\rho^+\pi^-$, $\rho^-\pi^+$ and $\rho^0\pi^0$, all of which interfere with one another. From a study of the Dalitz plot, one can extract $\sin 2\alpha$, as in $B_d^0(t) \to \pi^+\pi^-$. Even if there is penguin pollution, $\sin 2\alpha$ can be cleanly obtained with no discrete ambiguities. Furthermore — and more importantly for our purposes — there is enough information to extract in addition $\cos 2\alpha$. And the knowledge of $\cos 2\alpha$ removes the discrete ambiguity between α and $\pi/2 - \alpha$ (Eqs. 7 and 10) or between α and $-\pi/2 - \alpha$ (Eqs. 8 and 9).

There are other measurements which can remove the discrete ambiguity. The function $\cos 2\beta$ can be extracted in two different ways. First, it can be obtained from a study of the decay $B_d^0(t) \to \Psi + K \to \Psi + (\pi^- \ell^+ \nu)$. This decay, in which K mixing follows B mixing, is known as cascade mixing [17]. Second, $\cos 2\beta$ can also be measured by studying the time-dependent Dalitz plot for decays such as $B_d^0(t) \to D^+D^-K_s$ [18]. It is also possible to obtain $\sin 2\gamma$: if the two B_s mass eigenstates have a significant width difference [19], then the time-dependent study of the decay $B_s^0(t) \to D_s^\pm K^\mp$ allows one to extract $\sin 2\gamma$. Finally, the study of the Dalitz plot of $B_s^0(t) \to D^\pm \pi^\mp K_s$ allows one to measure $\sin 2(2\beta + \gamma)$ [18]. The knowledge of any of these functions — $\cos 2\alpha$, $\cos 2\beta$, $\sin 2\gamma$, or $\sin 2(2\beta + \gamma)$ — is sufficient to resolve the twofold discrete ambiguity.

Unfortunately, all of the above measurements are hard. Still, it will be necessary to perform such measurements in order to be sure that we have found new physics (or not). Such are the difficulties due to discrete ambiguities in the CP-violating asymmetries in the B system.

Acknowledgments

I would like to thank the organizers of MRST '98 for a very enjoyable conference. This research was financially supported by NSERC of Canada and FCAR du Québec.

REFERENCES

1. L. Wolfenstein, *Phys. Rev. Lett.* **51**, 1945 (1983).
2. R.M. Barnett et al., the Particle Data Group, *Phys. Rev.* **D54**, 1 (1996).
3. For discussions of the methodology, as well as previous updates of the constraints on the CKM matrix, see A. Ali and D. London, *Zeit. Phys.* **C65**, 431 (1995), *Nuovo Cim.* **109A**, 957 (1996), *Nucl. Phys. (Proc. Suppl.)* **54A**, 297 (1997).
4. For reviews, see, for example, Y. Nir and H.R. Quinn in *B Decays*, edited by S. Stone (World Scientific, Singapore, 1994), p. 520; I. Dunietz, *ibid.*, p. 550; M. Gronau, *Proceedings of Neutrino 94, XVI International Conference on Neutrino Physics and Astrophysics*, Eilat, Israel, May 29 - June 3, 1994, eds. A. Dar, G. Eilam and M. Gronau, *Nucl. Phys. (Proc. Suppl.)* **B38**, 136 (1995).
5. M. Gronau and D. Wyler, *Phys. Lett.* **265B**, 172 (1991). See also M. Gronau and D. London, *Phys. Lett.* **253B**, 483 (1991); I. Dunietz, *Phys. Lett.* **270B**, 75 (1991).

Improvements to this method have recently been discussed by D. Atwood, I. Dunietz and A. Soni, *Phys. Rev. Lett.* **78**, 3257 (1997).

6. R. Aleksan, I. Dunietz, B. Kayser and F. Le Diberder, *Nucl. Phys.* **B361**, 141 (1991); R. Aleksan, I. Dunietz and B. Kayser, *Zeit. Phys.* **C54**, 653 (1992).
7. For a review of new-physics effects in CP asymmetries in the B system, see M. Gronau and D. London, *Phys. Rev.* **D55**, 2845 (1997), and references therein.
8. Y. Nir and H.R. Quinn, *Phys. Rev.* **D42**, 1473 (1990); Y. Grossman, B. Kayser and Y. Nir, *Phys. Lett.* **415B**, 90 (1997).
9. For example, see Y. Nir and H.R. Quinn, Ref. [4]; Y. Grossman and M.P. Worah, *Phys. Lett.* **395B**, 241 (1997); D. London and A. Soni, *Phys. Lett.* **407B**, 61 (1997).
10. C.O. Dib, D. London and Y. Nir, *Int. J. Mod. Phys.* **A6**, 1253 (1991).
11. Y. Nir and D. Silverman, *Nucl. Phys.* **B345**, 301 (1990).
12. Other discussions of discrete ambiguities can be found in Y. Nir and H.R. Quinn, Ref. [8]; T. Goto, N. Kitazawa, Y. Okada and M. Tanaka, *Phys. Rev.* **D53**, 6662 (1996); L. Wolfenstein, presented at the second international conference on B physics and CP violation, Honolulu, Hawaii, 1997; Y. Grossman, Y. Nir and M.P. Worah, *Phys. Lett.* **407B**, 307 (1997); Y. Grossman and H.R. Quinn, *Phys. Rev.* **D56**, 7259 (1997).
13. B. Kayser and D. London, UdeM-GPP-TH-98-48, to be submitted to *Phys. Rev. D*.
14. D. London and R. Peccei, *Phys. Lett.* **223B**, 257 (1989); M. Gronau, *Phys. Rev. Lett.* **63**, 1451 (1989), *Phys. Lett.* **300B**, 163 (1993); B. Grinstein, *Phys. Lett.* **229B**, 280 (1989).
15. M. Gronau and D. London, *Phys. Rev. Lett.* **65**, 3381 (1990).
16. A.E. Snyder and H.R. Quinn, *Phys. Rev.* **D48**, 2139 (1993).
17. B. Kayser and L. Stodolsky, Max Planck Institute preprint MPI-PHT-96-112, hep-ph/9610522; B. Kayser, NSF preprint NSF-PT-97-2, hep-ph/9709382.
18. J. Charles, A. Le Yaouanc, L. Oliver, O. Pène and J.-C. Raynal, *Phys. Lett.* **425B**, 375 (1998).
19. See, for example, I. Dunietz, *Phys. Rev.* **D52**, 3048 (1995), and references therein.

Large CP Violation in $B \to K^{(*)}X$ Decays

T. E. Browder[1], A. Datta[2], X.-G. He[3] and S. Pakvasa[1]

[1] *Department of Physics and Astronomy University of Hawaii, Honolulu, Hawaii 96822*
[2] *Department of Physics and Astronomy University of Toronto, Toronto, Ontario M5S 1A7, Canada*
[3] *School of Physics University of Melbourne, Parkville, Victoria, 3052 Australia.*

Abstract. We consider the possibility of observing CP violation in quasi-inclusive decays of the type $B^- \to K^- X$, $B^- \to K^{*-} X$, $\bar{B}^0 \to K^- X$ and $\bar{B}^0 \to K^{*-} X$, where X does not contain strange quarks. We present estimates of rates and asymmetries for these decays in the Standard Model and comment on the experimental feasibility of observing CP violation in these decays at future B factories. We find the rate asymmetries can be quite sizeable.

I INTRODUCTION

The possibility of observing large CP violating asymmetries in the decay of B mesons motivates the construction of high luminosity B factories at several of the world's high energy physics laboratories. The theoretical and the experimental signatures of these asymmetries have been extensively discussed elsewhere [1–5]. At asymmetric B factories, it is possible to measure the time dependence of B decays and therefore time dependent rate asymmetries of neutral B decays due to $B - \bar{B}$ mixing. The measurement of time dependent asymmetries in the exclusive modes $\bar{B}^0 \to \psi K_s$ and $\bar{B}^0 \to \pi^+\pi^-$ will allow the determination of the angles in the Cabbibo-Kobayashi-Maskawa unitarity triangle. This type of CP violation has been studied extensively in the literature.

Another type of CP violation also exists in B decays, direct CP violation in the B decay amplitudes. This type of CP violation in B decays has also been discussed by several authors although not as extensively. For charged B decays calculation of the magnitudes of the effects for some exclusive modes and inclusive modes have been carried out [6–12]. In contrast to asymmetries induced by $B - \bar{B}$ mixing, the magnitudes have large hadronic uncertainties, especially for the exclusive modes. Observation of these asymmetries can be used to rule out the superweak class of models [13].

In this paper we describe several quasi-inclusive experimental signatures which could provide useful information on direct CP violation at the high luminosity facilities of the future. One of the goals is to increase the number of events available at experiments for observing a CP asymmetry. In particular we examine the inclusive decay of the neutral and the charged B to either a charged K or a charged K^* meson. By applying the appropriate cut on the kaon (or K^*) energy one can isolate a signal with little background from $b \to c$ transitions. Furthermore, these quasi-inclusive modes are expected to have less hadronic uncertainty than the exclusive modes, would have larger branching ratios and, compared to the purely inclusive modes they may have larger CP asymmetries. In this paper we will consider modes of the type $B \to K(K^*)X$ that have the strange quark only in the $K(K^*)$-meson.

In the sections which follow, we describe the experimental signature and method. We then calculate the rates and asymmetries for inclusive $B^- \to K^-(K^{*-})$ and $\bar{B}^0 \to K^-(K^{*-})$ decays.

II EXPERIMENTAL SIGNATURES FOR QUASI-INCLUSIVE $B \to SG^*$

In the $\Upsilon(4S)$ center of mass frame, the momentum of the $K^{(*)-}$ from quasi-two body B decays such as $B \to K^{(*)-}X$ may have momenta above the kinematic limit for $K^{(*)-}$ mesons from $b \to c$ transitions. This provides an experimental signature for $b \to sg^*$, $g^* \to u\bar{u}$ or $g^* \to d\bar{d}$ decays where g^* denotes a gluon. This kinematic separation between $b \to c$ and $b \to sg^*$ transitions is illustrated by a generator level Monte Carlo simulation in Figure 1 for the case of $B \to K^{*-}$. (The $B \to K^-$ spectrum will be similar). This experimental signature can be applied to the asymmetric energy B factories if one boosts backwards along the z axis into the $\Upsilon(4S)$ center of mass frame.

Since there is a large background ("continuum") from the non-resonant processes $e^+e^- \to q\bar{q}$ where $q = u, d, s, c$, experimental cuts on the event shape are also imposed. To provide additional continuum suppression, the "B reconstruction" technique has been employed. The requirement that the kaon and n other pions form a system consistent in beam constrained mass and energy with a B meson dramatically reduces the background. After these requirements are imposed, one searches for an excess in the kaon momentum spectrum above the $b \to c$ region. Only one combination per event is chosen. No effort is made to unfold the feed-across between submodes with different values of n.

Methods similar to these have been successfully used by the CLEO II experiment to isolate a signal in the inclusive single photon energy spectrum and measure the branching fraction for inclusive $b \to s\gamma$ transitions and to set upper limits on $b \to s\phi$ transitions [14,15]. It is clear from these studies that the B reconstruction method provides adequate continuum background suppression.

The decay modes that will be used here are listed below:

1. $B^- \to K^{(*)-}\pi^0$

2. $\bar{B}^0 \to K^{(*)-}\pi^+$

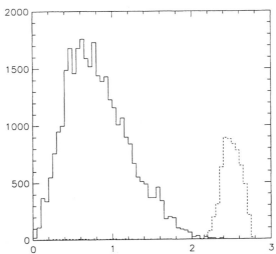

FIGURE 1. Generated Inclusive $B \to K^{*-}$ momentum spectrum. The component below 2.0 GeV/c is due to $b \to c$ decays while the component above 2.0 GeV/c arises from quasi-two body $b \to sg^*$ decay. The normalization of the $b \to c$ component is reduced by a factor of approximately 100 so that both components are visible.

3. $B^- \to K^{(*)-}\pi^-\pi^+$
4. $\bar{B}^0 \to K^{(*)-}\pi^+\pi^0$
5. $\bar{B}^0 \to K^{(*)-}\pi^+\pi^-\pi^+$
6. $B^- \to K^{(*)-}\pi^+\pi^-\pi^0$
7. $B^- \to K^{(*)-}\pi^+\pi^-\pi^+\pi^-$
8. $\bar{B}^0 \to K^{(*)-}\pi^+\pi^-\pi^+\pi^0$

In case of multiple entries for a decay mode, we choose the best entry on the basis of a χ^2 formed from the beam constrained mass and energy difference (i.e. $\chi^2 = (M_B/\delta M_B)^2 + (\Delta E/\delta \Delta E)^2$). In case of multiple decay modes per event, the best decay mode candidate is picked on the basis of the same χ^2.

Cross-feed between different $b \to sg$ decay modes (i.e. the misclassification of decay modes) provided the $K^{(*)-}$ is correctly identified, is not a concern as the goal is to extract an inclusive signal. The purpose of the B reconstruction method is to reduce continuum background. As the multiplicity of the decay mode increases, however, the probability of misreconstruction will increase.

The signal is isolated as excess $K^{(*)-}$ production in the high momentum signal region ($2.0 < p_{K^{(*)}} < 2.7$ GeV) above continuum background. To reduce contamination from high momentum $B \to \pi^-(\rho^-)$ production and residual $b \to c$

background, we assume the presence of a high momentum particle identification system as will be employed in the BABAR, BELLE, and CLEO III experiments.

We propose to measure the asymmetry $N(K^{(*)+}-K^{(*)-})/N(K^{(*)+}+K^{(*)-})$ where $K^{(*)\pm}$ originates from a partially reconstructed B decay such as $B \to K^{(*)-}(n\pi)^0$ where the additional pions have net charge 0 and $n \leq 4$ and one neutral pion is allowed and $2.7 > p(K^{(*)-}) > 2.0$ GeV. We assume that the contribution from $B \to K^- \eta' X$ decays has been removed by cutting on the η' region in X mass. It is possible that the anomalously large rate from this source [16,17] could dilute the asymmetry.

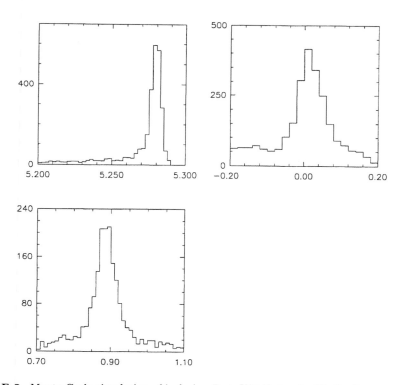

FIGURE 2. Monte Carlo simulation of inclusive $B \to K^{*-}X$ signal with the B reconstruction method: (a) The beam constrained mass distribution (b) The distribution of energy difference (c) The $K^-\pi^0$ invariant mass after selecting on energy difference and beam constrained mass

III EFFECTIVE HAMILTONIAN

In the Standard Model (SM) the amplitudes for hadronic B decays of the type $b \to q\bar{f}f$ are generated by the following effective Hamiltonian [18]:

$$H_{eff}^q = \frac{G_F}{\sqrt{2}}[V_{fb}V_{fq}^*(c_1 O_{1f}^q + c_2 O_{2f}^q) - \sum_{i=3}^{10}(V_{ub}V_{uq}^* c_i^u + V_{cb}V_{cq}^* c_i^c + V_{tb}V_{tq}^* c_i^t)O_i^q] + H.C.\,, \qquad (1)$$

where the superscript u, c, t indicates the internal quark, f can be u or c quark. q can be either a d or a s quark depending on whether the decay is a $\Delta S = 0$ or $\Delta S = -1$ process. The operators O_i^q are defined as

$$\begin{array}{ll} O_{f1}^q = \bar{q}_\alpha \gamma_\mu L f_\beta \bar{f}_\beta \gamma^\mu L b_\alpha, & O_{2f}^q = \bar{q}\gamma_\mu L f \bar{f} \gamma^\mu L b, \\ O_{3,5}^q = \bar{q}\gamma_\mu L b \bar{q}' \gamma_\mu L(R) q', & O_{4,6}^q = \bar{q}_\alpha \gamma_\mu L b_\beta \bar{q}'_\beta \gamma_\mu L(R) q'_\alpha, \\ O_{7,9}^q = \frac{3}{2}\bar{q}\gamma_\mu L b e_{q'} \bar{q}' \gamma^\mu R(L) q', & O_{8,10}^q = \frac{3}{2}\bar{q}_\alpha \gamma_\mu L b_\beta e_{q'} \bar{q}'_\beta \gamma_\mu R(L) q'_\alpha, \end{array} \qquad (2)$$

where $R(L) = 1 \pm \gamma_5$, and q' is summed over u, d, and s. O_1 are the tree level and QCD corrected operators. O_{3-6} are the strong gluon induced penguin operators, and operators O_{7-10} are due to γ and Z exchange (electroweak penguins), and "box" diagrams at loop level. The Wilson coefficients c_i^f are defined at the scale $\mu \approx m_b$ and have been evaluated to next-to-leading order in QCD. The c_i^t are the regularization scheme independent values obtained in Ref. [9]. We give the non-zero c_i^f below for $m_t = 176$ GeV, $\alpha_s(m_Z) = 0.117$, and $\mu = m_b = 5$ GeV,

$$\begin{array}{lll} c_1 = -0.307, & c_2 = 1.147, & c_3^t = 0.017, \\ c_4^t = -0.037, & c_5^t = 0.010, & c_6^t = -0.045, \\ c_7^t = -1.24 \times 10^{-5}, & c_8^t = 3.77 \times 10^{-4}, & c_9^t = -0.010, \\ c_{10}^t = 2.06 \times 10^{-3}, & c_{3,5}^{u,c} = -c_{4,6}^{u,c}/N_c = P_s^{u,c}/N_c, & \\ c_{7,9}^{u,c} = P_e^{u,c}, & c_{8,10}^{u,c} = 0, & \end{array} \qquad (3)$$

where N_c is the number of colors. The leading contributions to $P_{s,e}^i$ are given by: $P_s^i = (\frac{\alpha_s}{8\pi})c_2(\frac{10}{9} + G(m_i,\mu,q^2))$ and $P_e^i = (\frac{\alpha_{em}}{9\pi})(N_c c_1 + c_2)(\frac{10}{9} + G(m_i,\mu,q^2))$. The function $G(m,\mu,q^2)$ is given by

$$G(m,\mu,q^2) = 4 \int_0^1 x(1-x) \ln \frac{m^2 - x(1-x)q^2}{\mu^2}\, dx\,. \qquad (4)$$

All the above coefficients are obtained up to one loop order in electroweak interactions. The momentum q is the momentum carried by the virtual gluon in the penguin diagram. When $q^2 > 4m^2$, $G(m,\mu,q^2)$ becomes imaginary. In our calculation, we use $m_u = 5$ MeV, $m_d = 7$ MeV, $m_s = 200$ MeV, $m_c = 1.35$ GeV [19,20].

We assume that the final state phases calculated at the quark level will be a good approximation to the sizes and the signs of the FSI phases at the hadronic level for quasi-inclusive decays when the final state particles are quite energetic as is the case for the B decays in the kinematic range of experimental interest [6].

IV MATRIX ELEMENTS FOR $B^- \to K^- X$ AND $\bar{B}^0 \to K^- X$

We proceed to calculate the matrix elements of the form $<KX|H_{eff}|B>$ which represents the process $B \to KX$ and where H_{eff} has been described above. The effective Hamiltonian consists of operators with a current × current structure. Pairs of such operators can be expressed in terms of color singlet and color octet structures which lead to color singlet and color octet matrix elements. In the factorization approximation, one separates out the currents in the operators by inserting the vacuum state and neglecting any QCD interactions between the two currents. The basis for this approximation is that, if the quark pair created by one of the currents carries large energy then it will not have significant QCD interactions. In this approximation the color octet matrix element does not contribute because it cannot be expressed in a factorizable color singlet form. In our case, since the energy of the quark pairs that either creates the K or the X state is rather large, factorization is likely to be a good first approximation. To accommodate some deviation from this approximation we treat N_c, the number of colors that enter in the calculation of the matrix elements, as a free parameter. In our calculation we will see how our results vary with different choices of N_c. The value of $N_c \sim 2$ is suggested by experimental data on low multiplicity hadronic B decays [3]. The amplitude for $B \to K^{(*)} X$ can in general be split into a three body and a two body part. Detailed expressions for the matrix elements, decay distributions and asymmetries can be found in [21]

V RESULTS AND DISCUSSION

In this section we discuss the results of our calculations. We find that there can be significant asymmetries in $B \to K(K^*)X$ decays especially in the region $E_K > 2$ GeV which is also the region where an experimental signal for such decays can be isolated. The branching ratios are of order $O(10^{-4})$ which are within reach for future B factories. The contribution of the amplitude with the top quark in the loop accounts for 60-75% of the inclusive branching fraction. However, since the top quark amplitude is large and has no absorptive part in contrast to the c quark amplitude, the top quark contribution reduces the net CP asymmetry from 30-50% to about 10%. This calculation includes the contribution from electroweak penguins. We find that the electroweak penguin contributions increase the decay rates by 10-20% but reduce the overall asymmetry by 20-30%. The main sources of uncertainties in our calculation are discussed extensively in [21].

The asymmetries are sensitive to the values of the Wolfenstein parameters ρ and η. The existing constraints on the values of ρ and η come from measurements of $|V_{ub}|/|V_{cb}|$, ϵ_K in the K system and ΔM_{B_d}. (See Ref. [22] for a recent review). In our calculation we will use $f_B = 170$ MeV and choose $(\rho = -0.15, \eta = 0.33)$.

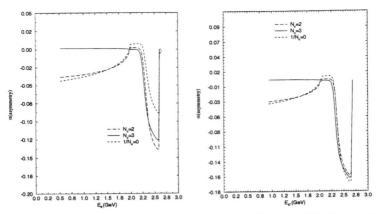

FIGURE 3. Predicted Asymmetries for $B^- \to K^- X$ and $B^- \to K^{*-} X$ as a function of the kaon energy. The three sets of curves indicate the sensitivity of the asymmetry to the value of N_c. The values $N_c = 2, 3, \infty$ are considered.

TABLE 1. Integrated decay rates and asymmetries for $B \to K^{(*)} X$ Decay

Process	Branching Ratio (1.65×10^{-4})	Integrated Asymmetry
$B^- \to K^- X$	$1.02, 0.79, 1.20$	$-0.10, -0.11, -0.050$
$B^- \to K^- X (E_K \geq 2.1 \text{GeV})$	$0.81, 0.74, 0.77$	$-0.12, -0.12, -0.07$
$\overline{B}^0 \to K^- X$	$0.6, 0.7, 0.8$	$-0.12, -0.12, -0.13$
$B^- \to K^{*-} X$	$1.37, 1.24, 2.30$	$-0.11, -0.14, -0.11$
$B^- \to K^{*-} X (E_{K^*} \geq 2.1 \text{GeV})$	$1.05, 1.16, 1.67$	$-0.14, -0.15, -0.14$
$\overline{B}^0 \to K^{*-} X$	$1.05, 1.16, 1.39$	$-0.15, -0.15, -0.16$

In Fig. 3 we show the asymmetries for K and K^* in the final state in charged B decays for different values of N_c. Variation of the asymmetries with the different inputs in our calculation are presented in detail in [21].

In Table. 1 we give the branching fractions and the integrated asymmetries for the inclusive decays for different N_c, $q^2 = m_b^2/2$ (q is the gluon momentum in the two body part of the amplitude), $f_B = 170$ MeV, $\rho = -0.15, \eta = 0.33$. For the charged B decays we also show the decay rates and asymmetries for $E_K > 2$ (2.1) GeV as that is the region of the signal.

The above figures show that there can be significant asymmetries in $B \to K^{(*)} X$ decays, especially in the region $E_K > 2$ GeV which is the region of experimental sensitivity for such decays. As already mentioned, our calculation is not free of theoretical uncertainties. Two strong assumptions used in our calculation are the use of quark level strong phases for the FSI phases at the hadronic level and the choice of the value of the gluon momentum q^2 in the two body decays. Other uncertainties from the use of different heavy to light form factors, the use of factorization, the model of the B meson wavefunction, the value of the charm quark mass and the choice of the renormalization scale μ have smaller effects on the asymmetries [21].

VI CONCLUSION

We find significant direct CP violation in the inclusive decay $B \to K^- X$ and $B \to K^{*-} X$ for $2.7 > E_{K^{(*)}} > 2.0$ GeV. The branching fractions are in the 10^{-4} range and the CP asymmetries may be sizeable. These asymmetries should be observable at future B factories and could be used to rule out the superweak class of models.

VII ACKNOWLEDGMENTS

This work was supported in part by National Science and Engineering Research Council of Canada (A. Datta). A. Datta thanks the organizers of MRST for hospitality and an interesting conference.

REFERENCES

1. D. Boutigny et al., Letter of Intent for the Study of CP Violation and Heavy Flavor Physics at PEPII, SLAC Report-0443, The BABAR collaboration, 1994.
2. K. Abe et al., Physics and Detector of Asymmetric B Factory at KEK, KEK Report 90-23 (1991). Also see Letter of Intent for A Study of CP Violation in B Meson Decays, KEK Report 94-2, April 1994.
3. For a review see A. Ali, hep-ph/9610333 and Nucl. Instrum. Meth. **A384**, 8 (1996); M. Gronau, TECHNION-PH-96-39, hep-ph/9609430 and Nucl. Instrum. Meth. **A384**, 1 (1996); T.E. Browder and K. Honscheid, Progress in Nuclear and Particle Physics, Vol. 35, ed. K. Faessler, p. 81-220 (1995).
4. A. Buras, hep-ph/9509329 and Nucl. Instrum. Meth. **A368**, 1 (1995); J.L. Rosner, hep-ph/9506364 and Proceedings of the 1995 Rio de Janeiro School on Particles and Fields, 116; A. Ali and D. London, DESY 95-148, UDEM-GPP-TH-95-32, hep-ph/9508272 and Nuovo Cim. **109A**, 957 (1996).
5. N.G. Deshpande, X.-G. He and S. Oh, Z. Phys. **C74**, 359 (1997); M. Gronau, hep-ph/9609430 and Nucl. Instrum. Meth. **A384**, 1 (1996); R. Fleischer, Int. J. Mod. Phys. **A12**, 2459 (1997).
6. M. Bander, D. Silverman and A. Soni, Phys. Rev. Lett. **43**, 242 (1979).
7. J. Gerard and W.-S. Hou, Phys. Rev. Lett. **62**, 855 (1989); Phys. Rev. **D43**, 2909 (1991); H. Simma, G. Eilam and D. Wyler, Nucl. Phys. **B352**, 367 (1991).
8. L. Wolfenstein, Phys. Rev.**D43**, 151 (1991).

9. R. Fleischer, Z. Phys. **C58** 438; Z. Phys. **C62**, 81; G. Kramer, W.F. Palmer and H. Simma, Nucl. Phys. **B 428**, 77 (1994); Z. Phys. **C66**, 429 (1994); N.G. Deshpande and X.-G. He, Phys. Lett. **B336**, 471 (1994).
10. D.-S. Du and Z.-Z. Xing, Z. Phys. **C66**, 129 (1995).
11. A.N. Kamal and C.W. Luo, Phys. Lett. **B398**, 151 (1997).
12. G. Kramer, W.F. Palmer, and H. Simma, Z. Phys. **C66**, 429, (1995).
13. J.P. Silva and L. Wolfenstein, Phys. Rev. **D55**, 5331 (1997) and references therein.
14. M.S. Alam *et al.* (CLEO Collaboration), Phys. Rev. Lett. **74**, 2885 (1995).
15. K.W. Edwards *et al.* (CLEO Collaboration), CLEO CONF 95-8.
16. P.C. Kim (CLEO Collaboration), contribution to the Proceedings of the FCNC97 conference; J. Alexander and B. Behrens (CLEO Collaboration), contribution to the Proceedings of the BCONF97 conference.
17. For a theoretical study of $B \to \eta' X_s$ in the standard model and beyond see A. Datta, X.-G. He and S. Pakvasa, Phys. Lett. **B 419**, 369 (1998) and references therein.
18. M. Lautenbacher and P. Weisz, Nucl. Phys. **B400**, 37 (1993); A. Buras, M. Jamin and M. Lautenbacher, Nucl. Phys. **B400**, 75 (1993); M. Ciuchini, E. Franco, G. Martinelli and L. Reina, Nucl. Phys. **B415**, 403 (1994).
19. J. Gasser and H. Leutwyler, Phys. Rep. **87**, 77 (1982).
20. A. Manohar in *Review of Particle Physics* (Particle Data Group), Phys. Rev. **D54**, 303 (1996).
21. T. Browder, A. Datta, S. Pakvasa and X.-G. He, Phys. Rev. **D57**, 6829 (1998).
22. A. Ali and D. London, Nucl. Phys. Proc. Suppl. **54A**, 29, 1997, also hep-ph/9607392.

Vector Dilepton Production at Hadron Colliders in the 3-3-1 Model [1]

Thomas Grégoire

Laboratoire de René J.-A. Lévesque, Université de Montréal
C.P. 6128, succ. centre-ville, Montréal, QC, Canada

Abstract. One way to detect the presence of physics beyond the Standard Model is to look for new kinds of particles which are not predicted by the Standard Model. One example is a particle with lepton number ± 2, called a dilepton. We have studied the production of vector dileptons at hadron colliders in the context of a specific model: the 3-3-1 model.

INTRODUCTION

The Standard Model (SM) is quite successful in explaining the present experimental data. Despite this fact, many people think that there is something beyond the SM. A model of new physics could predict new types of particles, so one should look for them in order to detect physics beyond the SM. Over the years, many alternative models have been developed and each of them predicts new types of particles. Because we don't really have any experimental indication of which of these models is most likely to be true, the most effective way to look for new physics would be to study these exotic particles only from their general characteristics, without assuming the validity of any model. For example, we can imagine exotic particles having a lepton number of ± 2. These kinds of particles are called dileptons. The objective of our work was to study, in the most general way, dilepton production at hadron colliders. We shall see however that it impossible to do it in a completely general way for vector dileptons, so that one has to work in the context of particular model.

GENERAL FEATURES

As mentioned above, dileptons are exotic particles which have a lepton number of ± 2. They can have an electric charge of ± 2 or ± 1 and they can be scalar or vector. In this talk I will concentrate on vector dileptons only (Ref. [1] discusses in

---

[1] Talk based on work done in collaboration with B.Dion, D.London, L.Marleau, and H. Nadeau.

addition scalar dileptons). Dileptons always couple to leptons but they could also couple to quarks in some models.

The most striking feature of dileptons is the decay of a doubly-charged dilepton into two identical leptons, i.e.

$$Y^{--} \to 2e^- \ .$$

The signal of this decay would be a peak in the invariant mass spectrum of the e^-e^- at the mass of the dilepton. No process of the SM can produce such a peak, so if a doubly-charged dilepton could be produced, it should not be too hard to detect [2,3].

Ideally, we would like to do the calculation independently of any model. Thus, we computed the cross section from these 2 diagrams only, which should exist in any model:

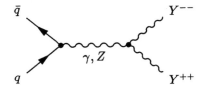

Unfortunately, the cross section calculated in this way violates unitarity when the dileptons are vectors. (For scalar dileptons though, it is not the case, so it is possible to do the calculation in a model-independent way, but I won't discuss it in this talk.) Thus, we had to choose a particular model in which to perform our calculations for the vector dileptons: we chose the 3-3-1 model [4,5].

THE 3-3-1 MODEL

General Features

The 3-3-1 model was introduced independently by Frampton [4], and by Pisano and Pleitez [5]. It has an $SU(3)_C \times SU(3)_L \times U(1)_X$ gauge symmetry, hence the name. One interesting feature of this model is that the anomaly cancellation takes place between the 3 families of fermions instead of taking place in each family separately as in the SM. So, in order to have a renormalizable theory, one needs to have 3 families. Another difference between this model and the SM is that in the latter, lepton number is conserved within each family while in the former, only the total lepton number is conserved. Finally, in this model, the third family of quark is treated differently from the first two and this could be related to the heaviness of the top quark.

Particle Content

In this model, the fermions are in triplets and anti-triplets of $SU(3)_L$ instead of being in doublets of $SU(2)_L$ as in the SM. Refs. [4] and [5] adopt two different representations for the quarks. These representations are equivalent [6] but we follow the convention adopted in Ref. [4]:

- Leptons: The 3 families of leptons are in anti-triplets of $SU(3)_L$ with $X = 0$. X is defined as $X = Q - \frac{1}{2}\lambda^3 - \frac{\sqrt{3}}{2}\lambda^8$, where the λ's are Gell-Mann matrices and Q the electric charge:

$$\begin{pmatrix} e^- \\ \nu_e \\ e^+ \end{pmatrix}_L, \begin{pmatrix} \mu^- \\ \nu_\mu \\ \mu^+ \end{pmatrix}_L, \begin{pmatrix} \tau^- \\ \nu_\tau \\ \tau^+ \end{pmatrix}_L \ : \ (1, 3^*, 0) \ .$$

- Quarks: The first two families of quarks are in triplets of $SU(3)_L$ with $X = -\frac{1}{3}$ and in singlets with $X = -\frac{2}{3}, \frac{1}{3}$ and $\frac{4}{3}$:

$$\begin{pmatrix} u \\ d \\ D^1 \end{pmatrix}_L, \begin{pmatrix} c \\ s \\ D^2 \end{pmatrix}_L \ : \ (3, 3, -\frac{1}{3}) \ ,$$

$$(\bar{u})_L, (\bar{c})_L \ : \ (3^*, 1, -\frac{2}{3}) \ ,$$

$$(\bar{d})_L, (\bar{s})_L \ : \ (3^*, 1, \frac{1}{3}) \ ,$$

$$(\bar{D}^1)_L, (\bar{D}^2)_L \ : \ (3^*, 1, \frac{4}{3}) \ .$$

The third family is in an anti-triplet of $SU(3)_L$ with $X = \frac{2}{3}$ and in singlets with $X = \frac{1}{3}, -\frac{2}{3}$ and $-\frac{5}{3}$:

$$\begin{pmatrix} b \\ t \\ T \end{pmatrix}_L \ : \ (3, 3^*, \frac{2}{3}) \ ,$$

$$(\bar{t})_L \ : \ (3^*, 1, -\frac{2}{3}) \ ,$$

$$(\bar{b})_L \ : \ (3^*, 1, \frac{1}{3}) \ ,$$

$$(\bar{T})_L \ : \ (3^*, 1, -\frac{5}{3}) \ .$$

D^1, D^2, T are new quarks with electric charge $-\frac{4}{3}, -\frac{4}{3}$ and $\frac{5}{3}$, respectively. They also have a lepton number of $+2, +2$ and -2, respectively.

The covariant derivative for triplets in this model is [2]:

$$D_\mu = \partial_\mu - ig\tfrac{\lambda^a}{2}W^a_\mu - igx\tfrac{X}{\sqrt{6}}X_\mu \ ,$$

where again the λ^a are the 8 Gell-Mann matrices, X is the $U(1)$ charge as defined earlier, and the W^a_μ and the X_μ are the gauge bosons fields. There are a total of 9 gauge bosons transmitting the electroweak interaction in the 3-3-1 model. They are:

- γ, W^+, W^- and Z: the usual SM gauge bosons,
- Z': a new neutral gauge boson,
- Y^{++}, Y^{--}, Y^+ and Y^-: the vector dileptons of this model.

PROCESSES CONSIDERED

The Processes

As mentioned earlier, the most striking feature of dileptons comes from the decay of a doubly-charged dilepton into 2 identical leptons, so we studied processes that could produce such dileptons:

$$q\bar{q} \to Y^{++}Y^{--} \text{ and } u\bar{d} \to Y^{++}Y^-$$

The first process produces two doubly-charged dileptons while the second one produces one doubly-charged dilepton along with a singly-charged one. I won't discuss the second process in this talk but it is examined in Ref. [1]. We also studied the following process:

$$q\bar{q} \to Y^{++} + 2e^-$$

In this process, 2 pairs of identical leptons will be produced as in the first process, but one of these pairs will be produced through a virtual dilepton instead of a real one.

Feynman Diagrams

Here are the various Feynman diagrams involved in the processes discussed in this talk:

- $q\bar{q} \to Y^{++}Y^{--}$:

- $q\bar{q} \to Y^{++} + 2e^-$:

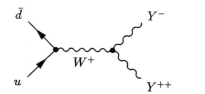 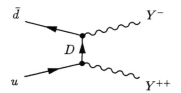

RESULTS

We computed the cross sections for these processes for different values of the masses of the dilepton and of the Z'. For this purpose, we had to evaluate the total width $\Gamma_{Z'}$ first. To do this, we assumed a mass of 600 GeV for the new quarks D and T. We found $\Gamma_{Z'} = 190.42$ for $M_{Z'} = 1$ TeV and $\Gamma_{Z'} = 971.44$ for $M_{Z'} = 2$ TeV. The big difference between the two values comes in large part from the fact that in the first case the $Z' \to Q\bar{Q}$ ($Q = D, T$) channel is closed while in the second it is open.

In Figs. 1 and 2, we plot the cross section for the production of two doubly-charged dileptons at the Tevatron ($\sqrt{s} = 1.8$ TeV) and at the LHC ($\sqrt{s} = 14$ TeV) respectively. Because we didn't take the detection efficiency into account in our analysis, we took a conservative number of 50 events as the minimum number of events needed to have an observable signal. The Tevatron has a luminosity of 100 pb^{-1}/yr, so the cross section would have to be at least 0.5 pb in order to have an observable signal. As we can see in Fig. 1, the Z' would then have to be lighter than 1.0 TeV in order to detect a signal at the Tevatron. The LHC has a luminosity of 10 fb^{-1}/yr, so a cross section of 0.005 pb would correspond to 50 events. Thus, as we can see in Fig. 2, the Z' could be as heavy as 3.0 TeV and the dilepton as heavy as 1.2 TeV and we would still be able to see a signal at the LHC.

As I mentioned in the previous section, at least three Feynman diagrams contribute to the $q\bar{q} \to Y^{++}Y^{--}$ process at tree level. One can then wonder if one of these diagrams dominates over the others. In particular, if $M_{Z'} > 2M_Y$, then the cross section could be dominated by the exchange of a real (or almost real) Z'. To see to what extent and for which masses of the Z' this is true, we plot in Figs. 3 and 4 the cross section for the production of two doubly-charged dileptons through a real Z' for various masses of the Z'. By comparing Fig. 1 with Fig. 3 and Fig. 2 with Fig. 4, we can see that if $M_{Z'} \lesssim 600$ GeV, it is a good approximation to say that the two dileptons would be mainly produced by the exchange of a real Z' at the Tevatron. The same is true at the LHC for $M_{Z'} \lesssim 1.4$ TeV.

On the other hand, we can see by looking at Figs. 1 and 2 that if $M_{Z'} \gtrsim 1.8$ TeV, the Z' contribution is negligible, both at the Tevatron and at the LHC. That means that for a heavy Z', the process $q\bar{q} \to Y^{++}Y^{--}$ is dominated by the other diagrams (the photon, Z and t-channel contributions). Thus, the curves labelled by $M_{Z'} = 1.8\text{-}3.0$ TeV in Figs. 1 and 2 correspond to the case where the Z' contribution to this process is essentially zero.

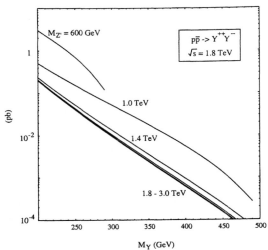

FIGURE 1. Cross section for $p\bar{p} \to Y^{++}Y^{--}$ at the Tevatron ($\sqrt{s} = 1.8$ TeV) as a function of M_Y, for various values of $M_{Z'}$.

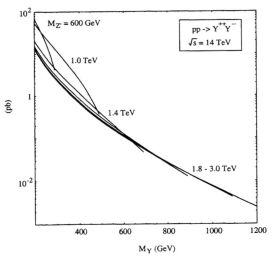

FIGURE 2. Cross section for $p\bar{p} \to Y^{++}Y^{--}$ at the LHC ($\sqrt{s} = 14$ TeV) as a function of M_Y, for various values of $M_{Z'}$.

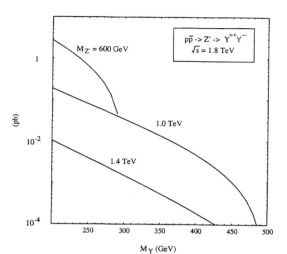

FIGURE 3. Cross section for $p\bar{p} \to Z' \to Y^{++}Y^{--}$ at the Tevatron ($\sqrt{s} = 1.8$ TeV) as a function of M_Y, for various values of $M_{Z'}$.

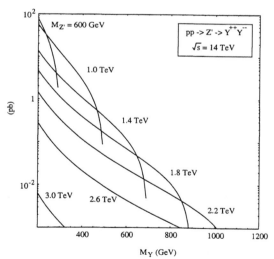

FIGURE 4. Cross section for $p\bar{p} \to Z' \to Y^{++}Y^{--}$ at the LHC ($\sqrt{s} = 14$ TeV) as a function of M_Y, for various values of $M_{Z'}$.

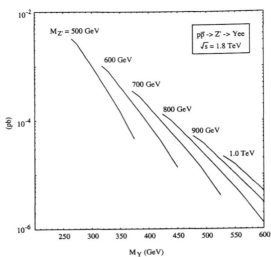

FIGURE 5. Cross section for $p\bar{p} \to Z' \to Yee$ at the Tevatron ($\sqrt{s} = 1.8$ TeV) as a function of M_Y, for various values of $M_{Z'}$.

Now, even for a real Z', if $M_{Z'} < 2M_Y$, it could not produce two real dileptons. On the other hand, it could produce one real dilepton along with two identical leptons ($q\bar{q} \to Yee$). So one can then ask the following question: if the Z' could be produced on shell but could not produce two real dileptons, is the cross section for $q\bar{q} \to Yee$ (which can be produced via an on-shell Z') bigger than that for $q\bar{q} \to Y^{++}Y^{--}$ (which cannot)?

To answer this question, we plot in Figs. 5 and 6 the cross section for $q\bar{q} \to Z' \to Yee$. This cross section is to be compared with the $M_{Z'} = 1.8$-3.0 TeV curves in Figs. 1 and 2, which essentially correspond to $q\bar{q} \to Y^{++}Y^{--}$ with no Z' contribution. If one looks at Figs. 5 and 6, one easily sees that, for a given dilepton mass, the cross section for $q\bar{q} \to Y^{++}Y^{--}$ with no Z' contribution is always bigger than the one for $q\bar{q} \to Z' \to Yee$. The answer to the above question is therefore no. In fact, the cross section for $q\bar{q} \to Z' \to Yee$ is so small that it couldn't be seen at all at the Tevatron. However, it could be seen at the LHC if $M_{Z'} < 800$ GeV. For such a mass range, as we saw earlier, the Z' would be produced on shell, so that $q\bar{q} \to Z' \to Yee \approx q\bar{q} \to Yee$ should be a good approximation.

Because the cross section for $q\bar{q} \to Y^{++}Y^{--}$ is always bigger than that for $q\bar{q} \to Yee$, the latter process would not be very useful as a probe for the existence of dileptons. On the other hand, it could be useful as a probe for the existence of the Z'. Indeed, if $M_{Z'} < 2M_Y$, the $q\bar{q} \to YY$ process could not occur through a real Z' while $q\bar{q} \to Yee$ could. So for some masses of the Z' and the Y, $q\bar{q} \to Yee$ could still be of some use.

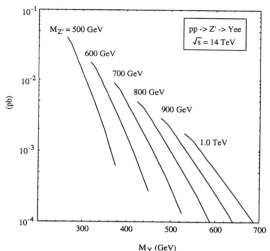

FIGURE 6. Cross section for $p\bar{p} \to Z' \to Yee$ at the LHC ($\sqrt{s} = 14$ TeV) as a function of M_Y, for various values of $M_{Z'}$.

CONCLUSION

If the dileptons of the 3-3-1 model exist, it should be possible to detect them at the LHC through the process $q\bar{q} \to Y^{++}Y^{--}$. At the Tevatron, the chances are not as good because the mass of the Z' would have to be in a more limited range.

The process $q\bar{q} \to Yee$ could be interesting because it could help to reveal the existence of the Z' if it is too light to produce two real dileptons. On the other hand, it would not be very helpful for detecting the dileptons because the cross section for this process will always be smaller than for $q\bar{q} \to Y^{++}Y^{--}$.

Acknowledgments

This research was financially supported by NSERC of Canada and FCAR du Québec.

REFERENCES

1. B. Dion, T. Grégoire, D. London, L. Marleau and H. Nadeau, to be submitted to *Phys. Rev. D*.
2. P.H. Frampton, J.T. Liu, B.C. Rasco and D. Ng, *Mod. Phys. Lett.* **A9**, 1975 (1994).
3. B. Dutta and S. Nandi, *Phys. Lett.* **340B**, 86 (1994).
4. P.H. Frampton, *Phys. Rev. Lett.* **69**, 2889 (1992).
5. F. Pisano and V. Pleitez, *Phys. Rev.* **D46**, 410 (1992).
6. D. Ng, *Phys. Rev.* **D49**, 4805 (1994).

Two-Higgs-Doublet-Models and Radiative CP Violation [1]

Otto C.W. Kong* and Feng-Li Lin[†]

*Department of Physics and Astronomy,
University of Rochester, Rochester NY 14627-0171.

† Department of Physics, and Institute for Particle Physics and Astrophysics,
Virginia Polytechnic Institute and State University, Blacksburg, VA 24061-0435.

Abstract. We discuss the feasibility of spontaneous CP violation being induced by radiative corrections in 2HDM's. Specifically, we analyze the cases of gaugino/higgsino effect on MSSM, and a new model with an extra exotic quark doublet. The new model, while demonstrating well the Georgi-Pais theorem, is also expected to be phenomenlogically interesting.

INTRODUCTION

The source of CP violation is one of the most important unsolved puzzles in particle physics. On the one hand, CP violation is observed experimentally only in the K^0-$\bar{K}^0$ system, with the corresponding weak CP phase compatible with the Kobayashi-Maskawa (KM) mechanism. On the other hand, the experimental bound on the neutron electric dipole moment indicates that the effective strong CP phase has to be exceedingly small,

$$\bar{\theta} < 10^{-9} \ . \tag{1}$$

The origin of this strong CP problem lies in the necessity of adding the so-called θ term to the effective QCD Lagrangian due to the contribution of instantons present in the topologically nontrivial QCD vacuum :

$$\mathcal{L}_{\text{eff}} = \frac{\theta \alpha_s}{8\pi} F^A_{\mu\nu} \tilde{F}^{A\mu\nu} \ , \tag{2}$$

where the dual field strength is given by $\tilde{F}_{\mu\nu} = \frac{1}{2}\epsilon_{\mu\nu\alpha\beta} F^{\alpha\beta}$. Through the anomaly in the QCD axial U(1) current, chiral U(1) transformations lead to shifts in θ, leaving

---

[1] Talk presented by O.K..

the physical combination $\bar{\theta} = \theta - \arg \det M_q$, where M_q is the quark mass matrix, as the effective strong CP phase. In the supersymmetrized version of the standard model, new CP phases from supersymmetry (SUSY) breaking terms also have to be small ($< 10^{-3}$). Spontaneously CP violation (SCPV) provides an elegant theory with a good control on the magnitude of the various CP phases. Hence models of SCPV keep generating new interest.

SCPV IN A TWO-HIGGS-DOUBLET MODEL

The most simple setting for achieving the SCPV scenario is given by a two-Higgs-doublet model (2HDM) [2]. The most general scalar potential for two Higgs doublets, ϕ_1 and ϕ_2, is given by

$$\begin{aligned} V(\phi_1, \phi_2) = & m_1^2 \phi_1^\dagger \phi_1 + m_2^2 \phi_2^\dagger \phi_2 - (m_3^2 \phi_1^\dagger \phi_2 + h.c.) \\ & + \lambda_1 (\phi_1^\dagger \phi_1)^2 + \lambda_2 (\phi_2^\dagger \phi_2)^2 + \lambda_3 (\phi_1^\dagger \phi_1)(\phi_2^\dagger \phi_2) \\ & + \lambda_4 (\phi_1^\dagger \phi_2)(\phi_2^\dagger \phi_1) + \frac{1}{2}[\lambda_5 (\phi_1^\dagger \phi_2)^2 + h.c.] \\ & + \frac{1}{2}\{\phi_1^\dagger \phi_2 [\lambda_6 (\phi_1^\dagger \phi_1) + \lambda_7 (\phi_2^\dagger \phi_2)] + h.c.\} \; . \end{aligned} \quad (3)$$

Assuming all the parameters in V being real, and denoting the vacuum expectation values (VEV's) of the neutral components of the Higgs doublets by

$$\langle \phi_1^0 \rangle = v_1 \quad \text{and} \quad \langle \phi_2^0 \rangle = v_2 e^{i\delta} \; ,$$

we have

$$\begin{aligned} \langle V \rangle = & m_1^2 v_1^2 + m_2^2 v_2^2 + \lambda_1 v_1^4 + \lambda_2 v_2^4 + (\lambda_3 + \lambda_4 - \lambda_5) v_1^2 v_2^2 \\ & + 2\lambda_5 v_1^2 v_2^2 \cos^2 \delta - (2m_3^2 - \lambda_6 v_1^2 - \lambda_7 v_2^2) v_1 v_2 \cos \delta \; , \\ = & M_1 v_1^2 + M_2 v_2^2 + (p v_1^4 + 2r v_1^2 v_2^2 + q v_2^4) \\ & + 2\lambda_5 v_1^2 v_2^2 (\cos \delta - \Omega)^2 - \frac{m_3^4}{2\lambda_5} \; ; \end{aligned} \quad (4)$$

where

$$\Omega = \frac{2m_3^2 - \lambda_6 v_1^2 - \lambda_7 v_2^2}{4\lambda_5 v_1 v_2} \; , \quad (5)$$

and

$$M_1 = m_1^2 + \frac{\lambda_6 m_3^2}{2\lambda_5} \; , \quad (6)$$

$$M_2 = m_2^2 + \frac{\lambda_7 m_3^2}{2\lambda_5} \; , \quad (7)$$

$$p = \lambda_1 - \frac{\lambda_6^2}{8\lambda_5}, \tag{8}$$

$$q = \lambda_2 - \frac{\lambda_7^2}{8\lambda_5}, \tag{9}$$

$$r = \frac{1}{2}(\lambda_3 + \lambda_4 - \lambda_5 - \frac{\lambda_6\lambda_7}{4\lambda_5}). \tag{10}$$

A nontrivial phase (δ) then indicates SCPV.

Let us look at the δ-dependence of $\langle V \rangle$. The extremal condition gives

$$-4\lambda_5 v_1^2 v_2^2 (\cos\delta - \Omega)\sin\delta = 0, \tag{11}$$

and the stability condition requires

$$\frac{\partial^2 V}{\partial \delta^2} = 4\lambda_5 v_1^2 v_2^2 [\cos\delta(\Omega - \cos\delta) + \sin^2\delta] > 0. \tag{12}$$

$\cos\delta = \Omega$ gives a SCPV solution, provided that $\lambda_5 > 0$ and $|\Omega| < 1$. Actually, Eq.(2) shows that this is the absolute minimum. In order for V to have a lower bound, we have the extra constraints

$$p > 0, \quad q > 0, \tag{13}$$

and

$$pq > [r + \lambda_5(\cos\delta - \Omega)^2]^2. \tag{14}$$

The latter reduced to

$$pq > r^2 \tag{15}$$

for the CP violating minimum.

However, in order to avoid flavor-changing-neutral-currents that could result, extra structure like natural flavor conservation (NFC) [3] has to be imposed on a 2HDM, which then forbids SCPV [4]. For instance, a natural way to impose NFC is to require that only one of the Higgs, say ϕ_1, transforms nontrivially under an extra discrete symmetry. This means that m_3^2, λ_6 and λ_7, and may be λ_5 too, all have to vanish. Similarly, a supersymmetric version of the standard model (SM) is naturally a 2HDM with NFC being imposed automatically by the holomorphy of the superpotential. The tree level scalar potential there has vanishing λ_5, λ_6, and λ_7, though the soft SUSY breaking B-term gives rise to a nonvanishing m_3^2. The interesting point of concern then is whether radiative corrections can modify the picture. Note that a positive $\Delta\lambda_5$ is needed for this radiative CP violation scenario.

RADIATIVE CP VIOLATION

For the case of the minimal supersymmetric standard model (MSSM), Maekawa [5] pointed out that there is a positive contributions to λ_5 from a finite 1-loop diagram (Fig. 1a) involving the gauginos and higgsinos, which could lead to SCPV. However, there are objections to the particular scenario. Maekawa [5] realized that the situation will not be able to give rise to sufficient weak CP vioation for the K^0-$\bar{K}^0$ system. Pomarol [6] argued that mass for the "psuedoscalar", m_A, is roughly proportional to $\sqrt{\lambda_5}$ and is at least more than a factor of three too small to be phenomenologically acceptable. Actually, we worked out the algebra to decoupled the Goldstone mode explicitly assuming the CP violating vacuum, and got the following results [1]: the 3×3 physical Higgs mass-squared matrix m^2_{Hij} gives

$$\det(m_H^2) = \lambda_5(pq - r^2)\sin^2 2\beta \sin^2 \delta ; \qquad (16)$$

as $p = q = -r > 0$ at tree level, when only a positive $\Delta\lambda_5$ is considered, $(pq - r^2)$ and hence $\det(m_H^2)$ becomes negative [see Eq.(8)]. In fact, with a negative r, $(pq < r^2)$ exactly violates the condition for the potential to be bounded from below. This inconsistency is also pointed out by Haba [7], who then suggested that it would be fixed when top/stop loop contributions to the other parameters, mainly λ_2, in the potential are included. From our perspective, in order to have a *consistent approximation* to the radiative corrections in whatever interesting region of the parameter space, loop contributions to *all* parameters in the scalar potential at the same order have to be considered. Corrections to all parameters in V do exists, as shown in Fig. 1.

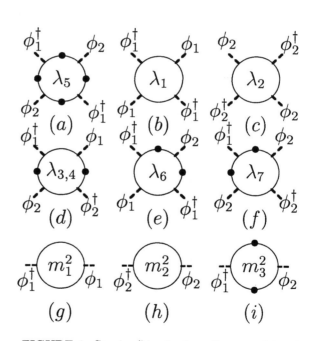

FIGURE 1. Gaugino/higgsino-loop diagrams giving rise to modifications to parameters in the potential. (Each dot indicates a helicity flip in the fermion propagator.)

A NEW MODEL WITH AN EXOTIC QUARK DOUBLET

Before going into discussion of a consistent 1-loop treatment, we first present a new model introduced in Ref. [1] that we believe could be experimentally viable. Our new model has an extra pair of vectorlike quark doublets, Q and $\bar{Q}$, with the following couplings

$$\mathcal{L}_Q = M_Q \bar{Q} Q + \lambda_Q \bar{t} \phi_1^\dagger Q \;, \tag{17}$$

as an addition to the two-Higgs-doublet SM or MSSM. Note that $\phi_1^\dagger$ is actually H_d, the Higgs (super)multiplet that gives masses to the down-type quarks; and ϕ_2 is H_u. So, $T_3 = -1/2$ component of Q, denoted by T, has the same charge as the top quark and mixes with it after electroweak (EW) symmetry breaking. The other part of the doublet is a quark of electric 5/3. The 1-loop diagram, now with the gaugino/higgsino propagators replaced by that of the quarks, leads to $\Delta\lambda_5 (\sim 3\lambda_Q^2 \lambda_t^2 / 16\pi^2)$ and could be very substantial for large Yukawa couplings. (Note that ϕ_1 and ϕ_2 vertices now have λ_Q and λ_t couplings respectively).

The mass matrix of the t-T system is given by

$$\mathcal{M}_t = \begin{pmatrix} \lambda_t \langle \phi_2 \rangle & \lambda_Q \langle \phi_1^\dagger \rangle \\ 0 & M_Q \end{pmatrix} \;. \tag{18}$$

Notice that the model actually has an effective KM phase to account for weak CP phenomenology. Assuming M_Q to be roughly around the same order as the EW scale, the model can also easily get around the "small m_A" objection. Moreover, its modification to top quark phenomenology would be very interesting, and will provide an experimental check on its viability. The Q-$\bar{Q}$ exotic quarks could naturally arise, for example, as the only extra quarks from some interesting models with a SM-like chiral fermion spectra embedding the three SM families in a intriguing way [8].

A CONSISTENT 1-LOOP TREATMENT AND THE GEORGI-PAIS THEOREM

Here we give a brief discussion of our consistent 1-loop treatment for both MSSM and the new model, particularly in relation to the Georgi-Pais theorem [9]. Readers are referred to our original paper [1] for more details.

The Georgi-Pais theorem states that radiative CP violation can occur if and only if there exists spinless bosons which are massless in the tree approximation. Naively looking, the above radiative CP violation pictures violate the theorem. There are some confusing statement in the literature in relation to the situation. However, the theorem has a presumption, that no fine tuning be allowed. Results from our

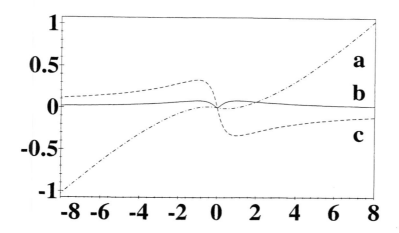

FIGURE 2. Plots of radiative correction from chargino loop verses $m(= M_{\tilde{g}}/\mu)$, with mass mixing from EW symmetry neglected. (a)$\Delta m_3^2(\overline{MS})$ in $25 \times g^2\mu^2/16\pi^2$; (b)$\Delta\lambda_5$; (c)$\Delta\lambda_6$ $(= \Delta\lambda_7)$; both of the latter curves with values in $g^4/16\pi^2$.

---

consistent treatment of the two models discussed to be sketched below show exactly that some form of fine tuning is unavoidable for radiative CP violation to occur in both cases. There has also been statements about the smallness of m_A being a necessary consequence of the Georgi-Pais theorem. Comparison between the two models dicsussed here clearly illustrated that is not true. Smallness of m_A in the MSSM cases is rather the result of the small gauge couplings used to produce $\Delta\lambda_5$.

A positive $\Delta\lambda_5$ is a necessary but *not sufficient* condition for radiative CP violation. This is clearly illustrated in our discussion of the scalar potential and the problem of the Maekawa picture. In our opinion, it is at least of theoretical interest to see what the 1-loop gaugino/higgsino effect alone could do to the vacuum solution of a supersymmetric 2HDM. A consistent treatment of the 1-loop effect should of course take into consideration contributions to all the 10 parameters in the potential V. Recall that another essential condition for the existence of the CP violating vacuum solution is $|\Omega| < 1$.

We plot the numerical results of major interest in Fig.2. The plots are for the chargino contributions only, as functions of $m = M_{\tilde{g}}/\mu$, the gaugino-higgsino mass ratio. Our results here presented give the 1-loop effect before EW symmetry breaking, i.e. mass mixing between the gaugino and higgsino were not considered. Further modifications due to the symmetry breaking are not expected to change the general features. Here, the neutralino contributions can simply be inferred from symmetry.

Taking the renormalized value of m_3^2 as a free parameter, it has a magnitude that increases fast with that of m for $|m| > 1$. Obviously, some fine tuning is needed to get $|\Omega| < 1$, though a small window on m always exists not too far from $|m| = 1$, for each sign of μ, for a not too small μ. With EW-scale μ, $\Delta m/m$ of the admissible regions are of order 10^{-2}, though the severe fine tuning can be tamed by having small μ.

In the our new model, though the new quark doublet Q has mass before EW symmetry breaking, a similar set of 1-loop diagrams, as those given in Fig.1, can only be completed when the EW breaking mass of the top and its mixing with T are taken into consideration. But then the plausibly large Yukawa couplings give

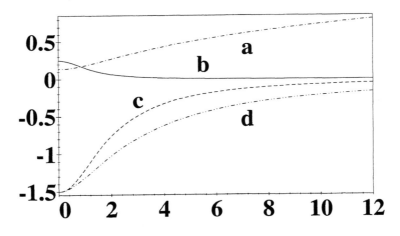

FIGURE 3. Plots of radiative correction from t-T loop verses $m(= M_Q/v_2)$. (a)$\Delta m_3^2(\overline{MS})$ in $10 \times 3\lambda_t^2 v_2^2/16\pi^2$; (b)$\Delta\lambda_5$; (c)$\Delta\lambda_6$; (d)$\Delta\lambda_7$; all of the latter three curves with values in $3\lambda_t^4/16\pi^2$; ($\lambda_t = \lambda_Q$, $\tan\beta = 1$ assumed).

rise to substantial results. In Fig.3, we presented some numerical results. The plots use again m as parameter which in this case denotes M_Q/v_2. For simplicity, we assume $\lambda_Q = \lambda_t$.

Here, Δm_3^2 ($\overline{MS}$) always has the opposite sign to that of $\Delta\lambda_6$ or $\Delta\lambda_7$, making a naive use of the value to fit in the $|\Omega| < 1$ condition impossible. Hence, for SCPV to occur, a tree level m_3^2 value is needed. While the $m_{3(tree)}^2$ then has to be chosen to roughly match the t-T 1-loop effect, the large Yukawa couplings make this relatively natural, as a value of the order v_2 is all required. Accepting that, $\Delta m_{3(tree)}^2/m_{3(tree)}^2$ of the solution region is not quite small, $\gtrsim .35$ for $m \leq 1$, representing only a moderate fine tuning. The small m_3^2 required is natural, as its zero limit provides the scalar potential with an extra Peccei-Quinn type symmetry.

CONCLUSION

We have performed a consistent 1-loop analysis of the feasibility of radiatively induced SCPV, for both the MSSM and our proposed new model with a pair of vector-like exotic quark doublets. Our results clarified some issues concerned. The new model, especially a SUSY version, is expected to be phenomenologically interesting. Though a complete studies of its various features and experimental viability still have to be performed, we have illustrated at least how it can easily overcome the objections to the corresponding scenario within MSSM. Further studies of the model is under progress.

REFERENCES

1. Kong, O.C.W., and Lin, F.-L., *Phys. Lett.* **B 419**, 217 (1998).
2. Lee, T.D., *Phys. Rev.* **D 8**, 1226 (1973).
3. Glashow, S.L., and Weinberg, S., *Phys. Rev.* **D 15**, 1958 (1977).
4. Branco, G.C., *Phys. Rev.* **D 22**, 2901 (1980).
5. Maekawa, N., *Phys. Lett.* **B 282**, 387 (1992).
6. Pomarol, A., *Phys. Lett.* **B 287**, 331 (1992).
7. Haba, N., *Phys. Lett.* **B 398**, 305 (1997).
8. Kong, O.C.W., *Mod. Phys. Lett.* **A11**, 2547 (1996); *Phys. Rev.* **D 55**, 383 (1997).
9. Georgi, H., and Pais, A., *Phys. Rev.* **D10**, 1246 (1974).

Minimal Ten-parameter Hermitian Texture Zeroes Mass Matrices and the CKM Matrix

M. Baillargeon[a], F. Boudjema[b], C. Hamzaoui[a] and J. Lindig[c]

[a] *Département de physique, Université du Québec à Montréal*
C.P. 8888, Succ. Centre Ville, Montréal, Québec, Canada, H3C 3P8
[b] *Laboratoire de Physique Théorique LAPTH*
Chemin de Bellevue, B.P. 110, F-74941 Annecy-le-Vieux, Cedex, France.
[c] *Institute for Theoretical Physics, Leipzig Univertity*
Augustusplatz 10, 04109 Leipzig, Germany

Abstract. Hermitian mass matrices for the up and down quarks with texture zeroes but with the minimum number of parameters, ten, are investigated. We show how these *minimum parameter* forms can be obtained from a general set of hermitian matrices through weak basis transformations. For the most simple forms we show that one can derive exact and compact parameterizations of the CKM mixing matrix in terms of the elements of these mass matrices (and the quark masses).

INTRODUCTION

Within the three-family standard model, the Yukawa interaction provides ten physical measurable parameters, the six quark masses and the four parameters of the CKM mixing matrices. Although ten is certainly a large number if the model is to be viewed as a fundamental theory, this number of parameters in fact emerges from two 3×3 Yukawa matrices which in total amount to as many as 36 parameters! Clearly there is a large number of redundant parameters and therefore in order to better corner the mechanism of symmetry breaking it could be advantageous to work with mass matrices that exhibit only the least number of parameters, *i.e.* ten. Having to deal with a minimal set not only eases the computational task, like going from the mass matrices to the mixing matrix, but when confronting the set with the data this may even help in better exhibiting patterns of mass matrices that hint towards further relations between some of the elements of the set. In this case one can entertain the existence of symmetries and models beyond the standard model that can explain the approximate relations. At the same time this general minimum approach could also reveal whether some relations, like those that relate some mixing angles to ratios of masses, are in fact not specific to a particular constrained model but are rather generic in a much wider class of models.

Most of the *constrained* matrices that aim at relating the mixing angles to the quark masses, and hence reduce the number of parameters to much less than the needed 10, are based on so called texture zeroes mass matrices. These are based either on some specific "beyond the standard model" scenario or by postulating some ad-hoc ansatz. In ansätze were these zeroes are not related to any symmetry only the non-zero elements are counted as parameters, although it is clear that there are numerous ways of keeping to ten independent parameters. For instance instead of zeroes one can take some elements to be equal or have any other definite relations between them. Democratic mass matrices with all elements equal are a case in mind, these are a one-parameter model which when written in an appropriate basis can be turned into matrices with all elements but one being zero. In our approach we keep within the popular textures zeroes paradigm and look for those bases were only the minimum number of (non-zero) independent parameters appears explicitly.

Obviously, dealing with less than ten parameters invariably leads to relations between masses and mixings. In many cases and for a certain range of masses the less-than-ten parameter descriptions may turn out not to be supported by data if the textures are over-predictive. On the other hand if one works with, at least, ten independent parameters then one should always reproduce the data since there should be possible to make a one-to-one mapping. A general classification of symmetric textures zeroes with a number of parameters less than ten has been given in [1], while Branco Lavoura and Mota [2] (BLM) have been the first to point out that for non hermitian matrices some textures zero à la Fritzsch [3] were just a rewriting of the mass matrices in a special basis and thus the zeroes of the much celebrated Fritzsch ansatz were " void" of any physical content. In fact the BLM approach for non hermitian matrices still involves twelve parameters, the extra two being related to the phase conventions taken for the CKM[1].

Recently an approach based on BLM has been pursued by some authors [4–6], taking a specific pattern of non hermitian matrices and in some cases re-expressing one of the mass matrices with the help of the phenomenological parameterization of the CKM matrix. In this talk we will concentrate on hermitian texture zero matrices having zeroes in the non-diagonal entries, the case with zeroes on the diagonal will be presented in a longer communication [7]. It is known that within the standard model one can always express the mass matrices in a basis were they are hermitian [2,8]. Also, in the case of non hermitian matrices our results should be understood as applying to the hermitian square matrices, $H = MM^\dagger$. We supply a systematic list of all possible texture zeroes that contain the minimal set of ten parameters and show how these textures can be reached from a general set of two hermitian matrices for up and down quarks, through specific weak basis transformation which we construct explicitly. Among all the patterns that we list,

---

[1] One should be fair and say that, sometimes, keeping one or two of the redundant parameters may prove useful. However we will stick with the minimalist description

one shows a particularly very simple and appealing structure which has a direct connection to the Wolfestein parameterization [9]. For this we have been able to analytically construct a compact exact formulae for the mixing matrix.

SIMPLE TEXTURE ZEROES QUARK MASS MATRICES AND THE CHOICE OF BASIS

The key observation as concerns the search of a suitable basis, ideally one with the maximum number of zeroes, is that starting from any set of matrices for the up and down quark, the physics is invariant if one performs a weak basis transformation on the fields. In the case of the standard model, one can choose any right-handed basis for both the up quark fields (u_R) and the down quark fields (d_R), as well as any basis for the doublets of left-handed fields (Q_L). All these bases are related to each other through unitary transformations, $u_R \to V_u u_R$; $d_R \to V_d u_R$; $Q_L \to U_L Q_L$. Therefore all sets of mass matrices related to each other through

$$M'_u = U_L^\dagger M_u V_u, \qquad M'_d = U_L^\dagger M_u V_d \tag{1}$$

give rise to the same physics (same masses and mixing angles in the charged current).

For hermitian matrices this means that weak basis transformations involve only a single unitary transformation, *i.e.*, $U_L = V_u = V_d = U$ and therefore one can use either the set M_u, M_d or the set M'_u, M'_d with $M'_f = U^\dagger M_f U$. In the case of hermitian matrices, one is starting with a set of 18 parameters and the task is to find a unitary matrix U which can absorb 8 redundant parameters. This should always be possible since a 3×3 unitary matrix has nine real parameters, but since an overall phase transformation $U = e^{i\phi} \mathbf{1}$ does not affect weak bases transformations, a unitary matrix provides the required number of variables to absorb the redundant parameters.

Phase transformations
One special case of this type of unitary transformations which always proves useful, even in the case of non-hermitian matrices, is the one provided by unitary phase transformations $U_{ij} = e^{i\phi_i} \delta_{ij}$. Because a global phase does not affect the transformation, we set $\phi_1 = 0$ without loss of generality. This type of matrix is therefore a two-parameter matrix. Applying this type of transformation on both M_u, M_d one has the freedom to choose $\phi_{2,3}$ such that two phases out of the six contained in the hermitian M_u, M_d can be set to zero. The only restriction is that one can not, in general, simultaneously remove the phases of both $M_u(ij)$ and $M_d(ij)$ (*i.e.* for the same (ij)). In any case, two parameters, or rather phases, out of the 18 can always be removed this way.

The simple case of a basis where one matrix is diagonal
It is always possible to take $U = U_d^D$ (U_u^D), that is the unitary matrix that diagonalises the down (up) matrix. In these specific bases where one matrix is diagonal,

TABLE 1. Location of the zeroes for the 18 different forms.

	M_u	M_d			M_u	M_d
1	(1,2)	(1,3) and (2,3)		10	(1,3) and (2,3)	(1,2)
2	(1,2)	(1,2) and (2,3)		11	(1,2) and (2,3)	(1,2)
3	(1,2)	(1,2) and (1,3)		12	(1,2) and (1,3)	(1,2)
4	(1,3)	(1,3) and (2,3)		13	(1,3) and (2,3)	(1,3)
5	(1,3)	(1,2) and (2,3)		14	(1,2) and (2,3)	(1,3)
6	(1,3)	(1,2) and (1,3)		15	(1,2) and (1,3)	(1,3)
7	(2,3)	(1,3) and (2,3)		16	(1,3) and (2,3)	(2,3)
8	(2,3)	(1,2) and (2,3)		17	(1,2) and (2,3)	(2,3)
9	(2,3)	(1,2) and (1,3)		18	(1,2) and (1,3)	(2,3)

the other, non-diagonal matrix, will then have no zero in general but 9 real parameters (of which 3 can be taken as phases in the non-diagonal entries). Applying an extra phase transformation removes two phases and therefore one does indeed end up with 3 parameters in M_d (the masses) and 7 in M_u making up a total of ten which is the minimal number.

It is worth mentioning that similar bases (where one of the matrices is diagonal) have been studied in the literature but for the case of non-hermitian matrices [4-6]. It is easy to see that one can easily recover these bases. Indeed, one can apply on our hermitian matrices, the following transformations: assuming one is starting with a diagonal M_u take $U = V_u$ as a phase transformation or simply just the unit matrix, then it is always possible to choose V_d such that M_d turns into a non-hermitian matrix but with extra zeroes. We leave the proof and a discussion of these kind of (diagonal) bases to our longer communication [7].

Non trivial cases: Non diagonal matrices with no diagonal zero

In the above simple case one had three zeroes[2]. In fact requiring that one maintains 10 parameters, and in the case of hermitian matrices where the zeroes are set on the off-diagonal elements, three is the maximum number of zeroes. The non trivial cases are when these three zeroes are shared between the up-quark and down-quark matrices, that is one off-diagonal zero in one matrix and two off-diagonal zeroes in the other. Indeed, having more than three off-diagonal zeroes, four say, one is left with the six real parameters on the diagonals plus two complex numbers which reduce to two real numbers after a phase transformation has been applied and thus leading to only 8 real parameters. Therefore, by requiring off-diagonal zeroes the problem is rather simple: one only has to combine a matrix with one off-diagonal zero with a matrix with two off-diagonal zeroes. For each of these matrices there are three possibilities of where to put the zero. All in all, one counts 18 such possibilities or patterns. These are displayed in

---

[2] Since one is dealing with hermitian matrices, the number of zeroes is that contained on one side of the diagonal.

All of these combinations can in fact be classified in only two distinct cases which can not be obtained from each other by a simple relabeling of the axes. Denoting the two arbitrary hermitian mass matrices in those bases by $M_u = A'$ and $M_d = B'$, these cases are explicitly:

$$A' = \begin{pmatrix} A'_{11} & 0 & 0 \\ 0 & A'_{22} & A'_{23} \\ 0 & A'^*_{23} & A'_{33} \end{pmatrix}, \quad B' = \begin{pmatrix} B'_{11} & 0 & B'_{13} \\ 0 & B'_{22} & B'_{23} \\ B'^*_{13} & B'^*_{23} & B'_{33} \end{pmatrix} \text{ case I} \quad (2)$$

and

$$A' = \begin{pmatrix} A'_{11} & 0 & 0 \\ 0 & A'_{22} & A'_{23} \\ 0 & A'^*_{23} & A'_{33} \end{pmatrix}, \quad B' = \begin{pmatrix} B'_{11} & B'_{12} & B'_{13} \\ B'^*_{12} & B'_{22} & 0 \\ B'^*_{13} & 0 & B'_{33} \end{pmatrix} \text{ case II}, \quad (3)$$

where A'_{11} is an eigenvalue. Of course, one can exchange the role of A' and B' so that $A' = M_d$ and $B' = M_u$.

To prove the existence of these bases and show how they are reached, it is easiest to first move to the basis where A is diagonal.

Denoting the eigenvalues of A by λ_i, $(i = 1, 2, 3)$ and $A'_{11} = \lambda_1$, we have in the eigenbasis of A

$$A = \begin{pmatrix} \lambda_1 & 0 & 0 \\ 0 & \lambda_2 & 0 \\ 0 & 0 & \lambda_3 \end{pmatrix}, \quad B = \begin{pmatrix} B_{11} & B_{12} & B_{13} \\ B^*_{12} & B_{22} & B_{23} \\ B^*_{13} & B^*_{23} & B_{33} \end{pmatrix}. \quad (4)$$

The unitary matrix which leads to the form for A' in both Eq. 2 (Case I) and Eq. 3 (caseII) is simply

$$U = \begin{pmatrix} 1 & 0 & 0 \\ 0 & x_2 & x_3 \\ 0 & y_2 & y_3 \end{pmatrix} \quad (5)$$

with the complex numbers x_2, x_3, y_2, y_3 subject to the orthonormality conditions. It is then trivial (and always possible) to find the appropriate combinations of x_2, x_3, y_2, y_3 that lead to either B' in the above two cases [7]. For instance in the first case, requiring $B'_{12} = x_2 B_{12} + x_3 B_{13} = 0$ gives the appropriate U. All other cases with two off-diagonal zeroes in one matrix and one in the other are treated in an analogous way. The proofs are obtained from case I and case II just by relabeling the indices. Of course, the case where the two quark matrices make up between them only two-zeroes being much less constrained is always easier to construct.

CKM MATRICES FROM OFF-DIAGONAL TEXTURE ZEROES HERMITIAN MATRICES

The advantage of texture zeroes matrices yet accommodating all the ten physical parameters is that they allow to easily express the mixing matrix solely in term of the elements describing the mass matrices. One could then work backward and use the hierarchy observed in a particular parameterization of the CKM mixing matrix, together with the hierarchy in the masses, to exhibit further correlations in the elements of the mass matrices expressed in a simple basis that already exhibits zeroes.

Recently, Rašin [10] has devised a procedure to express the CKM matrix as a function of the mass matrices in the general case where no zero element is found in neither M_u nor M_d. He uses a product of rotation matrices and phase matrices to diagonalise a general 3×3 matrix. However, even when we require M_u to be diagonal, which is a special case of [10], we are still left with large formulae which include sines and cosines of angles for which only the tangent is explicitly known. These results do not give compact expressions for the CKM matrix elements. Only when more zeroes are imposed do the results simplify. Even with the simple textures that are displayed in Table 1, the recipe given in [10] leads to tedious and complicated formulae [7] which moreover come with an ambiguity in determining the signs of the sines and cosines. We will show that, with the textures that are displayed in Table 1, there exists a more compact way of expressing the CKM that does not make use of any sines or cosines but exhibits the masses and the elements of the mass matrices explicitly.

Each combinations in Table 1 will lead to a particular parameterization of the mixing matrix. We concentrate on parameterization 14 not only to illustrate how the diagonalisation of the matrices is carried out exactly, and hence how one expresses the CKM, but also because it leads to a parameterization of the Kobayashi-Maskawa matrix which is directly related to the Wolfenstein parameterization [9].

To achieve this, we first apply a weak basis phase transformation to the form 14, such that the only remaining phase is located in the up quark matrix. Thus one is dealing with

$$M_u = \begin{pmatrix} u & 0 & ye^{i\phi} \\ 0 & \lambda_c & 0 \\ ye^{-i\phi} & 0 & t \end{pmatrix}, \quad M_d = \begin{pmatrix} d & x & 0 \\ x & s & z \\ 0 & z & b \end{pmatrix}. \quad (6)$$

Note that this parameterization allows to have as input, at the level of the mass matrices, the physical mass of the charm quark, λ_c. In what follows all physical masses will be denoted by λ_i, the index i being a flavour index.

These mass matrices are diagonalised through the following unitary matrices,

$$U_u = \begin{pmatrix} \sqrt{\frac{u_t}{\lambda_{ut}}} & 0 & \sqrt{\frac{u_u}{\lambda_{ut}}}e^{i\phi} \\ 0 & 1 & 0 \\ -\sqrt{\frac{u_u}{\lambda_{ut}}}e^{-i\phi} & 0 & \sqrt{\frac{u_t}{\lambda_{ut}}} \end{pmatrix}, U_d = \begin{pmatrix} \sqrt{\frac{b_d d_s d_b}{\Delta \lambda_{ds} \lambda_{db}}} & \sqrt{\frac{d_d b_s d_b}{\Delta \lambda_{ds} \lambda_{sb}}}\sigma & \sqrt{\frac{d_d d_s b_b}{\Delta \lambda_{db} \lambda_{sb}}} \\ -\sqrt{\frac{d_d b_d}{\lambda_{ds} \lambda_{db}}} & \sqrt{\frac{d_s b_s}{\lambda_{ds} \lambda_{sb}}} & \sqrt{\frac{d_b b_b}{\lambda_{db} \lambda_{sb}}} \\ \sqrt{\frac{d_d b_s b_b}{\Delta \lambda_{ds} \lambda_{db}}} & -\sqrt{\frac{b_d d_s b_b}{\Delta \lambda_{ds} \lambda_{sb}}}\sigma & \sqrt{\frac{b_d b_s d_b}{\Delta \lambda_{db} \lambda_{sb}}} \end{pmatrix},$$

$$(7)$$

such that

$$U_{u,d}^\dagger M_{u,d} U_{u,d} = \begin{pmatrix} \lambda_{u,d} & 0 & 0 \\ 0 & \lambda_{c,s} & 0 \\ 0 & 0 & \lambda_{t,b} \end{pmatrix}, \qquad (8)$$

and

$$x_i = |x - \lambda_i| \quad (e.g. \ u_t = |u - \lambda_t|), \qquad (9)$$
$$\lambda_{ij} = |\lambda_i - \lambda_j|, \qquad (10)$$
$$\Delta = |b - d|, \qquad (11)$$
$$\sigma = \text{sign of } (b - d). \qquad (12)$$

Expressing the diagonalising matrices, U_u, U_d, with the help of the physical masses keeps the expressions of these matrices very compact. As $V = U_u^\dagger U_d$, we can now write the CKM matrix exactly:

$$V_{us} = \frac{\sigma\left(\sqrt{u_t d_d b_s d_b} + \sqrt{u_u b_d d_s b_b} e^{i\phi}\right)}{\sqrt{\Delta \lambda_{ut} \lambda_{ds} \lambda_{sb}}}, \qquad (13)$$

$$V_{ub} = \frac{\sqrt{u_t d_d d_s b_b} - \sqrt{u_u b_d b_s d_b} e^{i\phi}}{\sqrt{\Delta \lambda_{ut} \lambda_{db} \lambda_{sb}}}, \qquad (14)$$

$$V_{cd} = -\sqrt{\frac{d_d b_d}{\lambda_{ds} \lambda_{db}}}, \qquad (15)$$

$$V_{cb} = \sqrt{\frac{d_b b_b}{\lambda_{db} \lambda_{sb}}}, \qquad (16)$$

$$V_{td} = \frac{\left(\sqrt{u_u b_d d_s d_b} e^{-i\phi} + \sqrt{u_t d_d b_s b_b}\right)}{\sqrt{\Delta \lambda_{ut} \lambda_{ds} \lambda_{db}}}, \qquad (17)$$

$$V_{ts} = \frac{\sigma\left(\sqrt{u_u d_d b_s d_b} e^{-i\phi} - \sqrt{u_t b_d d_s b_b}\right)}{\sqrt{\Delta \lambda_{ut} \lambda_{ds} \lambda_{sb}}}, \qquad (18)$$

$$J \equiv \frac{\det[M_u, M_d]}{2i\lambda_{uc}\lambda_{ut}\lambda_{ct}\lambda_{ds}\lambda_{db}\lambda_{sb}} = \frac{\sqrt{u_u u_t d_d d_s d_b b_d b_s b_b}}{\lambda_{ut}\lambda_{ds}\lambda_{db}\lambda_{sb}} \sin\phi. \qquad (19)$$

We see that these expressions are surprisingly simple given that they come from the mass matrices. Moreover, contrary to some ansätze, this type of matrix can always be made to fit the data.

Nonetheless, we are now in a position to exploit the mass hierarchies. One can take x_x as a small perturbation, which means that in fact $\lambda_f \simeq f$ where f refers to a diagonal element. We then have from eq. 15 and 16,

$$d_d \simeq |V_{cd}|^2 \lambda_s, \qquad (20)$$
$$b_b \simeq |V_{cb}|^2 \lambda_b. \qquad (21)$$

We also have from eq.14

$$V_{ub} \simeq \frac{1}{\sqrt{\lambda_t}}\left(\sqrt{\frac{d_d b_b}{\lambda_b}} - \sqrt{u_u}\, e^{i\phi}\right) \simeq -\sqrt{\frac{u_u}{\lambda_t}}\, e^{i\phi}, \qquad (22)$$

where we have made the additional assumption that the terms involving the down-quarks are quadratic in the "perturbation" $d_d \times b_b$ compared to the term originating from the up quark matrix: u_u. This additional assumption is stronger than the previous ones since it also compares the strengths of the off-diagonal elements of the up *and* down quark matrices. In any case with these mild assumptions one can now trade d_d, b_b, u_u, i.e. d, b, u for the modulii of V_{cd}, V_{cb} and V_{ub} and physical masses (up to some signs).

Taking into account the size of d_d, b_b and u_u, we can now write

$$V \simeq \begin{pmatrix} V_{ud} & \sqrt{\frac{d_d}{\lambda_s}} & -\sqrt{\frac{u_u}{\lambda_t}}e^{i\phi} \\ -\sqrt{\frac{d_d}{\lambda_s}} & V_{cs} & \sqrt{\frac{b_b}{\lambda_b}} \\ \sqrt{\frac{d_d b_b}{\lambda_s \lambda_b}} + \sqrt{\frac{u_u}{\lambda_t}}e^{-i\phi} & -\sqrt{\frac{b_b}{\lambda_b}} & V_{tb} \end{pmatrix}, \qquad (23)$$

$$J = \sqrt{\frac{u_u d_d b_b}{\lambda_s \lambda_b \lambda_t}}\, \sin\phi. \qquad (24)$$

It is interesting to see that in this parameterization the V_{CKM} can be split into elements which originate either solely from the down-quark sector[3] or the up-quark sector. To recover a phenomenologically viable mixing matrix, one could thus concentrate on each sector separately. Moreover this parameterization is equivalent to the standard Wolfenstein parameterization

$$V_W = \begin{pmatrix} 1 - \frac{1}{2}\lambda^2 & \lambda & \lambda^3 A(\rho - i\eta) \\ -\lambda & 1 - \frac{1}{2}\lambda^2 & \lambda^2 A \\ \lambda^3 A(1 - \rho - i\eta) & -\lambda^2 A & 1 - \mathcal{O}(\lambda^4) \end{pmatrix}, \qquad (25)$$

with

$$\lambda = \sqrt{\frac{d_d}{\lambda_s}},$$

$$A = \frac{\lambda_s}{d_d}\sqrt{\frac{b_b}{\lambda_b}},$$

$$\rho = -\sqrt{\frac{u_u \lambda_s \lambda_b}{\lambda_t d_d b_b}}\, \cos\phi,$$

$$\eta = \sqrt{\frac{u_u \lambda_s \lambda_b}{\lambda_t d_d b_b}}\, \sin\phi. \qquad (26)$$

---

[3] A similar observation has also been made in [11].

Asking for maximal CP violation [12] sets $\phi = \pi/2$ and leads to $\rho = 0$.

From the form of the V_{CKM} matrix it is now an easy matter to find phenomenologically viable quark mass matrices. Most direct from our study is the *general* feature that if $d = (M_d)_{11} = 0$, then $d_d = \lambda_d = m_d$ and therefore one has the rather successful prediction [13,14]: $V_{us} \simeq \lambda \simeq \sqrt{\lambda_d/\lambda_s} = \sqrt{m_d/m_s}$ Moreover, introducing the perturbative parameter $\epsilon \ll 1$ and with all other parameters of order 1, we may write the hierarchical matrices:

$$M_u = \lambda_t \begin{pmatrix} 0 & 0 & c\,\epsilon^3\,e^{i\phi} \\ 0 & \lambda_c/\lambda_t & 0 \\ c\,\epsilon^3\,e^{-i\phi} & 0 & 1 \end{pmatrix}, \quad M_d = \lambda_b \begin{pmatrix} 0 & a\,\epsilon^3 & 0 \\ a\,\epsilon^3 & \epsilon^2 & b\,\epsilon^2 \\ 0 & b\,\epsilon^2 & 1 \end{pmatrix}. \quad (27)$$

This leads to $V_{us} = \sqrt{m_d/m_s} = a\,\epsilon$ whereas $|V_{ts}| = b\,\epsilon^2 = (b/a^2)\,|V_{us}|^2$ $(b, a \sim 1)$. Therefore, if one identifies $\lambda = a\epsilon$ then $A = b/a^2$. We could "adjust" a, b (and c) to better fit the data. The forms in Eq. 27 bear some resemblance to those presented in [11], but note that we arrive at these forms from a rather different approach.

Also in the down sector, the ansatz reproduces the correct ratio of masses. Note also that with the ansatz for the up-quark, copied somehow on that of the down quark, we get: $|V_{ub}| \simeq c\,\epsilon^3$.

CONCLUSIONS

We have shown that without any assumption on the mass matrices apart from hermiticity, it is always possible to find a quark basis such that 3 off-diagonal elements are vanishing, allowing to diagonalise unambiguously the mass matrices and obtain the mixing matrix. The case where either M_u or M_d is diagonal (and therefore all the 6 vanishing elements are contained in one single matrix) is of special interest but leads to lengthy formulae for the CKM matrix entries. In all other cases, we arrived at compact formulae for the mixing matrix. These compact formulae that express without any approximation the V_{CKM} matrix in terms of the masses and other elements of the mass matrices can be compared to popular parameterizations of the CKM matrix. The exact forms that we find make it transparent which further assumptions one can make (*i.e* more zeroes) to simplify the structure of the mass matrices and yet be compatible with the data. We have given one such example, and in passing we have shown how starting from the general 10 parameter bases, the mere assumption of one extra zero in $(M_d)_{11}$ gives the famous relation [13,14] $V_{us} = \sqrt{m_d/m_s}$ which is seen then to be rather generic to a large class of models and ansätze. From there one can add more constraints, for example we have presented an new ansatz which can be made to fit the data quite well.

ACKNOWLEDGEMENTS

The work of M. B. is supported by La Fondation de l'Université du Québec à Montréal and the work of C. H. is supported in part by N.S.E.R.C. of Canada.

REFERENCES

1. P. Ramond, R.G. Roberts and G.G. Ross, Nucl. Phys. **B406**, 19 (1993).
2. G.C. Branco, L. Lavoura and F. Mota, Phys. Rev. **D39**, 3443 (1989).
3. H. Fritzsch, Phys. Lett. **B73** (1978) 317.
4. D. Falcone, O. Pisanti, L. Rosa, Phys. Rev. **D57**, 195 (1998);
   L. Rosa, Talk given at the South European School on Elementary Particle Physics, Lisbon, October 6-15, 1997;
   D. Falcone and F. Tramontano, hep-ph/9806496.
5. Y. Koide, Mod. Phys. Lett. A **12** (1997) 2655.
   E. Takasugi, Prog. Theor. Phys. **98** (1997) 177.
6. K. Harayama and N. Okamura, Phys. Lett. **B387** 614 (1996).
7. M. Baillargeon, F. Boudjema, C. Hamzaoui and J. Lindig, in preparation.
8. E. Ma, Phy. Rev. **D43**, R2761 (1990).
9. L. Wolfenstein, Phys. Rev. Lett. **51** (1983) 1945.
10. A. Rašin, hep-ph/9802356.
11. W.S. Hou and G.G. Wong, Phy. Rev. **D52**, 5269 (1994).
12. G. Bélanger, C. Hamzaoui and Y. Koide, Phys. Rev. **D45**, 4186 (1992); B. Margolis, C. Hamzaoui and S. Punch, Talk given at the 17th Annual MRST 95 Meeting on High Energy Physics, Rochester, NY, May 8-9, 1995, hep-ph/9506214.
13. R. Gatto, G. Sartori and M. Tonin, Phys. Lett. **B28** (1968) 128.
    N. Cabbibo and L. Maiani, Phys. Lett. **B28** (1968) 131.
14. S. Weinberg, in " A Festschrift for I.I. Rabi" Trans. N.Y. Acad. Sci. Ser. II (1977), v. 38], p. 185; F. Wilczek and A. Zee, Phy. Lett. **70B**, 418 (1977); F. Wilczek and A. Zee, Phy. Rev. Lett. **42**, 421 (1979).

Power Counting in Non-Relativistic Effective Field Theories

Michael Luke

Department of Physics, University of Toronto
60 St. George St., Toronto, Canada
M5S 1A7
E-mail: luke@medb.physics.utoronto.ca

Abstract. The issue of a consistent power counting scheme in nonrelativistic theories is discussed, with NRQCD and Yukawa theory used as examples.

I INTRODUCTION

In this talk I want to discuss the issue of power counting in nonrelativistic effective field theories. In particular, I will be interested in the problem of scattering heavy particles near threshold in an effective field theory. This has been the subject of a great deal of interest in the recent literature, in a number of different contexts:

- non-relativistic QCD (NRQCD) and its applications to quarkonium production and decay [1]

- NRQED, and its applications to precision calculations in QED bound states [2]

- $e^+e^- \rightarrow$ hadrons near threshold, including $\bar{t}t$ production [3] and non-relativistic sum rules [4]

- nucleon-nucleon scattering from chiral perturbation theory [5].

I will focus on $\bar{Q}Q$ production in this talk, although the approach discussed can also be applied to chiral perturbation theory and nucleon-nucleon scattering [6].

The advantages of setting up such a calculation in an effective field theory are well-known, but let me reiterate them. Since we are interested in processes near threshold, nonrelativistic effects such as heavy pair production are suppressed, and it is simplest to work in an effective field theory where such effects have been integrated out. This has the advantage of calculational simplicity, but also the practical advantage that it gives one a simple way to organize a calculation, forcing one to concentrate on the relevant degrees of freedom of the problem. Symmetries and other simplifications of the nonrelativistic theory become manifest in the Lagrangian, and corrections to the nonrelativistic limit are straightforward to take into account systematically.

The last point is of crucial importance. An effective field theory is a nonrenormalizable theory, and so contains an infinite number of operators. In order to determine which operators to keep and which to neglect to a given order, it is necessary to have a power counting scheme. In some situations, such as the effective four-fermi theory obtained by integrating out the W and Z bosons in the Standard Model, the power counting is obvious, just given by the dimension of the operator. A nonrelativistic effective theory is somewhat more subtle, since there are two important low-energy scales characterizing the non-relativistic nature of the problem which are not independent, the typical energy $\frac{1}{2}mv^2$ and the typical three-momentum mv. However, once the power counting scheme is established, the effective theory provides a simple and straightforward way to calculate the corrections to the leading order nonrelativistic theory in a systematic way.

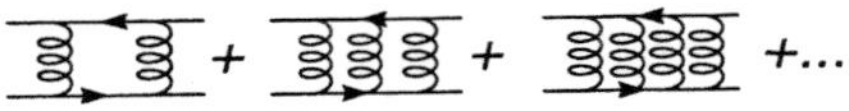

FIGURE 1. Ladder graphs in $\bar{Q}Q$ scattering.

As an example, consider the derivation of the Schrödinger equation from quantum field theory. As is well known, for heavy particle scattering near threshold perturbation theory breaks down because the ladder graphs in Fig. I are enhanced by kinematic factors. This is because for regions of the loop momentum both intermediate states in the one-loop box graph are almost on shell at the same time. This enhances it relative to other one loop graphs by a factor of $1/v$, where v is the relative three-velocity of the heavy particles. Similarly, an n-loop ladder graph is enhanced by $1/v^n$, so for $v \lesssim \alpha_s$, perturbation theory breaks down and the most singular parts of all the ladder graphs must be summed to all orders. This corresponds to solving the Schrödinger equation in a Coulomb potential. The appropriate power counting scheme for the nonrelativistic effective theory should therefore reduce to nonrelativistic quantum mechanics for two particles in an instantaneous Coulomb potential at the leading order, with corrections calculable order by order in the heavy particle three-velocity v (and other ratios of scales, such as $\Lambda_{\rm QCD}/m_Q v^2$, which characterizes non-perturbative effects).

Let us compare this expectation with the usual form of the NRQCD Lagrangian [1]:

$$\mathcal{L}_{\rm NRQCD} = -\frac{1}{2}(\mathbf{E}^2 - \mathbf{B}^2) + \psi^\dagger \left(iD_0 + \frac{\mathbf{D}^2}{2m_Q} + c_1 \frac{\mathbf{D}^4}{8m_Q^3} + \ldots \right)\psi$$
$$+\psi^\dagger \left(\frac{c_2 g}{2m_Q}\sigma \cdot \mathbf{B} + \frac{c_3 g}{8m_Q^2}(\mathbf{D}\cdot\mathbf{E} - \mathbf{E}\cdot\mathbf{D}) + \frac{c_4 g}{8m_Q^2}i(\mathbf{D}\times\mathbf{E} - \mathbf{E}\times\mathbf{D}) \right)\psi$$
$$+\ldots \tag{1}$$

where $\mathbf{E} = \boldsymbol{\nabla} A_0^a T^a$ and $\mathbf{B} = \boldsymbol{\nabla} \times \mathbf{A}^a T^a$ are the chromo-electric and -magnetic fields, and the coefficients $c_1 - c_4$ are all $1 + O(\alpha_s)$. By construction, this Lagrangian reproduces the physics of QCD order by order in v; in Ref. [7] a set of power counting rules were given which determine the v-scaling of any operator in $\mathcal{L}$ in quarkonium (that is, a $\bar{Q}Q$ state where the quarks have typical energies mv^2 and typical momenta mv). The first four operators in (1) are relative $O(1)$, while the next four are relative $O(v^2)$.

However, these velocity counting rules do not in themselves constitute a complete power counting scheme as discussed above. For example, the $1/v^n$ enhancement of ladder graphs is not apparent from these rules; nor is the fact that gluon exchange is instantaneous in the nonrelativistic theory[1]. The resulting theory is not manifestly nonrelativistic quantum mechanics at leading order.

The problem is simply that in this form, the effective Lagrangian has no small parameter - the expansion parameter v does not appear in $\mathcal{L}$, but rather characterizes the external states. Since the v dependence isn't manifest, there are still factors of v hidden in the graphs which can only be discovered upon closer inspection. It is the purpose of this talk to show how to make the v dependence explicit, giving a systematic approach to the nonrelativistic expansion in quantum field theory.

II MANIFEST VELOCITY SCALING

To make power counting manifest, we consider the relevant scales in the problem. For quarks, the typical energy is of order mv^2, whereas the typical three-momentum is of order mv. For gluons, things are a bit more complicated - the relevant scales depend on what one is interested in. The typical energies and momenta of gluons exchanged between scattering quarks scale like quark energies and momenta, of order mv^2 and mv respectively. Such gluons are far off shell and effectively instantaneous. On the other hand, on-shell gluons radiated from nonrelativistic quarks have energies and momenta of order mv^2. Operators in the effective theory containing spatial derivatives will therefore scale differently with v for these two modes of gluon, and so they must be treated separately in the effective theory if power counting is to be maintained. We will refer to the two modes as "potential" and "radiation" gluons.

A Quarks and Potential Gluons

It is simple to rewrite the theory so that the powers of v are manifest in the Lagrangian for nonrelativistic quarks and potential gluons [6]. Since the three-

---

[1] In Coulomb gauge, A_0 exchange is instantaneous; however this does not generalize to higher orders, since transverse gluons (which give a spin-dependent potential) are not manifestly instantaneous. Furthermore, in a theory such as Yukawa theory there is no gauge freedom to make scalar exchange instantaneous.

momenta of interest are of order mv and the energies of order mv^2, the v-scaling may be made explicit by introducing new coordinates

$$\vec{X} \equiv m_Q v \vec{x}, \quad T = m_Q v^2 t \tag{2}$$

in terms of which the rescaled momenta $\vec{P} = \vec{p}/m_Q v$ and $\varepsilon = E/m_Q v^2$ are both of order one. Thus, derivatives in the rescaled theory are all of order one, and the only place v enters is in the couplings; the v-scaling of operators is now given by their coefficients.

This also solves the problem of sensibly renormalizing the theory in dimensional regularization: by treating subleading operators as insertions, rather than including them at leading order, loop integrals are guaranteed not to introduce additional factors of v, and thus operators will not mix under renormalization with operators of lower order in v.

After performing the appropriate rescaling of the fields ψ and A^μ to ensure that their kinetic terms are correctly normalized, the effective Lagrangian (1) becomes (in Lorentz gauge with gauge parameter ξ)

$$\begin{aligned}\mathcal{L} = & \Psi^\dagger (i\partial_0 - g/\sqrt{v}\mathcal{A}_0)\Psi - \frac{1}{2}\Psi^\dagger (i\vec{\nabla} - g\sqrt{v}\vec{\mathcal{A}})^2 \Psi \\ & -\frac{1}{4}(\partial_i \mathcal{A}_j - \partial_j \mathcal{A}_i - g\sqrt{v} f_{abc}\mathcal{A}_i^b \mathcal{A}_j^c)^2 \\ & +\frac{1}{2}(\partial_i \mathcal{A}_0 - v\partial_0 \mathcal{A}_i - g\sqrt{v} f_{abc}\mathcal{A}_i^b \mathcal{A}_j^c)^2 - \frac{1}{2\xi}(v\partial_0 \mathcal{A}_0 - \vec{\nabla} \cdot \vec{\mathcal{A}})^2 \\ & + \ldots \end{aligned} \tag{3}$$

where Ψ and $\mathcal{A}$ are rescaled quark and potential gluon fields, respectively.

This simple rescaling gives the required results:

- $\bar{\Psi}\partial_0\Psi$ and $\frac{1}{2m_Q}\bar{\Psi}\vec{\partial}^2\Psi/(2m_Q)$ are the same order.

- The effective $\mathcal{A}_0$ coupling is $g/\sqrt{v}$, so for $v \lesssim \alpha_s$, the theory is strongly coupled and $\mathcal{A}_0$ exchange must be summed to all orders.

- $\partial_0 \mathcal{A}_i$ (and $\partial_0 \mathcal{A}_0$ from the gauge fixing term) are subleading in v, so the gauge field propagators are

$$D_{00} = \frac{i}{\vec{k}^2}, \quad D_{il} = \frac{i}{\vec{k}^2}\left[\delta_{ij} - (1-\xi)\frac{k_i k_j}{\vec{k}^2}\right].$$

Note that this corresponds to Lorentz, not Coulomb, gauge; the instantaneous potential is manifest in any gauge.

- Only ladder graphs survive, to all orders in v. This is simply because the Feynman rules for the nonrelativistic theory enforce time-ordering. The q_0 integral for a crossed-box loop integral, for example, vanishes because the poles are both on the same side of the axis.

Taken together, these results give the expected conclusion that for $v \lesssim \alpha_s$, all ladder graphs must be summed, using an instantaneous potential, reproducing the results of NRQM. Also note that the triple potential gluon vertex is $O(g\sqrt{v})$; thus, the appropriate coupling constant to use in the Schrödinger equation is $\alpha_s(mv)$ (note that to take into account the running of the coupling between m and mv correctly, one must be careful to match onto NRQCD at the scale mv).

Thus, this description works well for the potential gluons. Unfortunately, it completely misses the physics of the radiation gluons! This is simply because the gluon propagator doesn't have the right structure to describe real gluons, since there is no pole at $k_0^2 = \vec{k}^2$. Even if one is not interested in processes with external gluons, potential gluons cannot reproduce the analytic structure of full QCD - for example, the infrared divergence of the one-loop renormalization of the coupling is not reproduced, since the relevant graphs all vanish with potential gluons (they are not box graphs). Thus, without adding additional low-energy degrees of freedom (the radiation gluons), the effective theory cannot reproduce the infrared physics of full QCD, and therefore is not the correct low-energy theory.

B Radiation Gluons

The need for radiation gluons in addition to the potential was pointed out in Ref. [8]. Since both the energy and three-momentum are of order mv^2 for radiation gluons, the appropriate coordinate rescaling is

$$T = m_Q v^2 t, \quad \vec{Y} = m_Q v^2 \vec{x}. \tag{4}$$

The radiation gluons are much longer wavelength that the quarks and potential gluons, and so the quark-radiation gluon interaction must be expanded in powers of v to put all the powers of v in the couplings [8,9]:

$$\begin{aligned}\mathcal{L}_{int} &= -\frac{i}{2}gv\Psi_h^\dagger(\mathbf{X},T)\overset{\leftrightarrow}{\nabla}_i\Psi_h(\mathbf{X},T)\tilde{\mathcal{A}}_i(v\mathbf{X},T)\\ &= -\frac{i}{2}gv\Psi_h^\dagger(\mathbf{X},T)\overset{\leftrightarrow}{\nabla}_i\Psi_h(\mathbf{X},T)\left[1+vX\cdot\nabla+\ldots\right]\tilde{\mathcal{A}}_i(0,T).\end{aligned} \tag{5}$$

Note that since the theory is no longer translationally invariant, three-momentum is not conserved at the radiation gluon-quark vertex. This is in accord with the v power counting, since $mv^2 \ll mv$. Thus radiation gluons do not contribute to potential scattering to all orders in v, since they cannot transfer three-momentum between quarks.

Having established the velocity power counting, it is usually simplest just to use the unrescaled Lagrangian, but with both potential and radiation gluons included [10]:

$$\mathcal{L} = \psi_h^\dagger\left(i\partial_0 + \frac{\nabla^2}{2m} - gA_P^0 - gA_R^0(0,t)\right)\psi_h - \frac{1}{4}\left(\nabla^i A_P^j - \nabla^j A_P^i\right)^2$$

$$+\frac{1}{2}\left(\boldsymbol{\nabla} A_P^0\right)^2 - \frac{1}{4}G_R^{\mu\nu}G_{\mu\nu R} - \frac{1}{2\xi}\left(\boldsymbol{\nabla}\cdot\mathbf{A}_P\right)^2 - \frac{1}{2\xi}\left(\partial_\mu A_R^\mu\right)^2$$
$$+O(v, g\sqrt{v}), \tag{6}$$

where the subscripts P and R denote potential and radiation gluons, $G_R^{\mu\nu}$ is the field strength tensor for radiation gluons, and α is the gauge parameter. For practical calculations, this version of the Lagrangian is much more convenient than the rescaled version - the rescaling just gives a simple set of counting rules for each derivative and factor of each field in the effective theory. This set of rules differs from those in [7] in that it makes the complete v dependence in the theory manifest.

III MATCHING CONDITIONS

To see that all of this is working, it is instructive to calculate an explicit matching condition between the full and non-relativistic effective theories. Since the effective theory is required to reproduce the same infrared physics as the full theory, the matching conditions are only infrared safe if both gluon modes are included. This provides a nontrivial check (particularly at subleading orders) that the effective theory has the correct degrees of freedom.

In Ref. [10] this was shown explicitly in the case of an external current producing a fermion-antifermion pair in both QCD and a simple non-relativistic Yukawa theory (NRY). Consider first the nonrelativistic Yukawa theory, regulated in the infrared by the addition of a small scalar mass m_φ. The external current

$$J_\mu \bar\psi \gamma^\mu \psi. \tag{7}$$

matches at leading order in v onto the current in the effective theory

$$J_i\left(\chi_h^\dagger \sigma^i \psi_h + \psi_h^\dagger \sigma^i \chi_h\right) \tag{8}$$

(as well as other operators containing two fermions or two anti-fermions). To do the matching at one loop, we need the one-loop matrix element of the current in both theories. In the full theory, the infrared divergent part of the one-loop matrix element is

$$i\mathcal{A}_{IR} = \left[-\frac{ig^2}{8\pi\beta} + \frac{g^2}{2\pi^2} + O(\beta)\right] J_i u_h^\dagger \sigma^i v_h \ln m_\varphi + \ldots, \tag{9}$$

where u_h and v_h are two-component spinors,

$$\beta = \sqrt{1 - \frac{4m^2}{s}} \tag{10}$$

is the magnitude of the three-velocity of each fermion, and $s = (p_1 + p_2)^2$ is the invariant mass of the fermion-antifermion pair. The first term in Eq. (9) is singular as $\beta \to 0$, corresponding to the infinite complex anomalous dimension found in HQET at threshold. Since it is imaginary, it does not contribute to the decay rate at $O(g^2)$. The second term in Eq. (9) cancels in physical matrix elements with scalar bremsstrahlung.

In order to be able to match onto NRY at one loop, both of these divergences must be reproduced in the low energy theory (the singularities higher order in β will only be reproduced when operators higher order in the velocity are included in the effective theory). Diagrams with both potential φ_P and radiation φ_R scalars contribute to the amplitude in NRY. The wavefunction graphs with φ_P exchange vanish, while the vertex graph gives

$$i\mathcal{A}_P^V = -\frac{ig^2}{8\pi\beta} J_i u_h^\dagger \sigma^i v_h \ln m_\varphi + \ldots \qquad (11)$$

and reproduces the first term in Eq. (9). The sum of radiation scalar graphs gives

$$i\mathcal{A}_R = \frac{g^2}{2\pi^2} J_i u_h^\dagger \sigma^i v_h \ln m_\varphi + \ldots , \qquad (12)$$

reproducing the second term in Eq. (9).

Having demonstrated with an explicit scalar mass that the matching is infrared safe, it is simplest to regulate both the infrared and ultraviolet divergences in dimensional regularization. In this case, one-loop integrals containing radiation scalars are of the form

$$\int \frac{d^d k}{(2\pi)^d} f(k_0, k^2), \qquad (13)$$

which has no mass scale and so vanishes in dimensional regularization. Thus, radiation scalars (or gluons, in QCD) do not contribute to the one-loop matching conditions. In this scheme, the infrared safety of the matching conditions is demonstrated by the fact that the effective theory reproduces all terms which are non-analytic in the external momenta, whose effects cannot be mocked up via local operators. In Ref. [10] this was shown explicitly for an external vector current in QCD,

$$J^\mu = \bar\psi \gamma^\mu \psi. \qquad (14)$$

In the effective theory, this matches onto the following operators to relative $O(v^2)$,

$$J_\mu \bar\psi \gamma^\mu \psi \to c_1 \mathbf{O}_1 + c_2 \mathbf{O}_2 + c_3 \mathbf{O}_3 + \ldots + O(v^4) \qquad (15)$$

where

$$\begin{aligned}
\mathbf{O}_1 &= J_i \psi_h^\dagger \sigma^i \chi_h \\
\mathbf{O}_2 &= \frac{1}{4m^2} J_i \left[\psi_h^\dagger \left(\vec{\nabla} \cdot \boldsymbol{\sigma} \vec{\nabla}^i + \overleftarrow{\nabla} \cdot \boldsymbol{\sigma} \overleftarrow{\nabla}^i \right) \chi_h \right] \\
\mathbf{O}_3 &= \frac{1}{2m^2} J_i \left[\psi_h^\dagger \sigma^i \left(\vec{\nabla}^2 + \overleftarrow{\nabla}^2 \right) \chi_h \right]
\end{aligned} \qquad (16)$$

FIGURE 2. One loop contributions to quark-antiquark production in QCD.

The one-loop matrix element of this operator between heavy quark states near threshold is given by the diagrams in Fig. 2. Expanding the result in powers of three-momentum gives

$$i\mathcal{A}_{\text{QCD}} = u_h^\dagger \sigma^i v_h \left(1 - \frac{2g^2}{3\pi^2}\right) - \frac{1}{2m^2} u_h^\dagger \mathbf{p} \cdot \sigma \mathbf{p}^i v_h \left(1 - \frac{g^2}{3\pi^2}\right)$$
$$+ \frac{g^2}{12\pi^2} u_h^\dagger \sigma^i v_h \left[\frac{m}{|\mathbf{p}|}\left(\pi^2 + i\pi\left(\gamma_E + \frac{2}{d-4} + \ln\frac{\mathbf{p}^2}{\pi\mu^2}\right)\right)\right.$$
$$+ \frac{3|\mathbf{p}|}{2m}\left(\pi^2 + i\pi\left(\gamma_E - 2 + \frac{2}{d-4} + \ln\frac{\mathbf{p}^2}{\pi\mu^2}\right)\right) \quad (17)$$
$$\left. + \frac{\mathbf{p}^2}{3m^2}\left(\frac{2}{3} - 8\gamma_E - \frac{16}{d-4} - 8\ln\frac{m^2}{4\pi\mu^2}\right)\right]$$
$$+ \frac{g^2}{12\pi^2} \frac{u_h^\dagger \mathbf{p} \cdot \sigma \mathbf{p}^i v_h}{2m^2}\left[\frac{m}{|\mathbf{p}|}\left(-\pi^2 - i\pi\left(\gamma_E - 2 + \frac{2}{d-4} + \ln\frac{\mathbf{p}^2}{\pi\mu^2}\right)\right)\right]$$
$$+ O(v^3).$$

To reproduce this in NRQCD to $O(v^2)$, order, subleading terms in the NRQCD Lagrangian are required. The kinetic term is

$$\mathcal{L}_h = \psi_h^\dagger \left(i\partial_0 + \frac{\nabla^2}{2m}\right)\psi_h + \chi_h^\dagger \left(i\partial_0 - \frac{\nabla^2}{2m}\right)\chi_h$$
$$+ \frac{1}{8m^3}\left(\psi_h^\dagger \nabla^4 \psi_h - \chi_h^\dagger \nabla^4 \chi_h\right) + O(v^4). \quad (18)$$

At subleading order, the Coulomb potential gluons couple via the Darwin and spin-orbit terms

$$\mathcal{L}_{D,SO} = \frac{g}{8m^2}\left(\psi_h^\dagger T^a \psi_h + \chi_h^\dagger T^a \chi_h\right)\nabla^2 A_P^{0a} \quad (19)$$
$$+ i\frac{g}{4m^2}\epsilon^{ijk}\left(\psi_h^\dagger T^a \sigma^i \nabla^j \psi_h + \chi_h^\dagger T^a \sigma^i \nabla^j \chi_h\right)\nabla^k A_P^{0a} + O(g^3)$$

while transverse potential gluons couple through the operators

$$\mathcal{L}_{p \cdot A, F} = \frac{g}{2m}\left(\psi_h^\dagger (\mathbf{A}_P \cdot \nabla + \nabla \cdot \mathbf{A}_P)\psi_h - \chi_h^\dagger (\mathbf{A}_P \cdot \nabla + \nabla \cdot \mathbf{A}_P)\chi_h\right)$$
$$- \frac{g}{2m}\left(\psi_h^\dagger \sigma \cdot (\nabla \times \mathbf{A}_P)\psi_h - \chi_h^\dagger \sigma \cdot (\nabla \times \mathbf{A}_P)\chi_h\right) + O(g^3). \quad (20)$$

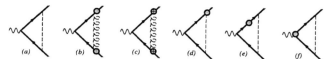

FIGURE 3. One loop contributions to quark-antiquark production in NRQCD. The dashed line corresponds to a potential A_0 gluon, the dashed gluon line to a potential A_i gluon. The shaded circles represent (b) the $\mathbf{p}\cdot\mathbf{A}$ vertex, (c) the Fermi vertex, (d) the Darwin vertex, (e) the relativistic kinematic correction to the fermion leg, and (f) $\mathbf{O}_2$. Implicit in both (d) and (e) are graphs with the same operator insertion on the antiquark line. The wavefunction graphs vanish.

The NRQCD amplitude is then given by the diagrams in Fig. 3,

$$
\begin{aligned}
i\mathcal{A}_{\text{NRQCD}} &= c_1 u_h^\dagger \sigma^i v_h - \frac{c_2}{2m^2} u_h^\dagger \mathbf{p}\cdot\sigma \mathbf{p}^i v_h \\
&+ \frac{g^2}{12\pi^2} u_h^\dagger \sigma^i v_h \left[c_1 \frac{m}{|\mathbf{p}|} \left(\pi^2 + i\pi \left(\gamma_E + \frac{2}{d-4} + \ln\frac{\mathbf{p}^2}{\pi\mu^2} \right) \right) \right. \\
&\left. + \frac{3|\mathbf{p}|}{2m} \left(c_1 \left(\pi^2 + i\pi \left(\gamma_E - \frac{5}{3} + \frac{2}{d-4} + \ln\frac{\mathbf{p}^2}{\pi\mu^2} \right) \right) - \frac{i\pi}{3} c_2 \right) \right] \\
&+ \frac{g^2}{12\pi^2} \frac{u_h^\dagger \mathbf{p}\cdot\sigma \mathbf{p}^i v_h}{2m^2} \left[\frac{m}{|\mathbf{p}|} \left(-c_2 \left(\pi^2 + i\pi \left(\gamma_E - 3 + \frac{2}{d-4} + \ln\frac{\mathbf{p}^2}{\pi\mu^2} \right) \right) \right.\right. \\
&\left.\left. - i\pi c_1 \right) \right].
\end{aligned}
\tag{21}
$$

As required, all the nonanalytic dependence on the external momentum cancels in the matching. This result is also gauge independent.

Comparing Eqs. (17) and (21) gives the coefficients $c_1 - c_3$ (regulating the low-energy theory as usual in $\overline{\text{MS}}$) to $O(g^2)$:

$$
\begin{aligned}
c_1 &= 1 - \frac{8\alpha_s}{3\pi} + O(\alpha_s^2) \\
c_2 &= 1 - \frac{4\alpha_s}{3\pi} + O(\alpha_s^2) \\
c_3 &= -\frac{\alpha_s}{9\pi}\left(\frac{2}{3} - 8\ln\frac{m^2}{\mu^2}\right) + O(\alpha_s^2).
\end{aligned}
\tag{22}
$$

The result for c_1 reproduces the familiar short-distance correction to $e^+e^- \to \bar{Q}Q$ near threshold [11], whereas c_2 and c_3 generalize this to $O(v^2)$. Note that the bare c_1 is finite while the bare c_3 is divergent. This reflects the fact that there are no infrared or ultraviolet divergences in the amplitude at $O(v^0)$ since the quarks are in a colour singlet state, and therefore cannot radiate a gluon at leading order in the multipole expansion.

IV CONCLUSIONS

We have introduced a power counting scheme which makes the velocity power counting in NRQCD manifest, corresponding to non-relativistic quantum mechanics with systematic calculation of subleading effects. This was accomplished by introducing two distinct gluon fields in the effective theory. Potential gluons propagate instantaneously and give rise to the QCD potential, whereas radiation gluons do not contribute to potential scattering, but correspond to on-shell gluons.

The infrared divergences arising in fermion-antifermion production in Yukawa theory at order v^{-1} and v^0 were shown to be reproduced in the nonrelativistic effective theory only when both potential and radiation scalars were included, and the matching conditions at that order were shown to be analytic in the external momentum. Finally, the matching conditions for quark-antiquark production by an external vector current were computed in NRQCD to $O(g^2 v^2)$.

ACKNOWLEDGMENTS

The work reported on here was done in collaboration with Aneesh Manohar and Martin J. Savage.

REFERENCES

1. W.E. Caswell and G.P. Lepage, Phys. Lett. **B167**, 437 (1986); G.T. Bodwin, E. Braaten, and G.P. Lepage, Phys. Rev. **D51**, 1125 (1995); Phys. Rev. **D55**, 5853 (1997) (E).
2. W. E. Caswell and G. P. Lepage, Phys. Rev. **A20**, 36 (1979); P. Labelle, G. P. Lepage and U. Magnea, Phys. Rev. Lett. **72**, 2006 (1994).
3. V. S. Fadin and V. A. Khoze, Yad. Phys. **48**, 487 (1988) [*Sov. J. Nucl. Phys.* **48**, 309 (1988)]; M. J. Strassler and M. E. Peskin, Phys. Rev. **D 43**, 1500 (1991); A. H. Hoang and T. Teubner, hep-ph/9801397.
4. V. A. Novikov et. al., Phys. Rept. **41**, 1 (1978); M. B. Voloshin, Int. J. Mod. Phys. **A10**, 2865 (1995).
5. S. Weinberg, Phys. Lett. **B51**, 288 (1990); Nucl. Phys. **B363**, 3 (1991); Phys. Lett. **B295**, 114 (1992); D.B. Kaplan, M.J. Savage and M.B. Wise, Nucl. Phys. **B478**, 629 (1996).
6. M. Luke and A. V. Manohar, Phys. Rev. **D55**, 4129 (1997).
7. G.P. Lepage et. al., Phys. Rev. **D46**, 4052 (1992).
8. B. Grinstein and I. Z. Rothstein, Phys. Rev. **D57**, 78 (1998).
9. P. Labelle, hep-ph/9608491.
10. M. Luke and M. J. Savage, Phys. Rev. **D57**, 413 (1998).
11. R. Karplus and A. Klein, Phys. Rev. **87**, 848 (1952); R. Barbieri, R. Gatto, R. Kögerler and Z. Kunzst, Phys. Lett. **B57**, 455 (1975).

Photoproduction of h_c

Sean Fleming* and Thomas Mehen[†]

Physics Department, The University of Toronto 60 St. George St., Toronto, Ontario M5S 1A7, Canada

[†]*California Institute of Technology, Pasadena, California 91125, U.S.A.*

Abstract. Using the NRQCD factorization formalism, we calculate the total cross section for the photoproduction of h_c mesons. We include color-octet and color-singlet mechanisms as well as next-to-leading order perturbative QCD corrections. The theoretical prediction depends on two nonperturbative matrix elements that are not well determined from existing data on charmonium production. For reasonable values of these matrix elements, the cross section is large enough that the h_c may be observable at the E831 experiment and at the HERA experiments.

Of all the charm-anticharm quark boundstates lying below the threshold of open charm production, the most elusive to experimental investigation is the h_c meson. To date this particle has been observed by only a few experiments, which measure an average mass of $M_{h_c} = 3.52614 \pm 0.00024$ GeV [1]. Studies of the h_c are difficult because it has quantum numbers $J^{PC} = 1^{+-}$, and thus cannot be produced resonantly in e^+e^- annihilation, or appear in the decay of a $J^{PC} = 1^{--}$ charmonium state via an electric dipole transition.

Since charm quarks are heavy compared to Λ_{QCD}, it is natural to view charmonium as a nonrelativistic system, where the h_c is a spin-singlet P-wave state of a c and $\bar{c}$. This approach is taken in the color-singlet model [2] (CSM) of quarkonium production and decay. In the CSM, it is assumed that the $c\bar{c}$ must be produced in a color-singlet state with the same angular-momentum quantum numbers as the charmonium meson which is eventually observed. However, the CSM has serious deficiencies. It is well known that perturbative QCD calculations of production and decay of P-wave quarkonia within the CSM are plagued by infrared divergences [3]. More recently, measurements made at the Fermilab Collider Detector Facility (CDF) show that the CSM also fails to accurately predict the production cross sections of S-wave quarkonia (J/ψ and ψ') [4].

The naive CSM has been supplanted by the NRQCD factorization formalism of Bodwin, Braaten, and Lepage [5]. This formalism allows the infrared safe calculation of inclusive charmonium production and decay rates. It also predicts new mechanisms in which a $c\bar{c}$ pair is produced at short distances in a color-octet state, and hadronizes into a final state charmonium nonperturbatively. These color-octet mechanisms can naturally account for the CDF data on J/ψ and ψ' production.

In this paper, we examine photoproduction of h_c within the NRQCD factorization formalism. Color-octet mechanisms play an essential role. By itself the CSM yields infrared divergent expressions for the h_c cross section; however, once color-octet contributions are included we obtain sensible predictions. The resulting expression then depends on two undetermined nonperturbative matrix elements. Using heavy-quark spin symmetry these can, in principle, be estimated from similar matrix elements extracted from data on the production and decay of χ_{cJ} mesons. Currently, theoretical and experimental uncertainties in the determination of the χ_{cJ} matrix elements preclude us from making definitive predictions for the h_c photoproduction cross section. This will be discussed in greater detail below. However, for some choices of the matrix elements which are consistent with current data, we find a large enough cross section that the h_c may be observed in current photoproduction experiments at the DESY HERA collider and at the Fermilab fixed target experiment E831.

In the NRQCD factorization formalism, inclusive quarkonium production cross sections have the form of a sum of products of short-distance coefficients and NRQCD matrix elements. The short-distance coefficients are associated with the production of a heavy quark-antiquark pair in specific color and angular-momentum states. They can be calculated using ordinary perturbative techniques, and are thus an expansion in the strong coupling constant α_s. The NRQCD matrix elements parameterize the hadronization of the quark-antiquark pair, and each scales as a power of the average relative velocity v of the heavy quark and antiquark as determined by the NRQCD velocity-scaling rules [6].

The NRQCD factorization formula for h_c photoproduction, at leading order in v, is

$$\sigma(\gamma + N \to h_c + X) = \int dx \sum_i f_{i/N}(x) \left[\hat{\sigma}(\gamma + i \to c\bar{c}(\mathbf{8}, {}^1S_0) + X) \langle \mathcal{O}_8^{h_c}({}^1S_0) \rangle^{(\mu)} + \right.$$
$$\left. \hat{\sigma}(\gamma + i \to c\bar{c}(\mathbf{1}, {}^1P_1) + X; \mu) \langle \mathcal{O}_1^{h_c}({}^1P_1) \rangle \right] , \quad (1)$$

where $f_{i/N}(x)$ is the probability of finding a parton i in the nucleon with a fraction x of the nucleon momentum. $\hat{\sigma}(\gamma + i \to c\bar{c}(\mathbf{8}, {}^1S_0) + X)$ is the short-distance coefficient for producing a $c\bar{c}$ pair in a color-octet ${}^1S_0$ configuration, and $\hat{\sigma}(\gamma + i \to c\bar{c}(\mathbf{1}, {}^1P_1) + X; \mu)$ is the short-distance coefficient for producing a $c\bar{c}$ pair in a color-singlet ${}^1P_1$ configuration. The matrix element $\langle \mathcal{O}_{1(8)}^{h_c}({}^{2S+1}L_J) \rangle$ describes the hadronization of a color-singlet (color-octet) ${}^{2S+1}L_J$ $c\bar{c}$ pair into an h_c. According to the v-scaling rules, both matrix elements in Eq. (1) scale as v^5. The scale μ appearing in Eq. (1) arises from the factorization of the cross section into long-distance and short-distance contributions. As we discuss below, the μ dependence of the color-octet matrix element is cancelled by the μ dependence of the color-singlet short-distance coefficient so that the expression for the cross section is μ

independent. Note that in addition to the NRQCD scale μ shown explicity in Eq. (1) there is a renormalization scale and a factorization scale (associated with the parton distribution function $f_{i/N}(x)$) which have been suppressed. These three scales do not have to be the same, however, in our numerical calculations we choose them to be equal.

The short-distance coefficients in Eq. (1) can be calculated using the techniques of Ref. [7]. At leading order in α_s, we obtain the following expression for the color-octet short-distance coefficient:

$$\frac{d\hat{\sigma}}{dz}(\gamma + g \to c\bar{c}(8, {}^1S_0)) = \frac{\pi^3 \alpha \alpha_s e_c^2}{4m_c^5} \delta(1-z), \qquad (2)$$

where $z = \rho/x$, $\rho = 4m_c^2/S_{\gamma N}$, and $S_{\gamma N}$ is the photon-nucleon center-of-mass energy squared. We calculate the color-singlet cross section, regulating the infrared divergences using dimensional regularization. The result in the $\overline{MS}$ scheme is:

$$\frac{d\hat{\sigma}}{dz}(\gamma + g \to c\bar{c}(1, {}^1P_1) + g; \mu) = \frac{8\pi^2 \alpha \alpha_s^2 e_c^2}{27 m_c^7} \left[f(z) + \frac{z^4(1+z^2)}{(1+z)^2} \frac{1}{(1-z)_\rho} \right.$$
$$\left. + 2\,\delta(1-z)\left(\ln(1-\rho) - \frac{5}{6}\right) \right] - \frac{\pi^3 \alpha \alpha_s e_c^2}{4m_c^5} \delta(1-z) \left[\frac{16\alpha_s}{27\pi m_c^2} \ln\left(\frac{\mu}{2m_c}\right) \right], \qquad (3)$$

where

$$f(z) = -\frac{z^5 \ln(z)}{(1+z)^3} + \frac{z^2(5 + 3z + 14z^2 + 2z^3 + 9z^4 - z^5)\ln(z)}{(1-z)^3(1+z)^5}$$
$$+ \frac{z^2(1 + z + 10z^2 + 4z^3 + 15z^4 - z^5 + 2z^6)}{(1-z)^2(1+z)^5}, \qquad (4)$$

and the functional distribution is defined by

$$\int_\rho^1 dz\, f(z) \left(\frac{1}{1-z}\right)_\rho = \int_\rho^1 dz\, \frac{f(z) - f(1)}{1-z}. \qquad (5)$$

The NRQCD expression for the cross section is obtained by substituting Eqs. (2) and (3) into Eq. (1). Note that a $1/\epsilon$ divergence in the color-singlet coefficient has been absorbed into the definition of the leading color-octet matrix element. The renormalized matrix element $\langle \mathcal{O}_8^{h_c}({}^1S_0)\rangle^{(\mu)}$ can be shown to obey the renormalization group equation [5],

$$\mu \frac{d}{d\mu} \langle \mathcal{O}_8^{h_c}({}^1S_0)\rangle^{(\mu)} = \frac{16\alpha_s}{27\pi m_c^2} \langle \mathcal{O}_1^{h_c}({}^1P_1)\rangle. \qquad (6)$$

Thus the logarithmic dependence on μ of the short-distance coefficient $\sigma(\gamma + g \to c\bar{c}(1, {}^1P_1 + X; \mu)$ is canceled to this order by the μ dependence of the renormalized matrix element $\langle \mathcal{O}_8^{h_c}({}^1S_0)\rangle^{(\mu)}$.

Note that the leading color-octet coefficient is $O(\alpha_s)$ while the leading color-singlet coefficient is $O(\alpha_s^2)$. Therefore, next-to-leading order QCD corrections to the color-octet coefficient are of the same order as the leading color-singlet coefficient, and must be included if we are to have a complete $O(\alpha_s^2)$ calculation of h_c photoproduction. The $O(\alpha_s^2)$ corrections to the color-octet contribution are computed in Ref. [8]. These corrections are included in our calculation of the cross section.

In order to make a prediction for the h_c photoproduction cross section, we must determine the values of the nonperturbative matrix elements $\langle \mathcal{O}_8^{h_c}(^1S_0)\rangle^{(\mu)}$ and $\langle \mathcal{O}_1^{h_c}(^1P_1)\rangle$. To eliminate large logarithms in the expression for the cross section we choose $\mu = 2m_c$. At this time there does not exist a direct measurement of these matrix elements. However, they are related to similar matrix elements for χ_{cJ} production and decay by the (approximate) heavy quark spin symmetry of NRQCD:

$$\langle \mathcal{O}_1^{\chi_{c1}}(^3P_1)\rangle = \langle \mathcal{O}_1^{h_c}(^1P_1)\rangle \left(1 + O(v^2)\right)$$
$$\langle \mathcal{O}_8^{\chi_{c1}}(^3S_1)\rangle = \langle \mathcal{O}_8^{h_c}(^1S_0)\rangle \left(1 + O(v^2)\right). \qquad (7)$$

The size of the $O(v^2)$ corrections in Eq (7) can be estimated by studying radiative χ_c and ψ' decays, where predictions based on heavy quark spin symmetry agree with experiment to 20% accuracy [9].

$\langle \mathcal{O}_1^{\chi_{c1}}(^3P_1)\rangle$ can be extracted from χ_{cJ} decays, or from the decay $B \to \chi_{cJ} + X$. The authors of Ref. [10] calculate inclusive hadronic χ_{cJ} decay including next to leading order α_s corrections. The result of their fit is:

$$\frac{\langle \mathcal{O}_1^{\chi_{c1}}(^3P_1)\rangle}{m_c^2} = 0.115 \pm 0.016 \text{ GeV}^3. \qquad (8)$$

The error includes only experimental uncertainties. It is particularly important to note that Ref. [10] does not include $O(v^2)$ relativistic corrections which are numerically of the same size as the $O(\alpha_s)$ perturbative corrections that have been included. Relativistic corrections to the J/ψ decay rate are large [11]. In the case of χ_{cJ} decay, relativistic corrections need to be analyzed before $\langle \mathcal{O}_1^{\chi_{c1}}(^3P_1)\rangle$ can be extracted with confidence.

The decay $B \to \chi_{cJ} + X$ is calculated in Ref. [12]. Measurements of B decay [13] allow an extraction of $\langle \mathcal{O}_1^{\chi_{c1}}(^3P_1)\rangle$:

$$\frac{\langle \mathcal{O}_1^{\chi_{c1}}(^3P_1)\rangle}{m_c^2} = 0.42 \pm 0.16 \text{ GeV}^3. \qquad (9)$$

Again, the error quoted above is only due to experimental uncertainties. The authors of Ref. [12] state that their calculation suffers from large theoretical uncertainties due to next to leading order QCD corrections to the Wilson coefficients and the subprocess $b \to c\bar{c}s$. Note that the two extractions of $\langle \mathcal{O}_1^{\chi_{c1}}(^3P_1)\rangle$ agree only at the 2σ level.

$\langle \mathcal{O}_8^{\chi_{c1}}(^3S_1)\rangle$ can be extracted from the decay $B \to \chi_{cJ} + X$, and from CDF data on χ_{cJ} production. The result of the fit to B decay, after running from the scale m_b to $2m_c$ and converting from a cutoff regularization scheme to dimensional regularization, is

$$\langle \mathcal{O}_8^{\chi_{c1}}(^3S_1)\rangle^{(2m_c)} = (2.9 \pm 2.0) \times 10^{-2} \text{ GeV}^3 \ . \tag{10}$$

As before only experimental errors are included. The result of two different fits to Tevatron data are

$$\langle \mathcal{O}_8^{\chi_{c1}}(^3S_1)\rangle = 0.98 \pm 0.13 \times 10^{-2} \text{ GeV}^3 \tag{11}$$

$$\langle \mathcal{O}_8^{\chi_{c1}}(^3S_1)\rangle^{(2m_c)} = 2.6 \times 10^{-2} \text{ GeV}^3 \ , \tag{12}$$

where the values are taken from Refs. [14] and [15] respectively. The error of Ref. [14] includes only experimental uncertainty; Ref. [15] does not quote errors. The central values of both fits lie within the 1σ error of the extraction from B-decays.

Note that in the χ_{cJ} production calculations of Ref. [14], there is no color-singlet contribution that gives an infrared divergence which needs to be absorbed into the definition of the color-octet matrix element, as is done in our calculation. Therefore, it is not possible to relate the the "bare" matrix element appearing in Eq. (11) with the renormalized matrix element needed for our calculation. However, the fragmentation calculation of χ_{cJ} production carried out in Ref. [15] makes use of the $g \to \chi_{cJ}$ fragmentation function. This fragmentation function includes both color-octet and color-singlet contributions, and has an infrared divergence in that is absorbed into the definition of the color-octet matrix element. This allows us to import the extracted value given in Eq. (12) into our calculation.

Not all values for $\langle \mathcal{O}_8^{h_c}(^1S_0)\rangle$ and $\langle \mathcal{O}_1^{h_c}(^1P_1)\rangle$ result in a physically sensible prediction for the cross section. Once the infrared divergence from the color-singlet contribution to the cross section is factorized, the remaining finite contribution is actually negative (for positive $\langle \mathcal{O}_1^{h_c}(^1P_1)\rangle/m_c^2$). If the ratio of the color-octet to color-singlet matrix elements is too small, it is possible to obtain physically meaningless results. This is the case if, for example, the central values of the color-singlet matrix element extracted from B-decays (Eqs. (9,10)) is used. Therefore it is impossible to put a lower bound on the cross section given our current state of ignorance concerning this matrix element. It is also important to point out that the NRQCD velocity scaling rules imply that $\langle \mathcal{O}_8^{h_c}(^1S_0)\rangle^{(2m_c)}$ and $\langle \mathcal{O}_1^{h_c}(^1P_1)\rangle/m_c^2$ should be roughly the same size. Thus, theory would prefer a smaller value for the matrix element $\langle \mathcal{O}_1^{h_c}(^1P_1)\rangle$, as suggested by the fit to χ_{cJ} decay. Clearly more accurate extractions from the Tevatron, from B-decays, and from χ_{cJ} decays are needed to clarify the situation. This may be possible once next-to-leading perturbative QCD corrections and leading relativistic corrections to these processes are calculated.

For the NRQCD matrix elements, we will use $\langle \mathcal{O}_1^{h_c}(^1P_1)\rangle/m_c^2 = 0.115$ GeV^3 as determined by the analysis of χ_{cJ} decays (Eq. (8)), and we choose $\langle \mathcal{O}_8^{h_c}(^1S_0)\rangle^{(2m_c)} =$

2.6×10^{-2} GeV^3. We use the CTEQ 3M parton distribution function with the factorization scale choosen to be equal to the NRQCD scale $\mu = 2m_c$. The resulting cross section is plotted as the solid line in Fig. 1. However, our results are extremely sensitive to the uncertainty in the determination of the matrix elements. To show this, we also plot, as the dashed line, the cross section with the choice $\langle \mathcal{O}_1^{h_c}(^1P_1)\rangle/m_c^2 = 0.2$ GeV^3. The cross section then drops by roughly a factor of four.

We conclude with a brief discussion of the possibility of observing the h_c. The E831 photoproduction experiment corresponds to roughly $\sqrt{S_{\gamma N}} = 20$ GeV, while the HERA experiments take photoproduction data at approximately $\sqrt{S_{\gamma N}} = 100$ GeV. At these energies the cross section for h_c production, assuming our first choice of NRQCD matrix elements, is 30 nb and 62 nb respectively. These cross sections are comparable to J/ψ production. The h_c can be detected either via the rare decay $h_c \to J/\psi + \pi$ with branching ratio on the order of 1% [16], or the radiative decay $h_c \to \eta_c + \gamma$ (BR = 50%) [17]. If it is possible to reconstruct these decay modes, then the possibility of observing the h_c is real, and should be experimentally investigated.

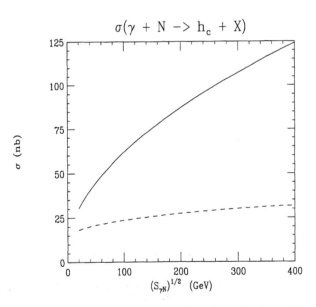

FIGURE 1. The total h_c photoproduction cross section $\sigma(\gamma + N \to h_c + X)$ as a function of the center of mass energy $\sqrt{S_{\gamma N}}$. The solid curve is for the choice $\langle \mathcal{O}_1^{h_c}(^1P_1)\rangle/m_c^2 = 0.115$ GeV^3 and $\langle \mathcal{O}_8^{h_c}(^1S_0)\rangle^{(2m_c)} = 2.6 \times 10^{-2}$ GeV^3. The dashed line is obtained by keeping the same value for the color-octet matrix element and by exit changing the value of the color-singlet matrix element to $\langle \mathcal{O}_1^{h_c}(^1P_1)\rangle/m_c^2 = 0.2$ GeV^3.

We would like to thank Christian Bauer, Adam Falk, Martin Gremm, Mike Luke, and Mark Wise for helpful discussions. T.M. would also like to thank the Fermilab summer visitors program for their hospitality while some of this work was being done. The work of S.F. is supported by NSERC. The work of T.M. is supported in part by the Department of Energy under grant number DE-FG03-ER 40701 and by a John A. McCone Fellowship.

REFERENCES

1. See The Review of Particle Physics, Phys. Rev. D**54**, 1 (1996) and references therein.
2. For an extensive review of the color singlet model, see G. A. Schuler, CERN-TH-7170-94, hep-ph/9403387 (unpublished).
3. R. Barbieri, R. Gatto, and E. Remiddi, Phys. Lett. **61B**, 465 (1976); R. Barbieri, M. Caffo, and E. Remiddi, Nucl. Phys. **B162**, 220 (1980); R. Barbieri, M. Caffo, R. Gatto, and E. Remiddi, Phys. Lett. **95B**, 93 (1980); ibid. **B192**, 61 (1981).
4. A. Sansoni (CDF Collaboration), Nucl. Phys. **A610**, 373c (1996).
5. G. T. Bodwin, E. Braaten, and G. P. Lepage, Phys. Rev. D **51** 1125 (1995).
6. G. P. Lepage, L. Magnea, C. Nakhleh, U. Magnea, and K. Hornbostle, Phys. Rev. D **46**, 4052 (1992).
7. E. Braaten and Y. Chen, Phys. Rev. D**54**, 3216 (1996); S. Fleming and I. Maksymyk, ibid. **54**, 3608 (1996); E. Braaten and Y. Chen, ibid. **55**, 2693 (1997).
8. F. Maltoni, M. L. Mangano, and A. Petrelli, CERN-TH/97-202, hep-ph/9708349 (unpublished).
9. P. Cho and M. Wise, Phys. Lett. **346B**, 129 (1995).
10. M. L. Mangano and A. Petrelli, Phys. Lett. **B352**, 445 (1995).
11. W. -Y. Keung and I. J. Muzinich, Phys. Rev. D **27**, 1518 (1983); M. Gremm and A. Kapustin, Phys. Lett. **B407**, 323 (1997).
12. G. T. Bodwin, E. Braaten, G. P. Lepage, and T. C. Yuan, Phys. Rev. D **46** R3703 (1992).
13. CLEO collaboration, R. Balest et al. in the proceedings of the Int. Conf. on High Energy Physics, Glasgow, July 1994 (Ref. GLS0248).
14. P. Cho and A. Leibovich, Phys. Rev. D **53**, 150 (1996); ibid. **53**, 6203 (1996).
15. M. Cacciari, M. Greco, M.L. Mangano, A. Petrelli, Phys. Lett. **356B**, 553 (1995); E. Braaten, S. Fleming, and T. C. Yuan, Ann. Rev. Nucl. Part. Sci. 46, 197 (1996).
16. M. B. Voloshin, Sov. J. Nucl. Phys. 43, 1011 (1986).
17. G. T. Bodwin, E. Braaten, and G. P. Lepage, Phys. Rev. D**46**, 1914 (1992).

$\bar{\Lambda}$ and λ_1 from inclusive B - decays

Christian Bauer

*Department of Physics, University of Toronto, 60 St. George Street,
Toronto, Ontario, M5S-1A7 Canada*

Abstract. In this talk I report on the possibility of extracting the heavy quark matrix elements $\bar{\Lambda}$ and λ_1 from inclusive B decays. In particular, I will focus on the extraction from moments of the rare decay $B \to X_s \gamma$ and discuss the associated theoretical uncertainties.

I INTRODUCTION

In the past few years our understanding of inclusive decays of heavy hadrons has increased dramatically. In [1] it was demonstrated that inclusive quantities in heavy quark decays can be expanded in terms of local operators. This is due to the presence of two widely separated scales, the hadronic scale Λ_{QCD} and the heavy quark pole mass m_h, leading to an expansion in Λ_{QCD}/m_h. At leading order the parton model result is reproduced and matrix elements of higher dimensional operators express the nonperturbative dynamics of the heavy quark inside the heavy hadron. The first corrections to the parton model result arise from matrix elements of dimension 5 operators. They can be parameterized by

$$\lambda_1 = \frac{1}{2M_B} \left\langle B \left| \bar{b}_v (iD)^2 b_v \right| B \right\rangle$$
$$\lambda_2 = \frac{1}{2M_B} \left\langle B \left| \frac{g}{2} \bar{b}_v \sigma_{\mu\nu} (iD^\mu)(iD^\nu) b_v \right| B \right\rangle, \qquad (1)$$

Since these operators are suppressed by a factor of $1/m_h^2$ they give an order 1% correction for B decays. The operators given in Eq. (1) together with the heavy quark pole mass m_h are the unknown parameters at this relative order. Note that usually one uses the constituent mass of the light degrees of freedom,

$$\bar{\Lambda} = M_B - m_b + \frac{(\lambda_1 + 3\lambda_2)}{2M_B} + \mathcal{O}\left(\frac{1}{M_B^2}\right), \qquad (2)$$

as a parameter instead of the pole quark mass.

Since the inclusive rate $\Gamma(B \to X_c e\bar{\nu})$, has been measured very accurately, a precise knowledge of the three parameters mentioned above should enable the determination of the CKM matrix element V_{cb} with a theoretical uncertainty of about 1%. This is one of the main reasons why there has been a tremendous interest in determining the parameters λ_1 and $\bar{\Lambda}$ [1]. QCD sum rules were first used by Ball and Braun [2] to obtain $\lambda_1 = -0.6 \pm 0.1\,\text{GeV}^2$ and later, in a more sophisticated way, by Neubert [3] who found $\lambda_1 = -0.1 \pm 0.05\,\text{GeV}^2$. There have also been lattice calculations [4] which yield $\lambda_1 = 0.09 \pm 0.14\,\text{GeV}^2$.

In this talk I want to focus on a method for determining these parameters in a model independent way directly from experiment through moments of inclusive B decays which can be reliably calculated. In the next section I will give a brief review of inclusive B decays. In section III I will discuss various ways to relate λ_1 and $\bar{\Lambda}$ to measurable quantities and explain how to estimate the nonperturbative uncertainties. I will show that the rare decay $B \to X_s \gamma$ can play an important role in obtaining the two parameters with small uncertainties. In section IV I will investigate the rare decay mode in some more detail, focusing on uncertainties from a cut on the photon energy that has to be imposed to perform the experimental measurement.

II INCLUSIVE B DECAYS

To discuss inclusive B decays, I will investigate the decay of a B meson via a current J_μ. For $B \to X_c e\bar{\nu}$, where X_c is any final state containing a c quark, this would be the weak current $J_\mu = \bar{c}\gamma_\mu \frac{1}{2}(1 - \gamma^5)b$, whereas for other decay modes the current can be more complicated. The quark level decay $b \to c e\bar{\nu}$ can be calculated perturbatively, but in order to compare this to the measured rate, one has to determine the matrix element $\langle X_c | \bar{c}\Gamma_\mu b | B \rangle$ which can not be calculated perturbatively. For inclusive modes, however, the optical theorem can be used to relate the hadronic tensor to a current - current correlator

$$W_{\mu\nu} = \sum_{X_c} d[P.S.]_{hadr} (2\pi)^4 \delta^4(P_B - P_X - q) \langle B | J_\mu^\dagger | X_c \rangle \langle X_c | J_\nu | B \rangle \qquad (3)$$

$$= 2\text{Im}(T_{\mu\nu}), \qquad (4)$$

where

$$T_{\mu\nu} = \int d^4x \langle B | T(J_\mu J_\nu^\dagger) | B \rangle. \qquad (5)$$

In a region far from the physical cuts the current - current correlator can be computed using an operator product expansion (OPE) as illustrated in Fig.1. The total rate (as well as all other inclusive quantities such as moments of a spectrum)

---

[1] The parameter λ_2 is given by the mass splitting of the vector and the pseudoscalar mesons $\lambda_2 = \frac{M_B^2 - M_{B^*}^2}{4} \approx 0.12\,\text{GeV}^2$.

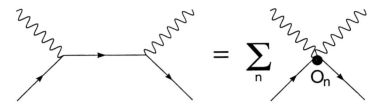

FIGURE 1. A diagrammatic representation of the OPE.

can therefore be expanded in inverse powers of m_b in addition to the perturbative expansion in α_s. As mentioned in the introduction, the leading order in the $1/m_b$ expansion reproduces the parton level result. In a very similar fashion, higher moments of inclusive spectra can be calculated as well.

III EXTRACTING λ_1 AND $\bar{\Lambda}$ AND THEORETICAL UNCERTAINTIES

To determine λ_1 and $\bar{\Lambda}$ from experiment one needs two independent quantities that can be determined from experiment and which can be calculated using the formalism explained in the previous section. There are three main suggestions in the literature. Two of them use moments of the weak decay $B \to X_c e \bar{\nu}$, whereas the third uses moments of the energy spectrum of the rare decay $B \to X_s \gamma$. To simplify the presentation, I will neglect the perturbative corrections to the moments in the following, which have of course all been included in the various analysis. The complete results to the order $\mathcal{O}\left(\alpha_s^2, \bar{\Lambda}\alpha_s, 1/m_b^3\right)$ can be found in the original papers.

A The decay $B \to X_c e \bar{\nu}$

Using the lepton energy spectrum one can calculate the two ratios [6]

$$R_1 = \frac{\int_{1.5\text{GeV}} E_l \frac{d\Gamma}{dE_l} dE_l}{\int_{1.5\text{GeV}} \frac{d\Gamma}{dE_l} dE_l} = 1.8 - 0.3\frac{\bar{\Lambda}}{M_B} - 0.4\frac{\bar{\Lambda}^2}{M_B^2} - 2.3\frac{\lambda_1}{M_B^2} - 4.0\frac{\lambda_1}{M_B^2} + \ldots \quad (6)$$

$$R_2 = \frac{\int_{1.7\text{GeV}} \frac{d\Gamma}{dE_l} dE_l}{\int_{1.5\text{GeV}} \frac{d\Gamma}{dE_l} dE_l} = 0.7 - 0.3\frac{\bar{\Lambda}}{M_B} - 0.7\frac{\bar{\Lambda}^2}{M_B^2} - 1.6\frac{\lambda_1}{M_B^2} - 4.9\frac{\lambda_1}{M_B^2} + \ldots, \quad (7)$$

which have been measured by CLEO [7]. Using their central values we show the two solutions in the $\lambda_1 - \bar{\Lambda}$ plane as the solid lines in Fig.2.

Another possibility is to use the hadron invariant mass spectrum of the same decay. This measurement has recently become feasible due to the development of the neutrino reconstruction technique, which requires a cut on the lepton energy.

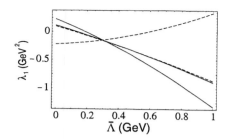

FIGURE 2. The two solutions for λ_1 and $\bar{\Lambda}$. The solid lines are for the lepton energy spectrum and the dashed lines correspond to the hadron invariant mass spectrum.

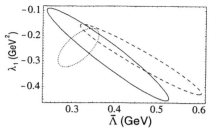

FIGURE 3. The 68 % confidence level ellipses. The solid and the dashed ellipse correspond to the lepton energy spectrum and the hadron invariant mass spectrum of $B \to X_c e \bar{\nu}$, respectively. The dotted ellipse corresponds to the $B \to X_s \gamma$ decay.

Using the preferred experimental cut of 1.5 GeV, the first two moments have been calculated to be [8]

$$M_1 = \langle s - \bar{M}_D^2 \rangle = M_B^2 \left[0.2 \frac{\bar{\Lambda}}{M_B} + 0.2 \frac{\bar{\Lambda}}{M_B} + 1.4 \frac{\lambda_1}{M_B^2} + 0.2 \frac{\lambda_1}{M_B^2} \right] + \ldots \quad (8)$$

$$M_2 = \langle (s - \bar{M}_D^2)^2 \rangle = M_B^4 \left[0.2 \frac{\bar{\Lambda}^2}{M_B^2} - 0.1 \frac{\lambda_1}{M_B^2} \right] + \ldots, \quad (9)$$

where $\bar{M}_D = (M_D + 3 M_{D^*})/4$ is the spin averaged D meson mass. The measurement of these two moments is currently in progress at CLEO. For the purpose of illustration we assume the hypothetical values $M_1 = 0.30 \, \mathrm{GeV}^2$ and $M_2 = 0.96 \, \mathrm{GeV}^4$ and plot the two solutions as dashed lines in Fig.2.

There are several uncertainties to the values extracted via these methods. There are, of course, the experimental errors to the measured quantities. There are also theoretical uncertainties, which can be divided into perturbative and nonperturbative. The size of the perturbative uncertainties can be estimated by calculating the $\alpha_s^2 \beta_0$ contributions, where $\beta_0 = 11 - 2n_f/3$ is the first term in the QCD beta function. The nonperturbative errors are estimated by calculating the moments to relative order $1/m_b^3$ [9]. The right hand sides of Eqs. (6) to (9) then receive additional contributions from six new and unknown matrix elements. Two of those, ρ_1 and ρ_2, are matrix elements of local operators and the other four, $\mathcal{T}_1$ to $\mathcal{T}_4$, are matrix elements of time ordered products The size of these uncertainties are estimated by varying these six unknown parameters in the range of values as expected by dimensional analysis [9]. Using the constraint obtained from a vacuum saturation approximation $\rho_1 \geq 0$ and a relation between ρ_2, $\mathcal{T}_3$ and $\mathcal{T}_4$ obtained from the mass splitting of the vector and pseudoscalar B and D mesons, a reasonable range for these parameters is $0 \leq \rho_1 \leq (0.5 \mathrm{GeV})^3$ and $-(0.5 \mathrm{GeV})^3 \leq \{\rho_2, \mathcal{T}_i\} \leq (0.5 \mathrm{GeV})^3$. The 68% confidence level ellipses in the $\bar{\Lambda} - \lambda_1$ plane are shown in Fig.3 by the solid and the dashed ellipse.

B   The decay $B \to X_s \gamma$

Unfortunately, the uncertainties on the two parameters extracted from these two methods are very large and, furthermore, the two error ellipses are parallel to each other. It would therefore be useful to have additional means of extracting the two parameters. It has been shown that moments of the photon spectrum of the rare decay $B \to X_s \gamma$ can do exactly that [10]. Since at the parton level this two body decay yields a monochromatic spectrum $d\Gamma/dE_\gamma \sim \delta(E_\gamma - m_b/2)$, the shape of the spectrum should contain interesting nonperturbative information. Calculating the mean photon energy and the variance, one finds (omitting perturbative contributions once again)

$$\langle E_\gamma \rangle = \frac{M_B - \bar{\Lambda}}{2} + \ldots$$
$$\mathrm{var}(E_\gamma) = -\frac{\lambda_1}{12} + \ldots, \qquad (10)$$

yielding two orthogonal lines in the $\bar{\Lambda} - \lambda_1$ plane. The nonperturbative uncertainties can be estimated in the same way as outlined in the previous paragraph [11] and one finds the error ellipse shown in Fig.3 by the dotted ellipse [2]. It is important to notice that the size of the uncertainties from this decay mode are considerably less than for the $b \to c$ modes and, moreover, the 68 % confidence level ellipse is orthogonal to the other two ellipses. Thus a combination of the two methods seems to be a promising method for extracting $\bar{\Lambda}$ and λ_1.

IV   UNCERTAINTIES FROM THE PHOTON ENERGY CUT

The signature for the rare decay $B \to X_s \gamma$ is a hard photon with energy of about 2.5 GeV. Unfortunately, bremsstrahlung photons from the weak decay $b \to c e \bar{\nu}$ can be as hard as 2.3 GeV. To eliminate this strong background, a cut on the photon energy at 2.2 GeV has been used in the CLEO analysis [12], which means that the moments measured are not fully inclusive quantities any more.

At present the cut implemented by the CLEO group reduces the available phase space to $\Delta \approx 500$ MeV, which is of order Λ_{QCD}. It can be shown that the convergence of the OPE breaks down for $\Delta \sim \Lambda_{QCD}$. However, it is only the most singular terms in the expansion that blow up [13] and they may be resummed into the so called shape function. In this section I will investigate the uncertainties on the theoretical quantities once a cut on the photon energy is included.

---

[2]   Note that since the two moments have not been measured accurately yet, the absolute position of this ellipse is unknown; only the relative size has meaning.

A Model independent bounds on the uncertainties

The moments with a cut on the photon energy, E_0, as measured by CLEO, are defined as

$$M_n^{E_0} = \frac{\int_0^{E_{max}} \theta(E-E_0) E^n \frac{d\Gamma}{dE} dE}{\Gamma^{E_0}}, \qquad (11)$$

where $\Gamma^{E_0} = \int_{E_0}^{E_{max}} \frac{d\Gamma}{dE} dE$. It is easy to see that $M_n^{E_0} \geq M_n$. It is possible to obtain bounds on these moments by replacing the θ function in (11) with a function $P(E, E_0)$ with the following properties [14]

$$\begin{aligned} P(E,E_0) &\leq 1 \quad \text{for} \quad E_0 \leq E \leq E_{max} \\ P(E,E_0) &\leq 0 \quad \text{for} \quad 0 \leq E \leq E_0, \end{aligned} \qquad (12)$$

which ensure that the function $P(E, E_0)$ is smaller than $\theta(E - E_0)$. Using this function we can now obtain the inequality

$$\int_0^{E_{max}} P(E,E_0) \frac{d\Gamma}{dE} dE = \Gamma_P^{E_0} \leq \Gamma^{E_0} \leq \Gamma. \qquad (13)$$

For a monotonically increasing function $P(E, E_0)$ it is also possible to obtain bounds on the moments as measured by experiment

$$\frac{\int_0^{E_{max}} P(E,E_0) E^n \frac{d\Gamma}{dE}}{\Gamma_P^{E_0}} = M_{n_P}^{E_0} \geq M_n^{E_0} \geq M_n. \qquad (14)$$

As discussed in the beginning of this section, the most singular terms in the OPE have to be resummed into a shape function

$$\frac{d\Gamma_s}{dE} = \Gamma \sum_{n=0}^{\infty} \left(\frac{m_b}{2}\right) A_n \delta^{(n)} \left(\frac{m_b}{2} - E\right), \qquad (15)$$

where the coefficient A_n now only contains the leading term in the $1/m_b$ expansion. Although the differential decay rate depends on an infinite number of unknown coefficients A_n (14), with the choice of a polynomial for the function $P(E, E_0)$ only a finite number of them contribute to the bounds. Since only the first 3 coefficients A_n are known, we have to use a polynomial of degree $s = 3 - n$ for the n'th moment. In Fig.4 and 5 the function

$$P(E,E_0) = 1 - \left(\frac{E_1 - E}{E_1 - E_0}\right)^s, \quad E_0 < E_1 \qquad (16)$$

has been used to obtain the bounds on the total rate and the first moment of the photon energy spectrum. [3] One can see that for the present cut at 2.2 GeV

---

[3] The bounds are only applicable if the shape function is positive definite. Models, such as the ACCMM model [15], do give a positive shape function.

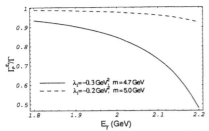
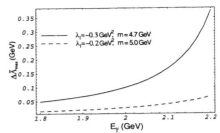

FIGURE 4. Bounds on the total rate for different values of λ_1 and m_b.

FIGURE 5. Bounds on the uncertainties on $\bar{\Lambda}$ for different values of λ_1 and m_b.

the bounds do not restrict the uncertainties on $\bar{\Lambda}$ to any satisfactory level. This of course does not necessarily mean that the uncertainties are large, it could just mean that the bounds are very weak. For a lower value of the cut, say 2 GeV, the uncertainties can be bound to a much smaller value.

For the second moment only a first degree polynomial can be used so that no strong bound can be obtained. In the next section we therefore use a model for the shape function to estimate the uncertainty on the parameters λ_1 and $\bar{\Lambda}$.

B Uncertainties using the ACCMM model

I will use a simplified version of the ACCMM model [15] to estimate the value of the moments as measured in the presence of a cut on the photon energy. Assuming a Gaussian distribution of the relative momentum of the b–quark inside the B meson, neglecting the mass of the s–quark and the momentum dependence of the b–quark mass, the spectral function is given by [16]

$$\frac{1}{\Gamma}\frac{d\Gamma}{dE} = \frac{1}{\sqrt{2\pi}\sigma_E} \exp\left\{-\frac{(E-\frac{m_b}{2})^2}{2\sigma_E^2}\right\}, \quad (17)$$

where $\sigma_E^2 = -\frac{\lambda_1}{12}$.

The difference between the first moment in the ACCMM model with and without a cut gives an estimate of the uncertainty on the value of $\bar{\Lambda}$ and the result is shown in Fig.6 for different values of σ_E.

For a cut at 2.2 GeV, the uncertainty on the parameter $\bar{\Lambda}$ in this model is between 20 MeV and 180 MeV, depending on the width of the spectrum. Since the real effect of the photon energy cut could easily exceed this estimate by a factor of two or three, a cut on the photon energy at 2.2 GeV could destroy the possibility of accurately determining the value of $\bar{\Lambda}$. If the cut could be lowered to $\approx$ 2 GeV, then an accurate extraction should be possible. This is in agreement with the model independent results obtained in the last section.

The result for a similar calculation for the variance of the spectrum is shown in Fig.7. For a cut of 2.2 GeV, the uncertainty on λ_1 in this model is between

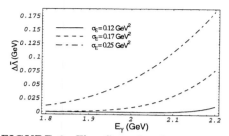
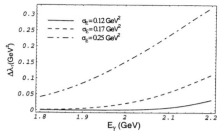

FIGURE 6. The effect of a photon energy cut on the extraction of $\bar{\Lambda}$ for different values of the parameters on the shape function F.

FIGURE 7. The effect of a photon energy cut on the extraction of λ_1 for different values of λ_1.

$0.05\,\text{GeV}^2$ and $0.3\,\text{GeV}^2$, Again, considering the fact that this is only a model calculation and it might underestimate the effect considerably, this indicates that an extraction of λ_1 from the present CLEO measurement might be unreliable. Lowering the cut to ≈ 1.9 GeV should enable a precise determination of this parameter from the decay $B \to X_s\gamma$.

V CONCLUSIONS

I have reviewed methods available for extracting the heavy quark matrix elements λ_1 and $\bar{\Lambda}$ from moments of inclusive B - decays. I have shown that moments of the photon spectrum in the rare decay $B \to X_s\gamma$ can provide extremely useful information on these two parameters, since the nonperturbative uncertainties in this decay mode are considerably less than for the weak decay $B \to X_c e\bar{\nu}$.

In the second part of the talk I have investigated the uncertainties due to a cut on the photon energy which has to be imposed in the measurement to reduce the strong background from bremsstrahlung photons from the decay $B \to X_c e\bar{\nu}$, which can be as energetic as 2.3 GeV. I have used an approach suggested in [14] to obtain a model independent bound on the uncertainty on $\bar{\Lambda}$, whereas no such bound could be derived for the uncertainty on λ_1. The bound indicates that an accurate extraction of $\bar{\Lambda}$ is definitely possible for a cut at 2 GeV, whereas for the present cut at 2.2 GeV the errors might be large. I have also used a simplified version of the ACCMM model to estimate the effect the photon energy cut. The uncertainties depend strongly on the width of the spectrum in the model. I again find that an accurate determination of $\bar{\Lambda}$ should be possible if the cut can be lowered to about 2 GeV. Depending on the width of the spectrum, the cut has to be lowered even further to allow a precise determination of λ_1.

REFERENCES

1. J. Chay, H. Georgi, B. Grinstein, Phys. Lett. B **247** (1990) 399.

2. P. Ball, M.V. Braun, Phys. Rev. D **49** (1994) 2472.
3. M. Neubert, Phys. Lett. B **389** (1996) 727.
4. V. Giménez, G. Martinelli, C.T. Sachrajda, Nucl. Phys. **B486** (1997) 227.
5. A.V. Manohar, M.B. Wise, Phys. Rev. D **49** (1994) 1310.
6. M. Gremm, A. Kapustin, Z. Ligeti, M.B. Wise, Phys. Rev. Lett. **77** (1996) 20.
7. J. Bartelt et al., CLEO Collaboration, CLEO/CONF 93-19; B. Barish et al., CLEO Collaboration, Phys. Rev. Lett. **76** (1996) 1570.
8. A.F. Falk, M. Luke, M. Savage, Phys. Rev. D **53** (1996) 24; A.F. Falk, M. Luke, Phys. Rev. D **57** (1998) 424.
9. M. Gremm, A. Kapustin, Phys. Rev. D **55** (1997) 6924.
10. A. Kapustin, Z. Ligeti, Phys. Lett. B **355** (1995) 318.
11. C. Bauer, Phys. Rev. D **57** (1998) 5611.
12. M.S. Alam et al., CLEO Collaboration, Phys. Rev. Lett. **74** (1995) 2885.
13. See for example M. Neubert, Phys. Rev. D **49** (1994) 3392 and 4623; I.I. Bigi, M.A. Shifman, N.G. Uraltsev, A.I. Vainshtein, Int. Jour. Mod. Phys. A**9** (1994) 2467.
14. A.F. Falk, Z. Ligeti, M.B. Wise, Phys. Lett. B **406** (1997) 225.
15. G. Altarelli et al., Nucl. Phys. **B208** (1982) 365.
16. M. Neubert, Phys. Rev. **D49** (1994) 4623.

Hyperfine Interactions in Charm and Bottom

Patrick J. O'Donnell

Physics Department, University of Toronto, Toronto, Ontario M5S 1A7, Canada

Abstract. Hyperfine interactions in the light meson and baryon sectors are generalized to the charm and bottom systems. It is pointed out that an attempt to increase the value of the wave function at the origin to account for the unusual ratio of Λ_b to the B^0 lifetimes could spoil the good agreement among the baryon and meson hyperfine mass–splittings. Including spin effects and taking phase space differences into account it is shown that the decay rate of the Λ_b can be increased relative to that of the B^0 meson by about 7%. The predominance of Λ_c, Λ and the nucleon in exclusive final states of Λ_b decays, relative to $\Sigma_c^{(*)}$, $\Sigma^{(*)}$ and Δ is predicted.

I INTRODUCTION

Before b physics there was charm. Initially there were predictions that the lifetimes of the charm particles would all be very similar. Now we know that the charm quark is not sufficiently heavy for these predictions to hold – not only can there be different quark–quark interactions but also final state interactions play an important part in charm decays. As the quark mass becomes heavier many differences among the properties of spin–1/2 and spin–3/2 baryons and also among pseudoscalar and vector mesons containing a heavy quark are expected to become less pronounced [1]. In the infinite mass limit there are model–independent predictions for all form factors in transitions from one heavy quark to another in terms of a single function of the momentum transfer [2]. As the quark mass increases it is expected that the lifetimes of particles containing one heavy quark will become very similar [3].

The expectation was that for the lifetimes, $\tau(\Lambda_b) \sim \tau(B^0)$, with just a slight difference when certain quark scattering processes that could occur in the Λ_b but not in the meson were included. These processes are (a) the "weak scattering" process, first invoked for the Λ_c^+ lifetime [4], and here of the form, $bu \to cd$, and (b), the so–called "Pauli interference" process $bd \to c\bar{u}dd$ [5,6]. Including these terms leads to a slight enhancement of the Λ_b decay rate giving the ratio of lifetimes, $\tau(\Lambda_b)/\tau(B^0)$, a value in the range of 0.9 to 0.95. The latest experimental results [7] are $\tau(\Lambda_b) = 1.11 \pm 0.13$ ps and $\tau(B^0) = 1.55 \pm 0.05$ ps. This gives a very much reduced fraction $\tau(\Lambda_b)/\tau(B^0) = 0.73 \pm 0.06$, or conversely a very much enhanced decay rate. (There is a recent CDF result [8] which would move this fraction higher than the world average to a value of $0.85 \pm 0.10 \pm 0.05$, in agreement with *both* values).

The enhancement of the decay width, $\Delta\Gamma(\Lambda_b)$ from the $q - q$ scattering involves replacing the usual flux factor by $|\psi(0)|^2$, the wave function at the origin of the pair of quarks bu in the Λ_b, (or the pair bd, for which the wave function is the same by isospin symmetry). This wave function at the origin naturally appears in hyperfine splitting in the quark model [9]. The usual expectation is that $|\psi(0)_{q\bar{q}}|^2 \sim |\psi(0)_{qq}|^2$ for the wave functions of the quark – antiquark pair in the meson and the quark – quark pair in the baryon, after taking into account a factor of two for the difference in color. Rosner [10] tried to account for the enhancement $\Delta\Gamma(\Lambda_b)$ by a change in the wave function $|\psi(0)_{bu}|^2$ of the quark – quark pair in the Λ_b; this might also help account for the surprisingly large hyperfine splitting suggested by the DELPHI group [11]. He was able to show that, under certain assumptions, there could be at most a $13 \pm 7\%$ increase of the total amount needed to explain the decay rate of the Λ_b.

In a more dramatic attempt to explain the lifetime problem it has been proposed [12] to allow the ratio $r = |\psi_{bq}^{\Lambda_b}(0)|^2/|\psi_{b\bar{q}}^{B_q}(0)|^2$ to vary between 1/4 and 4. Clearly such a large variation is definitely not conventional hyperfine relations.

Here I discuss the hyperfine relations among the mesons and baryons generalized to include the heavier particles. This leads to good predictions among the mass splittings and show that spin and phase space effects predict a significant enhancement of the Λ_B decay rate of about 7%. Furthermore, the spin, isospin and mass splittings lead to predictions about possible allowed final states in the decays, predictions which should be easily testable in the near future.

II HYPERFINE INTERACTIONS

First I shall begin with a review of the role of hyperfine interactions in the meson sector. A number of years ago it was pointed out [13] that for the ground state of mesons, $^3S_1$ and $^1S_0$ in the quark model spectroscopic notation, the difference $\delta m^2 = m_V^2 - m_P^2$ is approximately constant. This holds very well for states that contain at least one light quark and seems to be a consequence of the fact that the quantity $|\psi(0)_{q\bar{q}}|^2/\mu_{ij}$, where μ_{ij} is the reduced mass of the $q - \bar{q}$ system, is approximately constant; this is an exact result if the confining potential is a linear one. That is, for a linear confining potential the quantity $|\psi(0)_{q\bar{q}}|^2/\mu_{ij}$ is independent of the masses of the quarks; the quantity $(m_1 + m_2)\langle V_{hyp}\rangle$ is independent of the flavors of the quarks and the flavor dependence of hadron wave functions. This means that quantitative predictions for the hyperfine splitting could now be obtained for differing flavor combinations. The empirical flavor–independent regularity that meson hyperfine splitting seems to have when written as differences between the square of the masses can be obtained. From this, a simple formula in term of the quark masses and a single parameter gave predictions for the $(^3S_1)_{s\bar{c}}$, $(^3S_1)_{u\bar{b}}$, $(^3S_1)_{s\bar{c}}$ and $(^3S_1)_{s\bar{b}}$ mesons as well as the ones known at that time.

We can write the basic physics of the hyperfine interactions as

$$H = \sum_i m_i + V_{hyp} \tag{1}$$

with the expectation value of the hyperfine interaction being

$$\langle V_{hyp} \rangle = V \sum_{i>j} \frac{\langle \mathbf{s_i} \cdot \mathbf{s_j} \rangle}{m_i m_j} K_c |\psi(0)_{ij}|^2 \tag{2}$$

where V is the strength of the hyperfine interaction, $\mathbf{s_i}$ is the spin of the i-th quark and K_c is a color factor which is unity for a $q\bar{q}$ pair and 1/2 for two quarks in a baryon.

Now for mesons note that the hyperfine interaction can be written as

$$\langle V_{hyp} \rangle = V \frac{\langle \mathbf{s_1} \cdot \mathbf{s_2} \rangle}{m_1 + m_2} \frac{|\psi(0)_{12}|^2}{\mu_{12}} \tag{3}$$

where μ_{12} is the reduced mass of the quark–antiquark pair. For the sum of the two masses it is a good approximation to write

$$m_V + m_P \sim 2(m_1 + m_2) \tag{4}$$

so that

$$m_V^2 - m_P^2 \sim 2(m_1 + m_2) \langle V_{hyp} \rangle \tag{5}$$

With the latest values for the masses [7] this leads to the following approximate equalities,

$$\begin{array}{cccccc} m_\rho^2 - m_\pi^2 & \sim m_{K^*}^2 - m_K^2 & \sim m_{D^*}^2 - m_D^2 & \sim m_{D_s^*}^2 - m_{D_s}^2 & \sim m_{B^*}^2 - m_B^2 \\ 0.57 & \sim 0.55 & \sim 0.54 & \sim 0.58 & \sim 0.49. \end{array} \tag{6}$$

where the numbers are in units of GeV^2.

Lipkin [14] generalized this result to baryons by applying the identity used in atomic physics [15] which relates the wave function at the origin to the derivative of the potential. If $K_c W_{12}$ denotes the two–body attractive potential responsible for the quarks being bound, then, for the meson system,

$$\frac{|\psi(0)_{12}|^2}{K_c \mu_{12}} = \langle \frac{dW_{12}}{dr_{12}} \rangle \tag{7}$$

For baryons, W_{12} in Eq. (7) is replaced by the total potential for the three–body system and the derivative is that of the relative coordinate r_{12}. For the non–strange baryons, with a totally symmetric spatial wave function, this was used to obtain the constraint [14],

$$\frac{|\psi(0)_{12}|^2}{K_C \mu_{12}} = \langle \frac{dW_{12}}{dr_{12}} \rangle (17/12 \pm 1/12) \tag{8}$$

or, for spin-3/2 baryons and spin-1 mesons

$$< V_{hyp} >_{B(3/2)} = (17/16 \pm 1/16)\frac{q_M}{q_B} < V_{hyp} >_{M(1)} \tag{9}$$

where q_M and q_B are effective quark masses in the meson and baryon. This led to the relation

$$M_\Delta - M_N = (17/32 \pm 1/32)\frac{q_M}{q_B}(M_\rho - M_\pi)$$
$$293 \text{MeV} = 279 \pm 16 \text{MeV}. \tag{10}$$

For the strange hyperons, there is no overall spatial permutation symmetry and the quark masses are not equal, a situation which becomes even more obvious in both the charm and bottom systems. However, as the non equal-mass quark becomes heavier the approximation [14] $q_M/q_B \sim 1$ is expected to be more exact. Incidentally, this approximation gives the mass relations for the strange quark systems,

$$M_{\Sigma^*} - M_\Sigma \quad = M_{\Xi^*} - M_\Xi \quad = (17/32 \pm 1/32)\frac{q_M}{q_B}(M_{K^*} - M_K)$$
$$193.43 \pm 0.06 \text{MeV} = 213.68 \pm 0.086 \text{MeV} = 211 \pm 12 \text{MeV}. \tag{11}$$

The prediction is reasonably good even though it ignores wave function effects in the strange quark system of baryons [17]. In the heavy quark regime we might expect the analogous relations to be better.

A check on the charm sector seems to bear this out, for, in the same type of approximation we have the relation,

$$M_{\Sigma_c^*} - M_{\Sigma_c} \quad = M_{\Xi_c^*} - M_{\Xi_c'} \quad = (17/32 \pm 1/32)\frac{q_M}{q_B}(M_{D^*} - M_D)$$
$$65.7 \pm 0.06 \text{MeV} = 63.2 \pm 2.6 \text{MeV} = 75 \pm 4.4 \text{MeV}. \tag{12}$$

Here the recent precise measurement for the mass of Σ_c^* reported by the CLEO collaboration [16] has been used. The middle number comes from a theoretical estimate [18] based on an expansion in $1/m_Q$, $1/N_c$ and SU(3) flavor breaking. There is no measurement as yet of the mass of the Ξ_c^s. If the wave function of the bu pair in the baryon is enhanced using the prescription of Rosner then, in the charm sector, there would actually be a decrease in the wave function of the cu pair relative to that for the $q-\bar{q}$ pair in the mesons.

Since the attempt at enhancing the wave function did not succeed in explaining the Λ_b lifetime, and even leads to a decrease for the charm sector, it may be that the older symmetry results [13,14] should not be abandoned, especially since they are also capable of accounting for the ratios of the decay constants [19]; $f_B/f_D \sim 0.63$ [19] to be compared with the value 0.79 ± 0.21 deduced by Rosner [10]. How do these hyperfine relations apply to the b system? As mentioned above, a measurement has been reported by the DELPHI collaboration [11] of the large value 56 ± 16MeV for the mass difference $M_{\Sigma_b^*} - M_{\Sigma_b}$. Using this to check the comparison in the b system we end up with

$$M_{\Sigma_b^*} - M_{\Sigma_b} = (17/32 \pm 1/32)\frac{q_M}{q_B}(M_{B^*} - M_B)$$
$$56 \pm 16 \text{MeV} = 24.4 \pm 1.4 \text{MeV}. \quad (13)$$

which is out of line with all of the preceding comparisons. (No new analysis has been done since the conference report in 1995 [20]). This smaller value on the right hand side of Eq. (13) is more in line with a recent update of the baryon masses based on an expansion in $1/m_Q$, $1/N_c$ and SU(3) flavor breaking [18], which predicts an even smaller mass difference of 15.8 ± 3.3 MeV, with the errors being an estimate of the uncertainty in scaling up from the charm sector. The value of 56 MeV was part of the motivation for a new description [21] of the heavy baryons.

III EFFECTS OF HYPERFINE SPLITTING ON PHASE SPACE

Spin effects can also enhance the Λ_b lifetime relative to the B lifetime.

Assume a factorization assumption in which the b–quark decay is described for any given exclusive decay mode both in meson and baryon decays by

$$b \to c + X \quad (14)$$

where X denotes any hadron or multihadron state and the hadronic decay is described by combining the charmed quark with the spectator antiquark in the B decay and with the spectator diquark in the Λ_b decay. Now also assume that the spins of the spectator quarks are not changed during the transition.

Then the following hadronic decays are relevant if the c-quark combines with the spectators in the ground state configuration,

$$B \to D + X$$
$$B \to D^* + X$$
$$\Lambda_b \to \Lambda_c + X \quad (15)$$

The assumption of no change in spectator spin was invoked to make the result look straightforward but it is not crucial. The spin of the spectator diquark may be flipped by gluon exchange between the charmed quark and the diquark, but the isospin cannot be changed because the final state has isospin zero and cannot be changed by strong QCD interactions when the charmed quark combines with the diquark. The argument can be checked experimentally, as discussed below, as soon as data on Λ_b decays into different baryon final states are available.

This "spectator spin conservation" leads to spin factors that favor the baryons because the two quarks in the Λ_b are coupled to spin zero and can only combine with the c quark to make a Λ_c (i.e., $B \not\to \Sigma_c$ or $B \not\to \Sigma_c^*$), while the spins of the c quark and the spectator antiquark are uncorrelated in the B decay and favor the D^* over the D by a factor of 3:1. The resulting spectrum of final states in the meson case has the hyperfine energy averaged out, while in the baryon case our spectator assumption chooses the final state in the multiplet with the lowest hyperfine energy. The added hyperfine energy is available for the transition and leads to an enhancement in the phase space for the baryon transition over the meson transition. (A different argument [22] using the scaling of lifetimes as the inverse fifth power of hadronic rather than quark masses also gives a larger phase space for the baryon transition.)

A very rough calculation shows this hyperfine energy enhancement of the baryon phase space. A proper calculation would choose a mass for X and calculate the relative momentum for the three transitions, taking also into account the effect of the $D - \Lambda_c$ mass difference on the recoil energy which also favors the baryon transition. One might take the mass of X to be 1 GeV for a nonstrange-noncharmed transition and 2 GeV for the case where the W turns into a D_s.

Present B-decay data support this set of assumptions but are not sufficiently precise to be convincing. Decays into final states containing D and vector D^* modes are observed while those containing higher D^*'s are not. This supports the assumption that only the ground state configurations give appreciable contributions to phase space. The semileptonic partial widths show these spin factors in the final states clearly since the 3:1 factor favoring the D^* over the D seems to be present.

Unfortunately there are not enough data about the exclusive branching ratios, particularly for the baryons, to say more.

Another way to express this spin effect is to assume that the B and Λ_b decays would have the same phase space if hadron spectroscopy could be ignored, and the decays would sum over all final states without regard to spin form factors. This assumption is violated by spectator spin conservation, since it requires the u and d pair in the initial Λ_b to remain in a spin-zero state in the final state and not in spin-one. This immediately leads to a number of interesting predictions which can be checked by future experiments. Thus we have

(1) if the spin-zero diquark picks up a charm quark

$$\Lambda_b \to \Lambda_c + X, \quad \Lambda_b \not\to \Sigma_c^{(*)} + X \tag{16}$$

(2) if the spin-zero diquark picks up a strange quark

$$\Lambda_b \to \Lambda + X, \quad \Lambda_b \not\to \Sigma^{(*)} + X \tag{17}$$

(3) if the spin-zero diquark picks up a non–strange quark

$$\Lambda_b \to N + X, \quad \Lambda_b \not\to \Delta + X. \tag{18}$$

For a quantitative estimate here is a simple toy model for semileptonic decays. Assume that the Λ_b goes only to Λ_c, that the B goes to a statistical mixture (3/4) D* and (1/4) D and that all transitions to higher states are negligible. The phase space for the Λ_b decay is then given by $(m_{\Lambda_b} - m_{\Lambda_c})^5$. The phase space for the B decay is then given by $(3/4)(m_B - m_{D^*})^5 + (1/4)(m_B - m_D)^5$.

This well-defined model for semileptonic decays may be right or wrong, but it's predictions are easily calculated and the basic assumptions can be easily tested when exclusive branching ratios into baryon final states including spin-excited baryons become available. It gives the following result for the ratio of semileptonic partial widths:

$$\frac{\Gamma(\Lambda_b)}{\Gamma(B)} = 1.07 \tag{19}$$

or, in terms of the ratio of the lifetimes,

$$\frac{\tau(\Lambda_b)}{\tau(B)} = 0.938 \tag{20}$$

This shows a clear prediction of a significant enhancement of the Λ_b partial semileptonic width in comparison with the B. The Λ_b decay rate is enhanced by about 7%.

These results suggest (1) that phase space effects must be carefully taken into account using exclusive final states in lifetime calculations which compare the B and Λ_b decays; (2) that the validity of the spectator spin conservation model should be tested with new data and new analyzes to see whether the Λ_b decay branching ratios show the predicted predominance of Λ_c, Λ and nucleon in exclusive final states in comparison with Σ_c and Σ_c^*, Σ and Σ^*, and Δ, respectively. If only isospin is conserved, and not spectator spin, the Σ_c and Σ_c^* can appear only if accompanied by an appropriately charged π coming from the decay of a higher isoscalar resonance. Thus, in a decay $\Lambda_b \to \Lambda_c$, there should be two oppositely charged pions with the following mass constraints; $m_{\Lambda_c} + m_\pi = m_{\Sigma_c^{(*)}}$ while the mass of the Λ_c together with the mass of the two pions should add up to the mass of an isoscalar resonance.

Acknowledgment

This talk was based on work done in collaboration with H. Lipkin [23] which was supported in part by grant No. I-0304-120-.07/93 from The German-Israeli Foundation for Scientific Research and Development and by the Natural Sciences and Engineering Council of Canada. We thank J. Appel, A. Datta and Salam Tawfiq for helpful comments.

REFERENCES

1. For a review see M. Neubert, Phys. Rep. **245**, 259 (1994).
2. N. Isgur and M.B. Wise, Phys. Lett. **B232**, 113 (1989); B **237**, 527 (1990); M. Luke, Phys. Lett. **B252**, 447 (1990); M. Neubert, Phys. Lett. **B341**, 367 (1995); I.I. Bigi, M. Shifman, N.G. Uraltsev, and A. Vainshtein, Phys. Rev. Lett. **71**, 496 (1993); A. Manohar and M.B. Wise, Phys. Rev. **D49**, 1310 (1994); M. Luke and M. Savage, Phys. Lett. **B321**, 88 (1994).
3. A recent review is by I. Bigi, M. Shifman and N. Uraltsev, hep-ph/9703290.
4. V. Barger, J.P. Leveille and P.M. Stevenson, Phys. Rev. Lett. **44**, 226 (1980).
5. M.B. Voloshin and M.A. Shifman, Yad. Fiz. **41**, 187 (1985) [Sov. J. Nucl. Phys. **41**, 120 (1985)]; Zh. Eksp. Teor. Fiz. **91**, 1180 (1986) [Sov. Phys. − JETP **64**, 698 (1986)]; M.B. Voloshin, N.G. Uraltsev, V.A. Khoze and M.A. Shifman, Yad. Fiz. **46**, 181 (1987) [Sov. J. Nucl. Phys. **46**, 112 (1987)].
6. N. Bilić, B. Guberina and J. Trampetić, Nucl. Phys. **B248**, 261 (1984); B. Guberina, R. Ruckl and J. Trampetić, Zeit. Phys. **C33**, 297 (1986).
7. R.M. Barnett *et. al.*, Phys. Rev. **D54**, 1 (1996), and 1997 off-year partial update for the 1998 edition available on the PDG WWW pages (URL: http://pdg.lbl.gov).
8. J. Tseng, Fermilab preprint Fermilab-Conf-96/438-E, to appear in the proceedings of the Second International Conference on Hyperons, Charm and Beauty Hadrons, 1996.
9. A. De Rujula, H. Georgi and S.L. Glashow, Phys. Rev. **D12**, 147 (1975).
10. J.L. Rosner, Phys. Lett. **B379**, 267 (1996).
11. DELPHI Collaboration, DELPHI 95-107.
12. M. Neubert and C.T. Sachrajda, Nucl. Phys. **B483**, 339 (1997); M. Neubert, talk at B Physics Conference, Hawaii, March 1997.
13. M. Frank and P.J. O'Donnell, Phys. Lett. **B159**, 174 (1985); Zeit. Phys. **C34**, 39 (1987).
14. H.J. Lipkin, Phys. Lett. **B171**, 293 (1986); Phys. Lett. **B172**, 242 (1986).
15. J. Hiller, J. Sucher and G. Feinberg, Phys. Rev. **A18**, 2399 (1978).
16. CLEO Collaboration, G. Brandenburg *et. al.*, Phys. Rev. Lett. **78**, 2304 (1997).
17. I. Cohen and H.J. Lipkin, Phys. Lett. **B106**, 119 (1981).
18. E. Jenkins, Phys. Rev. **D55**, R10 (1997). See also, M. Savage, Phys. Lett **B359**, 189 (1995).
19. P. J. O'Donnell, Phys. Lett **B261**, 136 (1991).
20. M. Feindt, private communication.
21. Adam F. Falk, Phys. Rev. Lett. **77**, 223 (1996).
22. G. Altarelli *et. al.*, Phys. Lett. **B382**, 409 (1996).
23. H.J. Lipkin and P. J. O'Donnell, Phys. Lett. **B409**, 412 (1997).

Electromagnetic Interactions in Quantum Hall Ferromagnets

Rashmi Ray

Laboratoire de Physique Nucléaire
Université de Montréal
Montréal, Quebec H3C 3J7, Canada

Abstract. The $\nu = 1$ quantum Hall ground state in materials like GaAs is known to be ferromagnetic in nature. The exchange part of the Coulomb interaction provides the required attractive force to align the electronic spins spontaneously. The gapless Goldstone modes are the angular deviations of the magnetisation vector from its fixed ground state orientation. Furthermore, the system supports electrically charged spin skyrmion configurations. It has been claimed in the literature that these skyrmions have half-integral spin owing to the presence of a topological Hopf term in the effective action governing the spin excitations. However, it has also been claimed that the derivation leading to this term is somewhat flawed. In this article, we demonstrate the existence of this term unambiguously. Furthermore, we investigate the electromagnetic interactions of the spin excitations and obtain a compact expression for the leading nonminimal electromagnetic coupling of these degrees of freedom.

INTRODUCTION

It is well known that the soliton solutions that emerge in various field theoretic models play an important role in condensed matter physics [1]. Furthermore, there are situations where the solitons of purely bosonic theories acquire fermionic statistics due to topological reasons [2].

In 2+1 dimensions, O(3) nonlinear sigma models (NLSM) have baby skyrmions as static solitons [1]. Physically, these NLSM are known to provide a natural description of the dynamics of the Goldstone modes in a theory with spontaneous symmetry breaking (SSB). In the sequel, we shall discuss such a theory, that of the $\nu = 1$ Hall state in samples like GaAs, where the Coulomb interaction between electrons leads to the formation of a ferromagnetic ground state through SSB [3]. In this case, the Goldstone bosons are the neutral ferromagnetic magnons which are described by an effective O(3) NLSM action. The spin skyrmions in this model possess electric charge and are the lowest lying charged excitations in the system. Furthermore, as we shall see, the effective action also contains a topological term (the Hopf term) with an appropriate coefficient, such that these skyrmions have fermionic spin [4], [5].

In this article we also study the electromagnetic interactions of the spin excitations. As the magnons are neutral, the electromagnetic probes do not couple minimally, [6] but in a rather nontrivial manner.

NOTATION AND FORMULATION

The microscopic model for the quantum Hall system in samples like GaAs may be taken to be:

$$S = S_1 + S_2 \qquad (1)$$

with

$$S_1 \equiv \int dt \int d\vec{x}\ \psi^\dagger(\vec{x},t)[i\partial_t - \frac{1}{2m}(\vec{p} - \vec{A})^2 + \mu]\psi(\vec{x},t) \qquad (2)$$

$$S_2 \equiv -\frac{V_0}{2} \int dt \int d\vec{x}\ (\psi^\dagger \psi)^2 \qquad (3)$$

where $\partial_x A^y - \partial_y A^x = -B$, B being the external magnetic field and where μ is the chemical potential. $\psi(\vec{x},t)$ is a 2-component spinor satisfying the usual anticommutation relations.

Here, we have set the Pauli term in the action equal to zero by setting $g \to 0$. In this limit, the above action has an exact spin SU(2) symmetry.

More importantly, we have replaced the nonlocal Coulomb interaction with a local four-fermion interaction. While this has been done to simplify the analysis, it has been seen in [5] and in [7] that the SSB in the system may be attributed to the short-distance part of the Coulomb interaction. The long-distance part may be treated perturbatively.

The partition function of the system is immediately written as:

$$Z = \int D\psi \int D\bar\psi\ e^{i(S_1+S_2)}. \qquad (4)$$

Using the standard property of the normalised SU(2) generators, $t^a_{\alpha\beta}t^a_{\gamma\delta} = \frac{1}{2}\delta_{\alpha\delta}\delta_{\beta\gamma} - \frac{1}{4}\delta_{\alpha\beta}\delta_{\gamma\delta}$, we can reorganise the four-fermion term as [7]

$$S_2 = V_0 \int dt \int d\vec{x}\ [\vec{S}^2 + \frac{1}{4}\rho^2] \qquad (5)$$

where $\rho \equiv \bar\psi\psi$ is the density and $S^a \equiv \bar\psi t^a \psi$ the spin density.

Introducing the auxiliary fields ϕ and $\vec{h} \equiv h\hat{n}$, where $\hat{n}$ is a unit vector, we perform the standard Hubbard-Stratonovich transformation [7] to rewrite the partition function as

$$Z = \int D\phi \int D\vec{h} \; e^{-\frac{i}{V_0} \int dt d\vec{x} \; [\phi^2 + \frac{1}{4}h^2]} \int D\psi D\bar{\psi} \; e^{i \int dt d\vec{x} \; \bar{\psi}[i\partial_t + \mu + \phi - \frac{1}{2m}(\vec{p}-\vec{A})^2 + \frac{h}{2}\hat{n}\cdot\vec{\sigma}]}. \quad (6)$$

It may be shown quite straightforwardly that due to SSB, the ϕ integral and the h integral have non-zero saddle points. Evaluating them at those saddle points, we get

$$Z = \int D\hat{n} \int D\psi D\bar{\psi} \; e^{i \int dt d\vec{x} \; \bar{\psi}[i\partial_t + \mu - \frac{1}{2m}(\vec{p}-\vec{A})^2 + \zeta \hat{n}\cdot\vec{\sigma}]\psi} \quad (7)$$

where $\zeta \equiv \frac{\rho_0 V_0}{2}$ and $\vec{t} \equiv \frac{\vec{\sigma}}{2}$, $\vec{\sigma}$ being the Pauli matrices.

The ground state described by the saddle points of the ϕ and the h integrals is a ferromagnetic ground state. All orientations of the magnetisation are equivalent and conventionally we choose it to be along the $\hat{z}$ direction. The excited state describes the deviations of the magnetisation from this "up" configuration. It is described by the two angles required to specify the deviation of $\hat{n}$ from its ground state orientation along $\hat{z}$. Namely, these two angles describe the coset SU(2)/U(1) and correspond to the two Goldstone modes (there are two broken generators). The simplest way to extract these angular excitations from the fermionic fields is through a unitary transformation of these fields. This transformation is chosen to be such that the transformed fermionic fields describe locally spin up electrons. If $U(\vec{x},t) \in$ SU(2) be such a unitary matrix, we require that

$$U^\dagger \vec{\sigma} \cdot \hat{n} U = \sigma_z. \quad (8)$$

The spin up fermionic field is defined in terms of U as

$$\xi \equiv U^\dagger \psi$$
$$\bar{\xi} \equiv \bar{\psi} U \quad (9)$$

We further introduce the SU(2) valued pure gauge potentials

$$\Omega_\mu^a t^a \equiv U^\dagger i \partial_\mu U. \quad (10)$$

The corresponding field strengths vanish [8]. Thus

$$F_{\mu\nu}^a \equiv \partial_\mu \Omega_\nu^a - \partial_\nu \Omega_\mu^a + \epsilon^{abc} \Omega_\mu^b \Omega_\nu^c = 0. \quad (11)$$

Hence, the partition function is given by

$$Z = \int D\hat{n} \int D\chi D\bar{\chi} \; e^{i \int dt \; d\vec{x} \; \bar{\chi}[i\partial_t + \mu + \Omega_0^a t^a - \frac{1}{2m}(-i\partial_i - A^i - \Omega_i^a t^a)^2 + \zeta \sigma_z]\chi}. \quad (12)$$

If we now integrate the fields $\chi, \bar{\chi}$ out we obtain a standard functional determinant in terms of the SU(2) connections. If we had coupled U(1) valued electromagnetic perturbations a^μ minimally to the original electronic fields we would obtain

127

$$Z = \int D\hat{n} \; e^{iS_{eff}} \tag{13}$$

where

$$S_{eff} = -i \, \text{tr} \ln[i\partial_t + \mu - a_0 + \Omega_0^a t^a + \zeta\sigma_z - \frac{1}{2m}(-i\partial_i - A^i - a^i - \Omega_i^a)^2]. \tag{14}$$

We may organise the argument of the above in the form $i\partial_t + \mu - h_0 - V$, where

$$h_0 \equiv \frac{1}{2m}(\vec{p} - \vec{A})^2 - \zeta\sigma_z \tag{15}$$

and where V is taken to be perturbative.

It is quite obvious that h_0 is the hamiltonian describing the celebrated Landau problem. Thus we may immediately diagonalise it in the familiar Landau level (L.L.) basis, which is specified by the positive integer n (the L.L. index) and the continuous parameter X, which measures the degeneracy of the single particle spectrum ($\frac{B}{2\pi}$ states per unit area). $\pi^\dagger$ (π) is the L.L. raising (lowering) operator satisfying the standard commutator

$$[\pi, \pi^\dagger] = 2B. \tag{16}$$

The degeneracy of the single particle spectrum is exposed by the "guiding-centre" coordinates Z and $\bar{Z}$, satisfying the commutation relation

$$[Z, \bar{Z}] = \frac{2}{B}. \tag{17}$$

The coordinate operators (their holomorphic combinations) can be expressed in terms of these operators as

$$z = Z - \frac{i}{B}\pi^\dagger$$
$$\bar{z} = \bar{Z} + \frac{i}{B}\pi. \tag{18}$$

Thus, any function of the coordinates may be written as $f(Z - \frac{i}{B}\pi^\dagger, \bar{Z} + \frac{i}{B}\pi)$. This in turn may be Taylor expanded around $Z, \bar{Z}$ by adopting, for instance, a normal ordering prescription for $\pi, \pi^\dagger$. This automatically enforces an anti-normal ordering prescription on $Z, \bar{Z}$. We denote this through the symbol $\natural \cdots \natural$ in the sequel.

The basis that diagonalises h_0 can finally be written as $\{|n, X, \alpha\rangle\}$, where α specifies the electronic spin, namely $\alpha = \pm 1$. Thus,

$$\pi|n, X, \alpha\rangle = \sqrt{2Bn}|n-1, X, \alpha\rangle$$
$$\pi^\dagger|n, X, \alpha\rangle = \sqrt{2B(n+1)}|n+1, X, \alpha\rangle$$
$$\hat{X}|n, X, \alpha\rangle = X|n, X, \alpha\rangle$$
$$\sigma_z|n, X, \alpha\rangle = \alpha|n, X, \alpha\rangle. \tag{19}$$

Further,
$$h_0|n, X, \alpha\rangle = [(n + \frac{1}{2}) - \zeta\alpha]|n, X, \alpha\rangle. \quad (20)$$

Let us now define the operator $\hat{p}_0$ such that $[\hat{t}, \hat{p}_0] = -i$, with $\langle t|\hat{p}_0 = i\partial_t\langle t|$. Let us also introduce the frequency basis $\{|\omega\rangle\}$ with $\hat{p}_0|\omega\rangle = \omega|\omega\rangle$. Furthermore, $\langle\omega|t\rangle = \frac{1}{\sqrt{2\pi}}e^{i\omega t}$. We also introduce the spin projection operators $P_+ \equiv \frac{1}{2}(I + \sigma_z)$ and $P_- \equiv \frac{1}{2}(I - \sigma_z)$, which project onto $\alpha = \pm 1$ respectively.

The Green function is defined as $[i\partial_t + \mu - h_0]G \equiv I$. G is obviously diagonalised by the L.L. basis. The fact that the many body ground state if obtained by filling all the $n = 0$ states with $\alpha = 1$, fixes the pole structure of G.

Thus, if
$$G|n, X, \omega\rangle \equiv \Gamma^{(n)}(\omega)|n, X, \omega\rangle, \quad (21)$$

we have
$$\Gamma^{(0)}(\omega)P_+ = \frac{1}{\omega + \mu - \frac{\omega_c}{2} + \zeta - i\epsilon}P_+$$
$$\Gamma^{(0)}(\omega)P_- = \frac{1}{\omega + \mu - \frac{\omega_c}{2} - \zeta + i\epsilon}P_-$$
$$\Gamma^{(n)}(\omega)P_\pm = \frac{1}{\omega + \mu - (n + \frac{1}{2})\omega_c \pm \zeta + i\epsilon}P_\pm \quad (22)$$

where $n \neq 0$. Henceforth, we choose $\mu = \frac{\omega_c}{2} - \zeta$.

At this point, we may organise the perturbation in a power series in $\sqrt{B}$, as we have mentioned earlier. We first combine the U(1) and the SU(2) valued connections and write
$$\mathcal{A}_0 \equiv a_0 - \Omega_0^a$$
$$\mathcal{A}^i \equiv a^i + \Omega_i^a. \quad (23)$$

In terms of these, the perturbation is written as
$$V = \mathcal{A}^0 + \frac{1}{2m}\mathcal{A}^i\mathcal{A}^i - \frac{\mathcal{B}}{2m} - \frac{1}{2m}(A\pi + \pi^\dagger\bar{A}) \quad (24)$$

where $A \equiv \mathcal{A}^x + i\mathcal{A}^y, \bar{A} \equiv \mathcal{A}^x - i\mathcal{A}^y$ and $\mathcal{B} \equiv \partial_x\mathcal{A}^y - \partial_y\mathcal{A}^x$. Indicating the power of B, the external magnetic field, by a superscript, we can write
$$V = V^{(\frac{1}{2})} + V^{(0)} + V^{(-\frac{1}{2})} + \cdots \quad (25)$$

where the ellipses indicate terms subdominant in B. Here
$$V^{(\frac{1}{2})} = -\frac{1}{2m}(\natural A\natural + \pi^\dagger\natural\bar{A}\natural)$$
$$V^{(0)} = \natural[\mathcal{A}^0 - \frac{1}{2m}\mathcal{B} + \frac{1}{2m}(\mathcal{A}^i)^2]\natural$$
$$V^{(-\frac{1}{2})} = \frac{i}{B}[\natural\partial_{\bar{z}}\mathcal{A}^0\natural\pi - \pi^\dagger\natural\partial_z\mathcal{A}^0\natural]. \quad (26)$$

THE EFFECTIVE ACTION FOR THE GOLDSTONE MODES

Upon expanding the functional determinant in equation (14), we obtain

$$S_{eff} = -i \text{ tr } \ln[i\partial_t + \mu - h_0] + i \text{ tr } \sum_{l=1}^{\infty} \frac{1}{l}(GV)^l. \qquad (27)$$

We propose to compute the effective action in terms of the connections $\mathcal{A}^\mu$ to $O(1/B)$. To this order, we have

$$S_{eff} = i \text{ tr}[GV^{(0)} + \frac{1}{2}GV^{(\frac{1}{2})}GV^{(\frac{1}{2})} + GV^{(\frac{1}{2})}GV^{-(\frac{1}{2})} + GV^{(\frac{1}{2})}GV^{(\frac{1}{2})}GV^{(0)}] \qquad (28)$$

where we have used the cyclic property of the trace. Let us further define $S^{(1)} \equiv \text{ tr } GV^{(0)}$, $S^{(2)} \equiv \frac{i}{2}\text{tr } GV^{(\frac{1}{2})}GV^{(\frac{1}{2})} + i \text{ tr } GV^{(\frac{1}{2})}GV^{(-\frac{1}{2})}$ and $S^{(3)} \equiv -i \text{ tr } GV^{(\frac{1}{2})}GV^{(\frac{1}{2})}GV^{(0)}$. Then, using the result [9]

$$\int_{-\infty}^{\infty} \langle 0, X | \sharp f(\hat{Z}, \hat{\bar{Z}}) \sharp | 0, X \rangle = \frac{B}{2\pi} \int_{-\infty}^{\infty} dx \int_{-\infty}^{\infty} dy \, f(x, y) \qquad (29)$$

we obtain

$$S^{(1)} = \int dt \int d\vec{x} \, [-\rho_0 a^0 + \frac{1}{2}\rho_0 \Omega_0^z + \frac{\omega_c}{4\pi}(\partial_x \Omega_y^z - \partial_y \Omega_x^z) - \frac{\omega_c}{4\pi}\text{tr } P_+(\mathcal{A}^i)^2] \qquad (30)$$

$$S^{(2)} = \int dt \int d\vec{x} \, [\frac{\omega_c}{4\pi}\text{tr } P_+(\mathcal{A}^i)^2 - \frac{\omega_c}{8\pi}(\partial_x \Omega_y^z - \partial_y \Omega_x^z) - \frac{\zeta}{8\pi}((\vec{\Omega}_i)^2 - (\Omega_i^z)^2)$$
$$- \frac{\zeta}{4\pi}(\vec{\Omega}_x \times \vec{\Omega}_y)^z - \frac{1}{4\pi}\epsilon^{\mu\nu\rho}a_\mu \partial_\nu a_\rho + \frac{1}{4\pi}\epsilon^{\mu\nu\rho}a_\mu \partial_\nu \Omega_\rho^z$$
$$+ \frac{3}{16}\vec{\Omega}_0 \cdot (\vec{\Omega}_x \times \vec{\Omega}_y)] \qquad (31)$$

$$S^{(3)} \simeq \int dt \int d\vec{x} \, [-\frac{1}{16\pi}\vec{\Omega}_0 \cdot (\vec{\Omega}_x \times \vec{\Omega}_y)]. \qquad (32)$$

Here, $\rho_0 \equiv \frac{B}{2\pi}$. Thus combining the terms together, we obtain

$$S_{eff} = \int dt \int d\vec{x} \, [\frac{1}{2}\rho_0 \Omega_0^z - \frac{\zeta}{8\pi}((\vec{\Omega}_i)^2 - (\Omega_i^z)^2) - \frac{\zeta}{4\pi}(\vec{\Omega}_x \times \vec{\Omega}_y)^z$$
$$+ \frac{1}{8\pi}\vec{\Omega}_0 \cdot (\vec{\Omega}_x \times \vec{\Omega}_y) - \rho_0 a^0 + \frac{\omega_c}{4\pi}b - \frac{1}{4\pi}\epsilon^{\mu\nu\rho}a_\mu \partial_\nu a_\rho$$
$$+ \frac{1}{4\pi}\epsilon^{\mu\nu\rho}a_\mu \partial_\nu \Omega_\rho^z] \qquad (33)$$

where $b \equiv \partial_x a^y - \partial_y a^x$.

We can express the unitary matrix U introduced in the previous section in terms of the three Euler angles θ, ϕ, χ. Namely,

$$U = e^{-i\frac{\phi}{2}\sigma_z} e^{-i\frac{\theta}{2}\sigma_y} e^{-i\frac{\chi}{2}\sigma_z}. \tag{34}$$

Again, as $U\sigma_z U^\dagger = \vec{\sigma}\cdot\hat{n}$, the unit vector $\hat{n}$ is given in terms of the Euler angles as

$$\hat{n} = (\sin\theta\cos\phi, \sin\theta\sin\phi, \cos\theta). \tag{35}$$

Also,

$$(\vec{\Omega}_x \times \vec{\Omega}_y)^z = \frac{1}{2}\epsilon^{ij}\hat{n}\cdot(\partial_i\hat{n}\times\partial_j\hat{n}) \tag{36}$$

and

$$(\vec{\Omega}_i)^2 - (\Omega_i^z)^2 = (\partial_i\hat{n})^2. \tag{37}$$

With this the effective action can be written as

$$S_{eff} = \int dt \int d\vec{x}\, [\frac{B}{4\pi}\cos\theta\partial_t\phi - \frac{\zeta}{8\pi}(\partial_i\hat{n})^2 - \zeta\rho_p + \frac{1}{48\pi}\epsilon^{\mu\nu\rho}\vec{\Omega}_\mu\cdot(\vec{\Omega}_\nu\times\vec{\Omega}_\rho)$$
$$+\frac{1}{4\pi}\epsilon^{\mu\nu\rho}a_\mu\partial_\nu\Omega_\rho^z - \rho_0 a^0 + \frac{\omega_c}{4\pi}b - \frac{1}{4\pi}\epsilon^{\mu\nu\rho}a_\mu\partial_\nu a_\rho] \tag{38}$$

where $\rho_p \equiv \frac{1}{8\pi}\epsilon^{ij}\hat{n}\cdot(\partial_i\hat{n}\times\partial_j\hat{n})$ is a topological density (the Pontryagin density), such that $\int d\vec{x}\, \rho_p(\vec{x}) =$ integer.

Let us look at the various terms in the effective action. The first term, with a single time derivative is a so called Wess-Zumino term which is ubiquitous in the theory of the path integral representation of spin systems. It is actually a Berry phase and in this derivation has emerged through a derivative expansion [10]. The second term is the standard kinetic energy term of a NLSM. These two terms together tell us that the dispersion of the Goldstone bosons is that of ferromagnetic magnons. Setting the electromagnetic perturbations to zero momentarily, we see that the coefficient of the fourth term, the so called Hopf term is such that the skyrmionic solitons of the NLSM have half-integral spin [2]. When the electromagnetic perturbations are present, there is a mixing between the angular degrees of freedom representing the Goldstone modes and the electromagnetic fields. This is seen in the fifth term in the effective action and can be rewritten in terms of the Euler angles as $-\frac{1}{4\pi}\sin\theta\epsilon^{\mu\nu\rho}a_\mu\partial_\nu\theta\partial_\rho\phi$. This provides a rather elegant expression for the electromagnetic coupling of the angular variables parametrising the coset $SU(2)/U(1)$. Since only two independent angles are required to describe this coset, it is gratifying that the angle χ has dropped out of the expression.

The mean electromagnetic currents can be obtained by functional differentiation of the effective action. For example, upon setting the electromagnetic perturbations to zero, the mean density in the spin-textured (excited) state is

$$\langle j_0 \rangle = \frac{B}{2\pi} + \frac{1}{8\pi}\epsilon^{ij}\hat{n}\cdot(\partial_i\hat{n}\times\partial_j\hat{n}). \tag{39}$$

The deviation from the ground state value is thus precisely equal to a topological density. Now as $e=1$, the electrical charge density is also equal to the Pontryagin density. Now, there are configurations $\hat{n}$ known as skyrmions such that $\int d\vec{x}\,\rho_p = 1$. Hence, creating a spin skyrmion in the excited state is equivalent to adding one unit of electrical charge to the system. Thus spin skyrmions are carriers of charge in these Hall systems.

CONCLUSIONS

In this article, we have derived some new results and have rederived some well-known ones on the subject of quantum Hall ferromagnets.

Previously as in [7] the $\nu=1$ system in GaAs has been studied, but exclusively in the L.L.L. approximation. However, the Coulomb interaction [5] as well as the electromagnetic perturbations are non-diagonal in the L.L. basis and lead to a mixing of the L.L.. Thus, as we have done here, it is important to have a procedure where explicit projection to the L.L.L. is avoided. In our method, the effects due to L.L. mixing appear as subleading terms in the derivative expansion scheme that we have adopted.

In this manner, we have shown how a NLSM describing the electronic spin excitations, emerges naturally in terms of the angular variables parametrising the coset $SU(2)/U(1)$. We have further shown that in terms of these angles, a Hopf term emerges naturally in the effective action. It is this term that is responsible for rendering the charged skyrmions in the system fermionic.

Apart from an economical derivation of these known results, we have for the first time, obtained a compact expression for the nonminimal electromagnetic coupling [6] of the spin excitations of the system. This tells us precisely how the electromagnetic currents are affected by the spin excitations. This, we believe, could have interesting experimental ramifications.

REFERENCES

1. R. Rajaraman, *Solitons and Instantons* (North Holland, Amsterdam, 1982).
2. F. Wilczek & A. Zee, Phys. Rev. Lett. **51**, 2250 (1983).
3. S.L. Sondhi et. al., Phys. Rev. **B47**, 16419 (1993).
4. W. Apel & Yu.A. Bychkov, Phys. Rev. Lett. **78**, 2188 (1997).
5. R. Ray & J. Soto, cond-mat/9708067.
6. J.M. Roman & J. Soto, cond-mat/9709298.
7. K. Moon et. al., Phys. Rev. **B51**, 5138 (1995).
8. G.E. Volovik & V.M. Yakovenko, J. Phys.: Condens. Matter **1**, 5263 (1989).
9. R. Ray & J. Soto, Phys. Rev. **B54**, 10709 (1995).
10. R. Jackiw, Int. Jour. Mod. Phys. **A3**, 285 (1988). P. de Sousa Gerbert, Ann. Phys. **189**, 155 (1989). D. Düsedau, Phys. Lett. **B205**, 312 (1988).

The Quantum Hall effect, Skyrmions and Anomalies

A. Travesset

*Department of Physics, Syracuse University,
Syracuse, New York 13244-1130, USA*

Abstract.
We discuss the properties of Skyrmions in the Fractional Quantum Hall effect (FQHE). We begin with a brief description of the Chern-Simons-Landau-Ginzburg description of the FQHE, which provides the framework in which to understand a new derivation of the properties of FQHE Skyrmions (S. Baez, A.P. Balachandran, A. Stern and A. Travesset *cond-mat 9712151*) from anomaly and edge considerations.

I THE QUANTUM HALL EFFECT

Experiments carried out starting at the end of the 70's (see [2]) revealed that in some samples, the Hall effect presents fascinating and completely unexpected properties, the most outstanding being the broad plateaus for the transverse conductivity σ_H (or for the transverse resistivity $R_H = 1/\sigma_H$) and the vanishing of the longitudinal resistivity at this plateaus, see Fig. 1. There are other important properties, such as the incompressibility of the Hall liquid and the presence of quantum mechanically induced currents if the sample has edges. It is fair to say that, in spite that there are still open questions, theorist have succeeded in giving successful approaches to the problem. We give an introduction to the most common field theory formulation [3], but we will necessarily be very descriptive. We refer to for example [4] for a pedagogical review and [5] for a recent and more detailed analysis. The second part of the talk is devoted to the properties of FQHE Skyrmions, a problem which has also been addressed in this MRST meeting in [6].

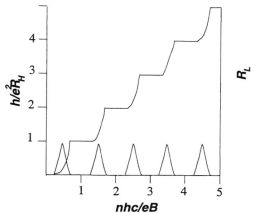

FIGURE 1. The plateaus in the integer Hall effect resistivity plotted versus the filling fraction. The left vertical axis is the inverse transverse resistivity in dimensionless units, and the right one the longitudinal in arbitrary units. With higher resolution settings smaller plateaus at fractional filling fractions may be observed as well.

A The CSLG Field theory of the FQHE

A very intuitive picture to understand the FQHE is to consider the total flux in the sample $\Phi = BA$, where B is the external magnetic field and A the total area, as being carried by N_f particles, the fluxons, each one carrying an elementary flux unit $\Phi_0 = \frac{2\pi\hbar c}{e}$ *. The physical electrons are represented as composite objects, a bare bosonic particle with an odd number of fluxons attached to it restoring its fermionic nature. The filling fraction is defined as

$$\nu = \frac{N_{el}}{N_f}, \qquad (1)$$

where N_{el} is the number of physical electrons in the sample. If $\nu = \nu_k \equiv \frac{1}{2k+1}$, $k \in \mathbf{N}$, there are exactly $2k+1$ fluxons per electron. It is, therefore, an "optimal" situation. In any other filling fraction [†] there will be an unbalanced number of fluxons with respect to electrons. How the system responds to this situation is the key for understanding the properties of the QH liquid.

The previous picture may be described in a Landau-Ginzburg formalism, the Chern-Simons-Landau-Ginzburg (CSLG) Lagrangian density being [3]

$$\mathcal{L} = \phi^\dagger(\mathbf{r})[i\partial_t - ea_0]\phi(\mathbf{r}) - \frac{1}{2m^*}\left|\phi^\dagger(\mathbf{r})\left[-i\nabla - e(\mathbf{A}(\mathbf{r}) + \mathbf{a}(\mathbf{r}))\right]\phi(\mathbf{r})\right|^2 -$$

$$-\frac{1}{2}\int d\mathbf{r}' V(\mathbf{r}-\mathbf{r}')[|\phi(\mathbf{r})|^2 - \rho_0][|\phi(\mathbf{r}')|^2 - \rho_0] - \frac{1}{2}g\mu_B B\phi^\dagger(\mathbf{r})\sigma_z\phi(\mathbf{r}) -$$

---

*) unless explicitly stated, $\hbar = c = 1$ units are used.
†) We do not consider generalized filling fractions.

$$-\frac{e^2}{4\vartheta}\epsilon^{\mu\nu\sigma}a_\mu(\mathbf{r})\partial_\nu a_\sigma(\mathbf{r}) \ . \tag{2}$$

First of all $\phi = (\phi_1, \phi_2)$ is a two-component complex bosonic field (the "bare electron"). The field $\mathbf{a}(\mathbf{r})$ is a gauge field that attaches $\frac{\vartheta}{\pi} = (2k+1)$ fluxons to the boson, which makes the physical electron. ρ_0 is the actual density of electrons (which obviously coincides with the density of bosons). The external magnetic field is described by $\mathbf{A}$ ($\nabla \times \mathbf{A} = -Be_z$). There are two extra terms, $V(\mathbf{r}) = \frac{e^2}{\epsilon|\mathbf{r}|}$ accounts for the Coulomb repulsion between electrons, and the coupling between the spin and the magnetic field, which involves g^s, the gyromagnetic ratio, the Bohr magneton μ_B and the Pauli matrix σ_z.

The Landau-Ginzburg ground state is the lowest energy solution of Eq. 2. Let us try the ansatz $\phi = \sqrt{\rho_0}(1,0)$. The equations of motion imply

$$\frac{N_{el}}{A} = \rho_0 = \frac{B}{\Phi_0}\frac{\pi}{\vartheta} = \frac{1}{2k+1}\frac{N_f}{A} \ . \tag{3}$$

So, the optimal situation described above, in which each electron captures an odd number fluxons ($2k+1$ of them) is the ground state. The filling fraction is then $\nu = \nu_k = \frac{1}{2k+1}$, the longitudinal resistivity is zero and the transverse conductivity $\sigma_H = \frac{\nu_k e^2}{2\pi}$. The case $g^S \to 0$ is special, since the solution is not unique as $\phi = \sqrt{\rho_0}(1,0)U$, $U \in SU(2)$ is a solution of the equations of motion. However, the true ground state is ferromagnetic, i. e. corresponds to a unique choice of U. The $SU(2)$-symmetry is then spontaneously broken to $U(1)$. the ground state remains spin polarized at $g^S = 0$ as well (although spins do not necessarily point to the same direction of the magnetic field). As a consequence, there are Goldstone modes in the system manifesting in the form of neutral disipationless spin waves(magnons). But, remarkably, the properties of the ground state are essentially independ of g^S.

To go further, let us temporally simplify Eq. 2 and neglect the Coulomb repulsion. Let us add the term

$$-\frac{\lambda}{2}(\phi^\dagger(\mathbf{x})\phi(\mathbf{x}) - \rho_o)^2 \ . \tag{4}$$

If B is the external magnetic field at the ground state, we find solutions to the equations of motion in the form

$$\phi_1(r,\theta) = g_V(r)e^{-i\theta}$$
$$\phi_2(r,\theta) = 0$$
$$a_r(r,\theta) = 0$$
$$a_\theta(r,\theta) = \frac{eB}{2}r + n_V(r)$$
$$a_0(r,\theta) = m_V(r) \tag{5}$$

where $lim_{r\to\infty}g_V(r) = \sqrt{\rho_0}$, $lim_{r\to 0}g_V(r) = 0$ and $lim_{r\to\infty}n_V(r) = \frac{N_{vort}}{r}$. This solution has finite energy and corresponds to a vortex of vorticity N_{vort} located at $r = 0$ with a finite core $\xi \sim \sqrt{\lambda}$. The addition of the term Eq. 5 determines the size of the vortex. The vortex may be regarded as a particle on its own, the pseudo-particle, having electric charge and spin. As an example, we compute the total electric charge in the ground state of a QH liquid in the presence of a vortex. The equations of motions imply

$$J^0_{em} = \frac{e}{2\vartheta} \epsilon^{0\nu\sigma} \partial_\nu a_\sigma \qquad (6)$$

So, the total electric charge is

$$Q = \int_\Sigma d^2\mathbf{r} J^0_{em}(\mathbf{r}) = \frac{e}{2\vartheta} \oint_{\partial\Sigma} dl a_l = -e(N_{el} - \nu_k N_{vort}) , \qquad (7)$$

A vortex has charge $\frac{e}{2k+1}$. It is then a *fractionally* charged object. Its spin may be also be computed with result $\frac{N_{vort}}{2k+1}$. So, in general, FQHE vortices are *anyons*.

Now, already at this stage, we can explain some of the phenomenology of the FQHE. Let us assume that the QH liquid is at its ground state. Each fermion is a composite boson with an even number of fluxons attached. If the filling fraction is changed from $\nu_k = \frac{1}{2k+1}$ by tuning the external magnetic field for example, the statistical gauge field **a** screens the excess of magnetic field keeping Eq. 3 unaltered. That is, the density of electrons ρ_0 remains constant under small variations of the magnetic field. The system is therefore incompressible. Furthermore, as the number of electrons does not change, the area of the sample A remains constant as well, a property that is usually referred to as area-preserving. At some point, however, the magnetic field is enough to create/destroy a fluxon. The QH-liquid responds creating a vortex/anti-vortex. A vortex(or an antivortex) may change dramatically the properties of the liquid. Being a charged object, the FQHE properties may be destroyed because of the motion of those vortices. However, metals with a sufficient amount of disorder may pin down those vortices, and the properties of the ground state like the longitudinal and transverse conductivity are not changed. That explains the broad plateaus observed. Eventually, more vortices cannot be accommodated in the sample, and the longitudinal conductance grows from its zero value. This corresponds to a transition to the next ground state filling fraction. The CSLG description nicely accounts for the experimental situation in Fig. 1.

The case $g^s \sim 0$ is experimentally relevant in some cases such as $GaAs$ samples. We may try solutions to the equations of motion in the form

$$\begin{aligned}
\phi_1(r,\theta) &= g_S^{(1)}(r)e^{-i\theta} \\
\phi_2(r,\theta) &= g_S^{(2)}(r) \\
a_r(r,\theta) &= 0 \\
a_\theta(r,\theta) &= \frac{eB}{2}r + n_S(r) \\
a_0(r,\theta) &= m_S(r) ,
\end{aligned} \qquad (8)$$

where the assymptotics are the same as in Eq. 8, but in addition, we require $lim_{r\to 0} g_S^{(2)}(r) \neq 0$, $lim_{r\to\infty} g_S^{(2)}(r) = 0$. This solution minimizes the Coulomb energy; when the spin up component starts to decrease the spin down increases and the electron density deviations from the ground state value ρ_0 are much smaller than for the vortex. The spin contribution increases, but it is anyhow negligible since the assumed small gyromagnetic ratio. It should not come as a surprise then, that in this regime, fluctuations in spin have a lower energy than vortices, and become the relevant quasi-particles [7]. those solutions are of a topological nature as well, they are Skyrmions, as we will see.

Although the ground state is essentially independent of g^S, the properties of the pseudo-particles are very different. There is a critical value g_c^S such that for $g^S < g_c^S$ those are Skyrmions, while for $g^S > g_c^S$ they are vortices.

We have introduced the pseudo-particles and explained the crucial role they have in explaining the properties of the QH fluid. From now on, we concentrate on the study of its properties, and more specifically of FQHE Skyrmions. In the case of vortices, this may be accomplished by integrating out the bosonic field, i. e. going to the dual picture [4]. The properties of vortices, such as the statistics or its electric charge, are explicit.

In [1] we have generalized this description to the case of Skyrmions. As it is well known, the spin of Skyrmions may be determined from the coefficient of the Hopf term. We include this term and read the spin from its coefficient. To determine the couplings of the different terms in the Lagrangian we consider a sample with edges. In the presence of edges, there are quantum mechanically induced chiral gapless modes [8]. We shall determine all the properties of FQHE Skyrmions by imposing this chirality condition, together with the fact that Skyrmions may be created and destroyed.

B The properties of FQHE Skyrmions

Let us write the direction to which the spin field is pointing at in terms of a unitary vector $\vec{n}$. It is possible to everywhere identify an $SU(2)$ field degree of freedom g, associated with spin fluctuations from

$$n_i = \frac{1}{2}\text{Tr}(\sigma_i g^\dagger \sigma_3 g) \ . \qquad (9)$$

There is an $U(1)$ subgroup, which we arbitrarily take to be generated by the third Pauli matrix σ_3, which is gauged via the coupling to a statistical gauge field (the same gauge field mentioned above), so that the gauge invariant observables are defined on S^2 (parametrized by $\vec{n}$). Energy finiteness generally demands that g goes to the above $U(1)$ subgroup, i.e. $g \to \exp i\chi\sigma_3$, at spatial infinity. This corresponds to $\phi^\dagger\phi$ going to the ground state value at spatial infinity, and in effect this compactifies $\mathbf{R}^2$ to S^2. This shows the topological nature of the FQHE Skyrmions.

They are naturally associated with $\Pi_2(S^2)$, the elements of $\Pi_2(S^2)$ being labeled by the winding number

$$N_{Sky} = \int_{\mathbf{R}^2} d^2x \, T^0(g) = \frac{i}{4\pi} \int_{\mathbf{R}^2} d \, \text{Tr} \, \sigma_3 g^\dagger dg , \qquad (10)$$

where $T^0(g)$ is the time component of the topological current,

$$T^\mu(g) = \frac{i}{4\pi} \epsilon^{\mu\nu\lambda} \text{Tr} \sigma_3 \partial_\nu g^\dagger \partial_\lambda g . \qquad (11)$$

We first show what are the consequences of not including a Hopf term. Our starting point is the standard bulk action of the dual picture [8], including a coupling to the Skyrmion current

$$\mathcal{S}_H = \int_{\Sigma \times R^1} d^3x \left(\frac{\sigma_H}{2} \epsilon^{\mu\nu\lambda} A_\mu \partial_\nu A_\lambda - eA_\mu \mathcal{T}^\mu \right) , \qquad (12)$$

where Σ is the two dimensional spatial domain of the sample, R^1 accounts for time, and $e\mathcal{T}^\mu$ is the Skyrmion current. Additional terms may be added [1], but we just consider the topological sector that determines the charge and the spin. For us, A_μ is the external electromagnetic field which is not a dynamical variable. Its variations therefore just define the bulk current J^μ_{em} by $J^\mu_{em} = -\frac{\delta \mathcal{S}_H}{\delta A_\mu}$.
According to Eq. 12, the bulk electromagnetic current J^μ_{em} is

$$J^\mu_{em} = -\frac{\sigma_H}{2} \epsilon^{\mu\nu\lambda} F_{\nu\lambda} + e\mathcal{T}^\mu . \qquad (13)$$

For consistency the current J^μ_{em}, and consequently $\mathcal{T}^\mu$, must be conserved. This is the case for $\mathcal{T}^\mu$ proportional to the topological current T^μ:

$$\mathcal{T}^\mu = \kappa T^\mu . \qquad (14)$$

Eq. 13 implies that the electric charge density is $J^0_{em} = -\sigma_H F_{12} + e\kappa T^0$. Integrating it over the whole sample gives the total electric charge as $-eN_{el} + e\kappa N_{Sky}$, where N_{el} is the total number of electrons at the corresponding filling fraction $\nu = \frac{1}{2k+1}$. Thus, κ times e is the charge of a Skyrmion of unit winding number.

We now examine under what conditions the bulk action $\mathcal{S}_H$ is consistent with the existence of chiral edge currents. For the case of filling fraction $\nu = 1$, there is a single edge current on the boundary $\partial\Sigma$ of Σ, which may be represented by a 2d massless chiral relativistic Dirac fermion [8], while for fractional values of ν one gets a Luttinger liquid. Chirality implies that the electromagnetic current J^μ_{em} of the edge fermions satisfies $J^{em}_- = \frac{1}{\sqrt{2}}(J^0_{em} + J^1_{em}) = 0$. In the quantum theory this is known to lead to an anomaly, i.e. $\partial_\mu J^\mu_{em} \neq 0$.

It is convenient to bosonize the edge theory [9], and for this we shall introduce a scalar field ϕ on $\partial\Sigma$. In terms of this field, chirality will mean the following:

$$\mathcal{D}_- \phi = f(x^-) , \qquad (15)$$

where $x^- = \frac{1}{\sqrt{2}}(x^0 - x^1)$, $\mathcal{D}_\pm = \frac{1}{\sqrt{2}}(\mathcal{D}_0 \pm \mathcal{D}_1)$ and $\mathcal{D}_\mu$ denotes a covariant derivative. (Usually the more restrictive condition $f(x^-) = 0$ is assumed, but Eq. 15 seems enough for us.)

To proceed we shall pose an action principle for the edge field ϕ. The edge action $\mathcal{S}_{\partial\Sigma \times R^1}$ should be such that: i) The total action $\mathcal{S} = \mathcal{S}_H + \mathcal{S}_{\partial\Sigma \times R^1}$ is gauge invariant. ii) It is consistent with chirality, i.e. Eq. 15. We will show that these two conditions lead to a chiral electromagnetic current $J_-^{em} = 0$, which at the boundary is defined by $J_{em}^\mu = -\frac{\delta S}{\delta A_\mu}|_{\partial\Sigma \times R^1}$. Requirements i) and ii) also lead to the anomaly. For this recall that the one loop effects responsible for the anomaly in the fermionic theory appear at tree level in the bosonized theory. Thus we can expect to recover the anomaly from the classical equation of motion for ϕ [10].

We begin by addressing the issue i) of gauge invariance. If we ignore boundary effects, the bulk action is separately gauge invariant under transformations of the electromagnetic potentials A_μ,

$$A_\mu \to A_\mu + \partial_\mu \Lambda, \tag{16}$$

as well as under transformations of the fields g,

$$g \to g\, e^{i\lambda \sigma_3}, \tag{17}$$

where both Λ and λ are functions of space-time coordinates. On the other hand, taking into account the boundary $\partial\Sigma$, one finds instead that Eq. 16 gives the surface terms

$$\delta \mathcal{S}_H = -\frac{\sigma_H}{2}\int_{\partial\Sigma \times R^1} d\Lambda \wedge A + \frac{e\kappa i}{4\pi}\int_{\partial\Sigma \times R^1} d\Lambda \wedge \text{Tr}\sigma_3 g^\dagger dg, \tag{18}$$

while gauge invariance under transformations Eq. 17) persists. We now specify that under gauge transformations Eq. 16), the edge field ϕ transforms according to

$$\phi \to \phi + e\Lambda. \tag{19}$$

Then we can cancel both of the above boundary terms in Eq. 18 if we assume the following action for the scalar field ϕ:

$$\mathcal{S}_{\partial\Sigma \times R^1} = \frac{R^2}{8\pi}\int_{\partial\Sigma \times R^1} d^2 x\, (\mathcal{D}_\mu \phi)^2 + \frac{\sigma_H}{2e}\int_{\partial\Sigma \times R^1} d\phi \wedge A - \frac{\kappa i}{4\pi}\int_{\partial\Sigma \times R^1} d\phi \wedge \text{Tr}\sigma_3 g^\dagger dg. \tag{20}$$

In (20) we have added a kinetic energy term for ϕ, where the covariant derivative is defined by $\mathcal{D}_\mu \phi = \partial_\mu \phi - eA_\mu$. The coefficient R is real and is known to correspond to the square root of the filling fraction ν_k.

Concerning ii), extremizing Eq. 20) with respect to ϕ gives

$$R^2 \partial_\mu \mathcal{D}^\mu \phi = -\frac{2\pi\sigma_H}{e}F_{01} - 4\pi\mathcal{T}, \tag{21}$$

F_{01} being the electric field strength at the boundary and the index r denoting the direction normal to the surface. This equation can be rewritten as

$$2R^2 \partial_+ \mathcal{D}_- \phi = (eR^2 - \frac{2\pi\sigma_H}{e})F_{01} - 4\pi\mathcal{T}^r , \qquad (22)$$

using $\partial_+ = \frac{1}{\sqrt{2}}(\partial_0 + \partial_1)$ and $diag(1,-1)$ for the Lorenz metric. But the chirality condition Eq. 15) requires that the right hand side of Eq. 22) vanishes. For this we can set

$$\sigma_H = \frac{e^2 R^2}{2\pi} , \qquad (23)$$

which is the usual relation for the Hall conductivity (after identifying R^2 with the filling fraction ν_k). But we also need

$$\mathcal{T}^r = 0 \quad \text{at } \partial\Sigma . \qquad (24)$$

From Eq. 23) and Eq. 24, variations of A_μ give the following result for the edge current

$$J^\mu_{em} = -\frac{\delta\mathcal{S}}{\delta A_\mu}|_{\partial\Sigma \times R^1} = \frac{eR^2}{4\pi}(\mathcal{D}^\mu + \epsilon^{\mu\nu}\mathcal{D}_\nu)\phi , \qquad (25)$$

and thus it is chiral, i.e. $J^{em}_- = 0$. Here $\epsilon^{01} = -\epsilon^{10} = 1$. By taking its divergence we also recover the anomaly:

$$\partial_\mu J^\mu_{em} = \frac{eR^2}{4\pi}\partial_\mu(\mathcal{D}^\mu + \epsilon^{\mu\nu}\mathcal{D}_\nu)\phi = -\frac{e^2 R^2}{2\pi}F_{01} , \qquad (26)$$

where we again used Eq. 23 and Eq. 24.

In order to satisfy chirality in the above discussion, we needed not only to constrain the values of coefficients, but we also found it necessary to impose a boundary condition Eq. 24 on the topological current. As a result, the topological flux, and moreover Skyrmions, cannot penetrate the edge. Thus, provided g is everywhere defined in Σ, the total Skyrmion number within the bulk $\int_\Sigma d^2x\, T^0(g)$ is a conserved quantity, and for example, a nonzero value for the total topological charge cannot be adiabatically generated from the ground state.

Below, we generalize to the situation where the total Skyrmion number in the bulk is *not* restricted to being a constant. For this we need to drop the boundary condition Eq. 24, and thus allow for a nonzero topological flux into or out of the sample. One may interpret this as Skyrmions being created or destroyed at the edges. For this purpose, we consider an extension of the above description, where the Hopf term

$$\mathcal{S}_{WZ} = \frac{\Theta}{24\pi^2}\int_{\Sigma \times R^1}\text{Tr}(g^\dagger dg)^3 \qquad (27)$$

is added to the bulk action $\mathcal{S}_H$. [Note that Eq. 27 is a local version of the Hopf term]. This term does not affect the classical equations of motion since it is the integral of a closed three form. That term gives a nontrivial spin to the Skyrmion

$$\frac{\Theta N_{Sky}}{2\pi} . \tag{28}$$

We thus need the numerical value of Θ to determine the spin. For this purpose we now reexamine the boundary dynamics taking into account the Hopf term. We once again require i) gauge invariance and ii) chirality.

Concerning i), as before, the bulk action is not invariant under gauge transformations Eq. 16 of the electromagnetic potentials A_μ. In addition, unlike before, it is not invariant under gauge transformations Eq. 17 of the fields g. From $\mathcal{S}_{WZ}$ we pick up the surface term

$$\delta \mathcal{S}_{WZ} = \frac{i\Theta}{8\pi^2} \int_{\partial \Sigma \times R^1} d\lambda \wedge \text{Tr}\sigma_3 g^\dagger dg . \tag{29}$$

To cancel this variation along with Eq. 18, we once again assume the existence of an edge field ϕ which transforms like Eq. 19, simultaneously with the gauge transformations Eq. 16 of the electromagnetic potentials A_μ. We further specify that ϕ transforms according to

$$\phi \to \phi + \lambda , \tag{30}$$

simultaneously with the gauge transformations Eq. 17 of the fields g. Recall that this transformation is the one compatible with the covariant derivative in Eq. 2. Then we can cancel both of the boundary terms Eq. 18 and Eq. 29, making our theory anomaly free, if we assume the following action for the scalar field ϕ:

$$\mathcal{S}_{\partial\Sigma \times R^1} = \frac{R^2}{8\pi} \int_{\partial\Sigma \times R^1} d^2x \, (\mathcal{D}_\mu \phi)^2 + \frac{\sigma_H}{2e} \int_{\partial\Sigma \times R^1} d\phi \wedge A \tag{31}$$
$$- \frac{i}{4} \int_{\partial\Sigma \times R^1} \left(\frac{\Theta}{2\pi^2} d\phi + \frac{\sigma_H}{e} A \right) \wedge \text{Tr}\sigma_3 g^\dagger dg ,$$

provided we also impose that

$$\kappa = \frac{\pi \sigma_H}{e^2} + \frac{\Theta}{2\pi} . \tag{32}$$

Since ϕ admits gauge transformations Eq. 30, as well as Eq. 19, we must redefine the covariant derivative appearing in Eq. 31 according to

$$\mathcal{D}_\mu \phi = \partial_\mu \phi - \beta_\mu , \qquad \beta_\mu = eA_\mu - \frac{i}{2}\text{Tr}\sigma_3 g^\dagger \partial_\mu g . \tag{33}$$

With regard to ii), the equation of motion for ϕ is

$$R^2 \partial_\mu \mathcal{D}^\mu \phi = -\frac{2\pi\sigma_H}{e} F_{01} - 2\Theta T^r , \qquad (34)$$

which can be rewritten as

$$2R^2 \partial_+ \mathcal{D}_- \phi = (eR^2 - \frac{2\pi\sigma_H}{e}) F_{01} + 2(\pi R^2 - \Theta)\mathcal{T}^r . \qquad (35)$$

We recover the chirality condition Eq. 15 upon setting

$$\Theta = \pi R^2 , \qquad (36)$$

as well as (23). From Eq. 23) and Eq. 36), variations of A_μ i again give the edge current in the form of Eq. 25 (although the covariant derivative is now defined differently) and thus $J_-^{em} = 0$. By taking its divergence we get the anomaly equation:

$$\partial_\mu J_{em}^\mu = -\frac{eR^2}{2\pi} \epsilon^{\mu\nu} \partial_\mu \beta_\nu . \qquad (37)$$

Thus now we can satisfy the criterion of chirality without imposing any boundary conditions on the topological current. Substituting Eq. 36) into Eq. 32 (and using $R^2 = \nu_k$) also fixes κ to be the filling fraction. It follows that the Skyrmion charge is $e\nu_k N_{Sky}$. Eqs. 28, 36 then give the value for the spin to be $\frac{N_{Sky}\nu_k}{2}$. Therefore, within the above assumptions, a winding number one Skyrmion is a fermion when the filling fraction is one.

Acknowledgments

It is a pleasure to thank S. Baez, A.P. Balachandran and A. Stern for so many discussions. I also acknowledge interest and discussions with J. Soto and R. Ray. This research was supported by the U.S. Department of Energy under contract DE-FG02-85ER40237.

REFERENCES

1. S. Baez, A.P. Balachandran, A. Stern and A. Travesset, *cond-mat 9712151*.
2. R.E. Prange and S.M. Girvin, *The Quantum Hall effect*, Springer-Verlag, New York (1990) and references therein;
3. S. C. Zhang, H. Hansson and S. Kivelson, Phys. Rev. Lett. **62**, 82 (1989); **62**, 980 (1989); D. H. Lee and S. C. Zhang, Phys. Rev. Lett. **66**, 1220 (1991).
4. S. C. Zhang, Int. J. of Mod. Phys. **B 6**, 25 (1992) and references therein.
5. J.M. Leinaas and S. Viefers, *cond-mat 9712009*.
6. R. Ray, This proceedings.
7. S.L. Sondhi, A. Karlhede, S.A. Kivelson and E.H. Rezayi, Phys. Rev. **B 47**, 16419 (1993).
8. X.Wen, Adv. In Phys. 44 (1995) 405-473 and references therein.
9. A.P. Balachandran, L. Chandar and B. Sathiapalan, Nucl. Phys. **B 443**, 465 (1996); Int. J. of Mod. Phys. **A 11**, 3587 (1996).
10. C.G. Callan Jr. and J.A. Harvey, Nucl. Phys. B 250 (1985) 427; S.G. Naculich, Nucl. Phys. B 296 (1988) 837.

Integrability in Classical and Quantum Field Theory and the Bukhvostov-Lipatov Model

Bogomil Gerganov

Newman Laboratory of Nuclear Studies
Cornell University
Ithaca, NY 14853

Abstract. A brief overview of the notions of classical and quantum integrability in $1+1$ space-time dimensions is given with an emphasis on the existence of infinite number of conserved quantities in $(1+1)D$ field theories. Some methods for building such conserved quantities are outlined and are then applied to study the integrability of the Bukhvostov-Lipatov model, a 2-field generalization of the sine-Gordon/massive Thirring model. The integrability properties of both the classical/quantum and the bosonic/fermionic versions of the model are discussed. It is shown that the *classical fermionic* model possesses a conserved current of Lorentz spin 3. It is then established that the conservation law is spoiled at the quantum level – a fact that indicates that the *quantum* Bukhvostov-Lipatov model is not integrable.

I INTEGRABILITY IN $1+1$ SPACE-TIME DIMENSIONS

A Classical Integrability

Definition. Let the time evolution of a physical system be described by some (generally non-linear) partial differential equations. The system is said to be *classically integrable* if there exist *exact integrals* to its *classical equations of evolution*.

The pioneering work in the field of classically integrable non-linear systems was a 1967 paper by Gardner, Greene, Kruskal, and Miura [1]. The authors were able to introduce a non-linear change of variables that makes the Korteweg-de Vries equation linear and exactly solvable. In 1968 Lax generalized their work by developing a general procedure [2] for associating to a system of *non-linear* PDEs a *linear operator*, whose *eigenvalues* are *integrals* of the non-linear system.

In the following decades exact solutions to a number of non-linear classical systems have been found and the importance of the field has grown for several reasons:

The evolution of some very important *physical systems* is governed by non-linear equations that admit exact solutions. Their range of *applicability* includes fluid dynamics, non-linear optics, and condensed matter theory.

Most known classically integrable systems exhibit the presence of "particle-like" solutions to the non-linear equations of motion – solitary waves or *solitons*. Their properties present interest to many areas of physics, both applicable, such as optic fiber design, and purely theoretical, such as string theory.

The solutions of some integrable systems can be used as 0th-order approximations in perturbation theory to study non-integrable systems. Such refined perturbative techniques often are more powerful than conventional perturbation theory.

Some examples of integrable non-linear partial differential equations (PDEs) are:

– Korteweg-de Vries equation (KdV): $\phi_t + 6\phi\phi_x + \phi_{xxx} = 0$, [1, 2].

– Non-linear Schödinger equation: $i\phi_t = -\phi_{xx} + |\phi|^2 \phi$, [1, 2].

– sine-Gordon equation (sG): $\phi_{tt} - \phi_{xx} = \sin\phi$, [3].

A comprehensive introduction to the subject of classical integrability and the theory of solitons can be found in [4].

B Quantum Integrability in $1+1$ Space-Time Dimensions

Because of the specific (*strongly restrictive*) features of particle kinematics in $(1+1)$-dimensional space-time, there exist scattering theories which allow *precise computation* of all elements of the *exact* particle S-matrix. This is possible because the following 2 (related) properties, each of which can be taken as a *definition of quantum integrability* in $(1+1)D$:

Definition A. A QFT in 2 space-time dimensions is *integrable* if there exists an *infinite set* of independent, mutually commuting *integrals of motion*

$$[Q_r, H] = 0 \, , \quad [Q_r, Q_{r'}] = 0 \, , \quad r = 1, 2, 3, \ldots \, . \tag{I.1}$$

Definition B. A QFT in 2 space-time dimensions is *integrable* if the N-paticle S-matrix can be *factorized* into a product of $\frac{N(N-1)}{2}$ 2-particle S-matrices, as if the process of N-particle scattering were reduced to a *sequence of pair collisions*; thus, allowing all the matrix elements of S to be determined exactly.

It can be shown that **Property B** follows from **Property A**. Indeed, in $1+1$ dimensions the *only symmetry transformations* on the space-time are the *translations* and the *Lorentz boosts* That's why, in the limits $t \to \pm\infty$, any conserved tensorial quantity of rank r reduces to a polynomial in the asymptotic particle 2-momenta. In a diagonal basis this polynomial could be written as

$$\hat{Q}_r = \sum_{a \in IN} \hat{p}_a^r = \sum_{b \in OUT} \hat{p}_b^r \, . \tag{I.2}$$

The last result is important because it has 2 consequences:

1. **No particle production (the S-matrix is *elastic*).**
   Let $|p_1, p_2, \ldots, p_a, \ldots, p_{N_a}\rangle_{IN}$ and $|p'_1, p'_2, \ldots, p'_b, \ldots, p'_{N_b}\rangle_{OUT}$ be asymptotic particle states at $t \to \pm\infty$. If we act on these states with the conserved charge operator Q_r, we obtain, according to (I.2), that

$$\sum_{a=1}^{N_a} p_a^r = \sum_{b=1}^{N_b} p_b^r . \qquad (I.3)$$

Since there are infinite number of such conserved charges in the theory, the particle 2-momenta sholud satisfy *infinite number of conservation laws* like (I.3) with $r = 1, 2, 3, \ldots$. This is only possible if the *set of the ingoing 2-momenta is equal to the set of outgoing 2-momenta*:

$$\{p_1, p_2, \ldots, p_a, \ldots, p_{N_a}\}_{IN} = \{p'_1, p'_2, \ldots, p'_b, \ldots, p'_{N_b}\}_{OUT}, \quad N_a = N_b = N .$$

This implies, in particular, that the total particle number and the number of particles of given mass, $m^2 = p_\mu p^\mu$, is conserved. Therefore,

$$S_{fi} = {}_{OUT}\langle p'_1, p'_2, \ldots, p'_b, \ldots, p'_N | p_1, p_2, \ldots, p_a, \ldots, p_N\rangle_{IN} \text{ is } elastic.$$

2. **Multiparticle collisions reduce to a sequence of pair collisions (the S-matrix is *factorizable*).**
   Since there exists an infinite set of integrals of motion in our theory, it is natural to assume that there is a *particular symmetry transformation* that reduces at $t \to \pm\infty$ to *independent translations* of the individual particles. By performing such translations on can get *arbitrarily large space-time separations* of all the regions where pair collisions occur. Therefore, the pair collisioins can be treated as independent. The existence of infinite number of symmetries further implies that the time order of the collisions is irrelevant and, therefore, different permutations of the 2-particle S-matrices in the factorized product will give rise to the same multiparticle S-matrix (*Yang-Baxter equations*).

The above 2 properties, together with the Yang-Baxter equations and the requirements for *unitarity, analyticity*, and *crossing symmetry*, for many $(1+1)D$ QFTs provide enough constraints to compute the S-matrix *exactly*.

A more extensive review of the properties of the S-matrix in $(1+1)D$ can be found in [5].

II METHODS FOR BUILDING INTEGRALS OF MOTION OF HIGHER TENSORIAL RANK

A Classical Integrals of Motion

In a remarkable paper [2] Lax formulated a powerful technique for integrating exactly systems of non-linear evolution equations. Let

$$u_t = K(u) \qquad (II.1)$$

be some non-linear PDE and let $u(x,t)$ be a solution to (II.1). Lax showed that if we can associate some *selfadjoint* linear differential operator $\hat{L}_u$ with the function u, such that $\hat{L}_u$ is *unitary invariant* under the evolution of u according to (II.1), then the *eigenvalues of $\hat{L}_u$ are integrals of motion* of (II.1).

If we find the operator $\hat{L}_u$ for a certain PDE, it generates an *infinite series of integrable PDEs*. The higher order PDEs in the series can be used to build integrals of motion of higher rank and, thus, to construct and infinite series of conservation laws of the classical system. Using this and a variety of other methods, infinite sets of conserved quantities of increasing spin have been constructed for the classical MT, Korteweg-de Vries (KdV), and sG models [1–3, 6, 7].

B Quantum Integrals of Motion

A powerful tool for building conserved quantities for $2D$ *quantum* models is the technique of *perturbed conformal field theory* [8]. By treating a $2D$ QFT as a perturbed CFT, one can make use of conformal invariance to find an infinite number of conserved quantities in the conformal theory. It is then possible to study which (if any) of these conservation laws survive the perturbation. Zamolodchikov's paper [8] provides an easy way of computing the conserved current densities explicitly.

In CFT any conserved density T_{s+1} is a holomorphic function and $\partial_{\bar{z}} T_{s+1} = 0$. In the perturbed QFT that is no longer true and we can compute $\partial_{\bar{z}} T_{s+1}$, using Zamolodchikov's formula [8, eq.(3.14)]:

$$\partial_{\bar{z}} T_{s+1} = \lambda \oint_z \frac{d\zeta}{2\pi i} : \Phi(\zeta, \bar{z}) :: T_{s+1}(z) : \qquad (II.2)$$

If the RHS of (II.2) can be expressed as a total ∂_z-derivative of some local operator Θ_{s-1}, there exists a spin s conservation law surviving in the perturbed QFT:

$$\partial_{\bar{z}} T_{s+1} = \partial_z \Theta_{s-1}, \qquad \partial_z \overline{T}_{s+1} = \partial_{\bar{z}} \overline{\Theta}_{s-1} , \qquad (II.3)$$

T_{s+1} and Θ_{s-1} being the quantum conserved densities.

III THE BUKHVOSTOV-LIPATOV MODEL

The bosonic Bukhvostov-Lipatov model (BL) is a 2-field generalization of the sine-Gordon model (sG)[1]. It was first introduced by Bukhvostov and Lipatov [10] in a study of the $O(3)$-nonlinear σ-model and has drawn recent attention in works by Fateev [11] and Lesage, Saleur, Simonetti [12]. The model is defined by the action

$$S = \frac{1}{4\pi} \int dt dx \left[\frac{1}{2} (\partial_\mu \phi_1)^2 + \frac{1}{2} (\partial_\mu \phi_2)^2 + \lambda : \cos \hat{\beta}_1 \phi_1 :: \cos \hat{\beta}_2 \phi_2 : \right] , \qquad (III.1)$$

---

[1] Generalizations of the sine-Gordon model have been studied as early as 1981 [9].

where $\hat{\beta}_i \equiv \frac{\beta_i}{\sqrt{4\pi}}$, $i = 1, 2$. Using the and Coleman's bosonization prescription [13], Bukhvostov and Lipatov mapped[2] the quantum bosonic theory (III.1) to a dual theory of two Dirac fermions[3]:

$$S_{\text{full}} = \frac{1}{4\pi} \int dtdx \left\{ \frac{i}{2} \Psi^{\mathcal{D}} \gamma^\mu \partial_\mu \Psi - \frac{i}{2} \partial_\mu \Psi^{\mathcal{D}} \gamma^\mu \Psi + \frac{i}{2} X^{\mathcal{D}} \gamma^\mu \partial_\mu X - \frac{i}{2} \partial_\mu X^{\mathcal{D}} \gamma^\mu X \right.$$
$$\left. - m \left(\Psi^{\mathcal{D}} \Psi + X^{\mathcal{D}} X \right) - g \left(\Psi^{\mathcal{D}} \gamma^\mu \Psi \right) \left(X^{\mathcal{D}} \gamma^\mu X \right) - g' \left[\left(\Psi^{\mathcal{D}} \gamma^\mu \Psi \right)^2 + \left(X^{\mathcal{D}} \gamma_\mu X \right)^2 \right] \right\} .$$
(III.2)

where

$$g = \pi^2 \left(\frac{1}{\hat{\beta}_1^2} - \frac{1}{\hat{\beta}_2^2} \right) , \qquad g' = \frac{\pi^2}{2} \left(\frac{1}{\hat{\beta}_1^2} + \frac{1}{\hat{\beta}_2^2} - 4 \right) . \qquad \text{(III.3)}$$

The above model is obviously a 2-field generalization of the massive Thirring model (MTM). It involves 2 types of 4-fermion interactions: the interaction of the MT model with coupling g' and a new interaction term with coupling g.

In [10] Bukhvostov and Lipatov studied the integrability of the model (III.2), using Bethe ansatz [14]. They constructed the S-matrix for the *pseudoparticles* and showed that it satisfies Yang-Baxter equations in 2 cases:

1. $g = 0$, $\qquad\qquad\qquad g'$: unrestricted , $\qquad$ (III.4)
2. $g' = 0$, $\qquad\qquad\qquad g$: unrestricted . $\qquad$ (III.5)

In Case 1 (III.2) reduces to two decoupled copies of the MT model, known to be integrable both classically and quantum mechanically [6,15]. In Case 2 one obtains a new model, to which we will refer as *"the fermionic BL model"*:

$$S_{BL} = S_{\text{full}} |_{g'=0} . \qquad \text{(III.6)}$$

Based on their result for the pseudoparticle S-matrix, Bukhvostov and Lipatov conjectured that the quantum field theory (III.6) is integrable. The actual *physical* states, however, have *not* been constucted in their paper. To the best of our knowledge, computing the physical S-matrix and, thus carrying out Bethe ansatz for the model consistently to the end, still remains an open problem.

The **classical bosonic model** (III.1) was studied by Ameduri and Efthimiou. Using *Painlevé analysis*, they proved [16] that the model is *not* classically integrable. In the following sections we discuss in some details the integrability of the **classical fermionic model** (III.6) and the **quantum integrability** of the models (III.1) and (III.2), dual to each other in the quantum case. We focus, in particular, on constructing conserved quantities of higher tensorial rank in these theories.

---

[2]) After a rotation of the boson fields: $\phi_1 = \frac{1}{2\hat{\beta}_1} (\vartheta_1 + \vartheta_2)$ and $\phi_2 = \frac{1}{2\hat{\beta}_2} (\vartheta_1 - \vartheta_2)$.
[3]) Here $\Psi^{\mathcal{D}}$ denotes the *Dirac conjugate* of Ψ, $\Psi^{\mathcal{D}} \equiv \Psi^\dagger \gamma^0$. We use this unconventional notation rather than the usual "$\overline{\Psi}$" to avoid confusion with the *antichiral spinor component*, $\overline{\psi}$.

A Classical Fermionic Model

In Eucledian space-time, parametrized by the light cone coordinates $z = \frac{1}{2}(\tau + ix)$, $\bar{z} = \frac{1}{2}(\tau - ix)$, and in terms of spinor components[4], the action (III.6) becomes

$$S_E = \frac{1}{4\pi} \int d\tau dx \left[\frac{1}{2} \left(\psi_+ \partial_{\bar{z}} \psi_- + \psi_- \partial_z \psi_+ + \overline{\psi}_+ \partial_z \overline{\psi}_- + \overline{\psi}_+ \partial_z \overline{\psi}_- \right) + \right.$$
$$\left. - im \left(\overline{\psi}_+ \psi_- + \overline{\psi}_- \psi_+ \right) - g \left(\psi_+ \psi_- \overline{\chi}_+ \overline{\chi}_- + \overline{\psi}_+ \overline{\psi}_- \chi_+ \chi_- \right) + (\psi \leftrightarrow \chi) \right] , \quad \text{(III.7)}$$

which gives rise to the *classical* equations of motion

$$\partial_{\bar{z}} \psi_\pm (z,\bar{z}) = -im \overline{\psi}_\pm \pm 2g \psi_\pm \overline{\chi}_+ \overline{\chi}_- \ , \quad \partial_z \overline{\psi}_\pm (z,\bar{z}) = im \psi_\pm \pm 2g \overline{\psi}_\pm \chi_+ \chi_- \quad \text{(III.8)}$$

and the equations obtained with the substitutiuon $\psi \leftrightarrow \chi$.

Because of the space and time translational invariance, the theory possesses a trivial conserved charge of spin 1 — the energy-momentum tensor. Finding conserved quantitites of higher Lorentz spin is equivalent to finding non-trivial integrals of motion and is considered to be a strong indication for the classical integrability of the theory. We have been able to show [17] that the fermionic BL model (III.7) has a classically conserved charge of spin 3 in the bulk:

$$Q_3 = \int_{-\infty}^{+\infty} dx \, (T_4 - \Theta_2) \quad \text{and} \quad \overline{Q}_3 = \int_{-\infty}^{+\infty} dx \, (\overline{T}_4 - \overline{\Theta}_2) \ , \quad \text{(III.9)}$$

where the densities T_4 and Θ_2 are given by:

$$T_4 = \psi_+ \partial_z^3 \psi_- + 6g \partial_z \psi_+ \partial_z \psi_- \chi_+ \chi_- + (+ \leftrightarrow -) + (\psi \leftrightarrow \chi) \ , \quad \text{(III.10)}$$

$$\Theta_2 = im \overline{\psi}_+ \partial_z^2 \psi_- + 2mg \overline{\psi}_+ \partial_z \psi_- \chi_+ \chi_- + 4mg \overline{\psi}_+ \psi_- \chi_+ \partial_z \chi_- +$$
$$+ 4mg \overline{\psi}_+ \psi_- \partial_z \chi_+ \chi_- - 4im^2 g \psi_+ \psi_- \chi_+ \chi_- + (+ \leftrightarrow -) + (\psi \leftrightarrow \chi) \ . \quad \text{(III.11)}$$

(similarly, for $\overline{T}_4$ and $\overline{\Theta}_2$) and satisfy the continuity equation (II.3) with $s = 3$. The spin 3 charge (III.9) is conserved also in the theory on the half-line for specific types of boundary actions [17]. We intend to generalize our result to conserved quantities of arbitrary spin by using methods similar to the ones described in Refs. [1–3,6,7].

We have also found that the *more general* 2-fermion model (III.2) does *not* possess a classically conserved spin 3 current for arbitrary values of the couplings g and g'. Classical conservations laws of spin 3 exist only in the special cases $g = 0$ (2×MTM) and $g' = 0$ (BL model), i. e. the model (III.2) is classically integrable precisely in the cases (III.4) and (III.5), suggested by Bukhvostov and Lipatov.

---

[4] We use the following conventions:

$$\Psi \to \begin{pmatrix} \psi_- \\ \overline{\psi}_- \end{pmatrix}, \ \Psi^\dagger \to (\psi_+ \ \overline{\psi}_+); \ \Psi^D \equiv \Psi^\dagger \gamma^0; \ \gamma_0 = \begin{bmatrix} 0 & 1 \\ 1 & 0 \end{bmatrix}, \ \gamma_1 = \begin{bmatrix} 0 & 1 \\ -1 & 0 \end{bmatrix}$$

and, similarly, for X. The components of the metric tensor are $[g_{\mu\nu}] = [g^{\mu\nu}] = \text{diag}[1,-1]$.

B Quantum Integrability

In this section we will study the modifications to the classical conservation law due to quantum corrections, using *perturbed conformal field theory*[5]. Boson-fermion duality in 2 space-time dimensions allows us to write our expression for the classically conserved fermionic current (III.10) in bosonic form and then to study its quantum conservation in the model (III.1), treating the term $\lambda : \cos\hat{\beta}_1\phi_1 \cos\hat{\beta}_2\phi_2 :$ as a single relevant perturbation to the conformal theory of massless free bosons.

Using the substitutions $\psi_\pm =: e^{\pm i\vartheta_1}:$, $\chi_\pm =: e^{\pm i\vartheta_2}:$ and the vertex operator product expansions from CFT [18], we can map each term of T_4 to its dual operator in the bosonic theory. Since quantum corrections will, in principle, modify the coefficients of the classical T_4, we prefer to leave them as arbitrary functions of g and fix them later while computing the conserved current via perturbed CFT. Finally, using the relations in Footnote 2, we can write T_4 in terms of the boson fields ϕ_1 and ϕ_2 as

$$T_4 = \sum_{m+n=2} a_{mn} : (\partial_z^2\phi_1)^m (\partial_z^2\phi_2)^n : + \sum_{k+l=4} b_{kl} : (\partial_z\phi_1)^k (\partial_z\phi_2)^l :$$
$$+ c_1 : (\partial_z\phi_1)^2 \partial_z^2\phi_2 : + c_2 : \partial_z^2\phi_1 (\partial_z\phi_2)^2 : ; \quad m,n=0,1,2;\ k,l=0,1,2,3,4\ . \quad \text{(III.12)}$$

This is, in fact, the *most general ansatz* for T_4 for the double cosine model. It includes, up to addition of total ∂_z-derivatives, *all* operators of mass dimension 4 with arbitrary coefficients, functions of $(\hat{\beta}_1, \hat{\beta}_2)$ or, via (III.3), of g. As it turns out, the requirement that T_4 be conserved in the perturbed CFT is very restrictive and gives enough information to compute the exact form of these functions.

In starting this calculation, our goal was to find *all* the conditions on the couplings $\hat{\beta}_1$ and $\hat{\beta}_2$ for which the spin 3 charge is conserved. We expected that the 'BL manifold' (III.5) would be one of the integrable cases and then, by fermionizing back, we would be able to obtain an exact quantum expression for the fermionic T_4. At the end, we found that the spin 3 current is conserved only in 3 cases:

$$\hat{\beta}_1^2 - \hat{\beta}_2^2 = 0 \longrightarrow g = 0\ , \quad \text{(III.13)}$$

$$\hat{\beta}_1^2 + \hat{\beta}_2^2 = 1 \longrightarrow \left(\frac{2g'}{\pi^2} + 2\right)^2 - \left(\frac{g}{\pi^2}\right)^2 = 4\ , \quad \text{(III.14)}$$

$$\hat{\beta}_1^2 + \hat{\beta}_2^2 = 2 \longrightarrow \left(\frac{2g'}{\pi^2} + 3\right)^2 - \left(\frac{g}{\pi^2}\right)^2 = 1\ . \quad \text{(III.15)}$$

The first manifold is trivial: when $\hat{\beta}_1^2 = \hat{\beta}_2^2$ the double cosine model decouples into 2 sine-Gordon models, integrable both classically and quantum mechanically. The second two[6] quantum integrable[7] manifolds have been previously identified by

---

[5] See Section II.B and Ref. [8].
[6] We concentrate here on the first non-trivial manifold, (III.14), since this is one with physical significance (tunneling in quantum wires) [12].
[7] Another model that is *not integrable classically*, but is *quantum integrable* on certain submanifolds of the parameter space has been previously discussed by Shankar [9].

Fateev [11] and by Lesage, Saleur, and Simonetti [12]. On the BL manifold (III.5) the charge (III.9) is *not quantum conserved*, except in the trivial case when the manifolds (III.13), (III.5), and (III.14) intersect each other (free fermion point).

The above result becomes even more clear when we look at the fermionic parameter space. As we showed at the end of Section III.B, the general model has a *classically* conserved charge of spin 3 only if either $g = 0$ or $g' = 0$. Therefore, the *classical integrable manifolds* are simply the axes of the (g, g')-plane:

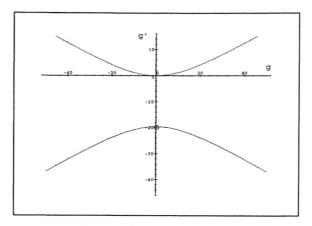

FIGURE 1. Fermionic parameter space.

We see that the manifold $g = 0$ is present both in the classical and in the quantum case. This merely reflects the fact that massive Thirring model is integrable both classically and quantum mechanically. In other words, if we start with the classical MT model, turning on quantum corrections will *not* take the classical conservation law away from the vertical axis. The coefficients of T_4 will be modified but they will still remain functions of g' only and will be independent of g.

In contrast, if we start with the classical fermionic BL model, with T_4 depending on g only (III.10), quantum corrections will not only modify the coefficients of the conserved current but will also introduce new terms, involving g'-dependence. Looking at Figure 1, we see that the g-axis, on which the model is integrable classically, is 'deformed' into the hyperbola (III.14) in the quantum case. The only point on the classical manifold where integrability is preserved by quantum mechanics is the origin, which, of course, is simply the the *free point* ($g = g' = 0$) of the fermionic theory.

IV   CONCLUSION

We have shown that the fermionic Bukhvostov-Lipatov model, given by the action (III.6), admits a nontrivial classical integral of motion of spin 3. This conservation law holds quantum mechanically only at the free fermion point $g = 0$ and is spoiled by quantum corrections for generic values of the coupling g. The more general fermionic model (III.2) admits a quantum conservation law of spin 3 for the specific relation (III.14) between the couplings g and g'. The study of the spin 3 conservation laws, therefore, suggests that the integrable manifold $(g : \text{free}, g' = 0)$ proposed by Bukhvostov and Lipatov does not survive in the quantum field theory.

ACKNOWLEDGEMENTS

I would like to thank my collaborators Costas Efthimiou and Marco Ameduri for completing this work. I would also like to thank André LeClair for many useful discussions and insights on the subject and to Frédéric Lesage for a discussion on the double-cosine model.

REFERENCES

1. Gardner, C. S., Greene, J. M., Kruskal, M. D., and Miura, R. M., *Phys. Rev. Lett.* **19**, 1095 (1967).
2. Lax, P. D., *Comm. Pure Appl. Math.* **21**, 467 (1968).
3. Takhtadzhyan, L. A., and Faddeev, L. D., *Theor. Math. Phys.* **21**, 1046 (1974).
4. Faddeev, L. D., Takhtajan, L. A., *Hamiltonian Methods in the Theory of Solitons*, Berlin/Heidelber: Springer-Verlag, 1987, and the references therein.
5. Zamolodchikov, A. B., in *Soviet Scientific Reviews*, Sec. **A**, Khalatnikov, I. M., ed., New York: Harwood Academic Publishers, 1980, vol. 2, pp. 1–40.
6. Berg, B., Karowski, M., and Thun, H. J., *Phys. Lett.* **B64**, 286 (1976); Fiume, R., Mitter, P. K., and Papanicolau, N., *Phys. Lett.* **B64**, 289 (1976).
7. Ferrara, S., Girardello, L., and Sciuto, S., *Phys. Lett.* **B76**, 303 (1978); Girardello, L., and Sciuto, S., *Phys. Lett.* **B77**, 267 (1978).
8. Zamolodchikov, A. B., *Adv. Studies in Pure Math.* **19**, 641 (1989).
9. Shankar, R., *Phys. Lett.* **B102**, 257 (1981).
10. Bukhvostov, A. P., and Lipatov, L. N., *Nucl. Phys.* **B180**, 116 (1983).
11. Fateev, V. A., *Nucl. Phys.* **B473**, 509 (1996).
12. F. Lesage, F., Saleur, H., and Simonetti, P., *Tunneling in Quantum Wires I: Exact Solution of the Spin Isotropic Case*, cond-mat/9703220; *Tunneling in Quantum Wires II: A New Line of IR Fixed Points*, cond-mat/9712019.
13. Coleman, S., *Phys. Rev.* **D11**, 2088 (1975).
14. Korepin, V. E., Bogoliubov, N. M., and Izergin, A. G., *Quantum Inverse Scattering Method and Correlation Functions*, Cambridge: Cambridge University Press, 1993.
15. Kulish, P. P., and Nisimov, E. R., *Pis'ma Zh. Eksp. Teor. Fiz.* **24**, 247 (1976).
16. Ameduri, M., Efthimiou, C., *J. Nonl. Math. Phys.* **5**, 132 (1998).
17. Ameduri, M., Efthimiou, C., and Gerganov, B., *The Bukhvostov-Lipatov Model*, work in progress.
18. Di Francesco, P., Mathieu, P., and Sénéchal, D., *Conformal Field Theory*, New York: Spinger, 1997.

Evidence for a Scalar $\kappa(900)$ Resonance in πK Scattering

Deirdre Black*

Department of Physics, Syracuse University, Syracuse, NY 13244-2210, USA

Abstract. Motivated by the $1/N_c$ expansion, we study a simple model in which the πK scattering amplitude is the sum of a *current − algebra* contact term and resonance pole exchanges. This phenomenological model is crossing symmetric and, when a putative light strange scalar meson κ is included, satisfies the unitarity bounds to well above 1 GeV. The model also features chiral dynamics, vector meson dominance and appropriate interference between the established $K_0^*(1430)$ resonance and its predicted background.

INTRODUCTION AND REVIEW OF MODEL

This presentation is based on research done with Amir H. Fariborz and Joseph Schechter from the Department of Physics at Syracuse University and Francesco Sannino from the Department of Physics at Yale University. It is a summary of the article given as reference [1] below and a generalization of ideas developed previously at Syracuse by Harada, Sannino and Schechter in the context of $\pi\pi$ scattering [2]. At that time evidence was found to support the existence of a low-mass relatively broad scalar resonance, denoted $\sigma[m_\sigma = 560\text{MeV}, \Gamma_\sigma = 370\text{MeV}$, with pole position $s = (0.585 - 0.178i)$ GeV], in addition to the well-established scalar $f_0(980)$ resonance. A number of other authors have also found similar or related results in different models (see references in [1]).

If one accepts a low-lying σ and notes the existence of the isovector scalar $a_0(980)$, as well as the $f_0(980)$, there would be three scalar resonances below 1GeV. A great deal of discussion and controversy over the years has surrounded the issue of the nature of such very low-mass scalars. The reason is that one expects the lowest-lying scalars in the quark model to be p-wave $q\bar{q}$ bound states and hence to have masses comparable to those of the axial and tensor mesons, already in the 1.2 - 1.6 GeV region (see for example [3]). As an example (see the discussion on page 355 of [4] under the "Note on Scalar Mesons") one might form a conventional scalar nonet from the $f_0(1370)$, $a_0(1450)$, $K_0^*(1430)$ and $f_J(1710)$. If an assignment like this is correct it raises the question of why the three scalar candidates σ, $f_0(980)$ and $a_0(980)$ are so light, and whether a general organizing principle for their dynamics can be found. From this point of view it is extremely interesting to see if a light strange scalar resonance, to be denoted κ, emerges in the study of πK scattering.

If the putative κ resembles the σ of [2] in that it is both broad and also only one of several comparable contributions to the pseudoscalar scattering amplitude, then the existence of such a scalar may not be immediately apparent from inspection of the phase shifts obtained from experiment. Clearly, the reliability of any prediction made about a κ meson depends on how accurately its "background" can be modeled. In the present approach we will use an effective chiral Lagrangian, since chiral symmetry works well near threshold. We will begin with the invariant amplitude which is manifestly crossing symmetric, thereby ensuring inclusion of cross-channel contributions from resonances known to exist in a given energy region.

This approach developed in [2] is inspired by the $\frac{1}{N_c}$ expansion [6] of QCD, in which the leading, order of $\frac{1}{N_c}$, contribution to the meson-meson scattering amplitude consists of tree diagrams - contact terms and resonance exchanges. Thus, away from the poles (which contain divergences of the theory in leading order since the resonance widths go as $\frac{1}{N_c}$) the leading order amplitudes are purely real. Hence we restrict ourselves to comparing the real parts of our computed amplitudes with the real parts of the amplitudes deduced from experiment. The imaginary parts may, assuming elasticity, be computed using dispersion relations.

A guiding physical principle in our analysis is to make sure that the amplitude obeys the very important unitarity bounds, which for our purposes means that:

$$\left|R_l^I\right| \leq \frac{1}{2}, \qquad (1)$$

where R_l^I is the isospin I and lth partial wave projection of the Real part of the invariant scattering amplitude.

In the case of $\pi\pi$ scattering it was found that the threshold region was well described by the current algebra contact term, but that it violated the positive unitarity bound at about 500MeV. Inclusion of the ρ-meson somewhat counteracted the rising current algebra term, suggesting that different contributions to the amplitude, which individually are not within the unitarity bounds, may "locally cancel" with each other to give the full, unitary amplitude. It was found that by also including the σ-meson unitarity could be restored. Furthermore the amplitude thus computed provided a background at the $f_0(980)$ pole. A unitary parameterisation of the amplitude, embodying interference between the resonance and its background, allowed a good fit to experiment up to about 1GeV.

EVIDENCE FOR THE SCALAR $\kappa(900)$ RESONANCE

Preliminary Study of the $I = \frac{1}{2}$ Channel up to 1GeV

Let us carry out an initial analysis of the $I = \frac{1}{2}$ and $J = 0$ projection of the real part of the πK scattering amplitude $R_0^{1/2}$. We give only a sketch of the calculations here and refer the reader to [1] for more details. We first define the $I = \frac{3}{2}$ invariant amplitude via $A^{3/2}(s,t,u) = A(\pi^+(p_1)K^+(p_2) \to \pi^+(p_3)K^+(p_4))$ where s, t and u are the Mandelstam variables. By crossing symmetry we have $A(\pi^+K^- \to \pi^+K^-) = A^{3/2}(u,t,s)$ which leads to

$$A^{1/2}(s,t,u) = \frac{3}{2}A^{3/2}(u,t,s) - \frac{1}{2}A^{3/2}(s,t,u). \tag{2}$$

We then define the usual partial wave isospin amplitudes according to [4,1].

As in the $\pi\pi$ case we start with the well-known "current algebra" amplitude. For this purpose, we can use the conventional chiral Lagrangian including only pseudoscalars:

$$\mathcal{L}_1 = -\frac{F_\pi^2}{8}\text{Tr}\left(\partial_\mu U \partial_\mu U^\dagger\right) + \text{Tr}\left[\mathcal{B}\left(U + U^\dagger\right)\right], \tag{3}$$

in which $U = e^{2i\frac{\phi}{F_\pi}}$, with ϕ the 3×3 matrix of pseudoscalar fields and $F_\pi = 132$ MeV the pion decay constant. $\mathcal{B}$ is a diagonal matrix (B_1, B_1, B_3) with $B_1 = m_\pi^2 F_\pi^2/8 = B_2$ and $B_3 = F_\pi^2(m_K^2 - m_\pi^2/2)/4$. This is the dominant minimal symmetry breaking term for the pseudoscalar mesons. We shall choose $m_\pi = 137$ MeV and $m_K = 496$ MeV. The $I = \frac{1}{2}$ invariant amplitude, which at tree level is simply a contact term as in Fig. 1(a), is

$$A_{CA}^{1/2}(s,t,u) = \frac{1}{2F_\pi^2}[2(s-u) + t], \tag{4}$$

and we will refer to this as the *current algebra* result. The $J = 0$ partial wave projection of the real part of (4) is shown in Fig. 2, indicating a severe violation of the unitarity bound (1) beyond approximately 900 MeV. As in the $\pi\pi$ case we will try to solve this problem by including resonance contributions to the scattering amplitude.

First consider the effect of the vector mesons whose interaction Lagrangian with pseudoscalars is given by [1]:

$$\mathcal{L} = -\frac{1}{2}m_v^2\text{Tr}\left[\left(\rho_\mu - \frac{v_\mu}{\tilde{g}}\right)^2\right], \tag{5}$$

where ρ_μ is the vector multiplet, $v_\mu = \frac{i}{2}\left(\xi\partial_\mu\xi^\dagger + \xi^\dagger\partial_\mu\xi\right)$ with $\xi = \sqrt{U}$, and $\tilde{g}$ is the $SU(3)$ gauge-type coupling constant.

There are ρ and K^* exchanges and a direct K^* pole as illustrated in Figs 1(b), 1(d) and 1(c). Also there is an additional contact term (which arises as a consequence of casting the Lagrangian in a chiral invariant form) which turns out to be most important in reversing the unitarity violation by the current algebra piece. We take $m_\rho = 0.77$GeV, $m_{K^*} = 0.89$GeV and the vector-pseudoscalar-pseudoscalar coupling constant to be $g_{\rho\pi\pi} = \frac{m_\rho^2}{\tilde{g}F_\pi^2} = 8.56$.

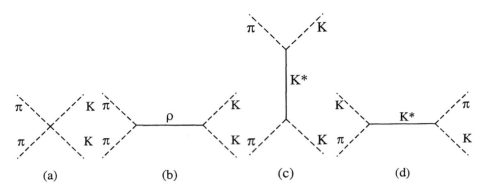

FIGURE 1. Examples of tree diagrams relevant for πK scattering in our model - we show a contact term a), s-channel diagram c) and t- and u-channel exchange diagrams b) and d). This is the complete set of diagrams for the vector contributions.

The effect of all the vector contributions, added to the current algebra piece is displayed in Fig. 2. It can be seen that, similarly to the $\pi\pi$ case, the introduction of vectors has pulled the curve down so that it almost lies within the unitarity bound.

So far we have not used any unknown parameters, but in the spirit of [2], we begin to include other resonances in this energy regime in order to offset the violation of the unitarity bound. There is the established $f_0(980)$ as well as the $\sigma(560)$ which should be included for self-consistency. Of course the role played by a possible strange scalar is of great interest.

In order to compute the scalar exchange diagrams we take the following scalar-pseudoscalar-pseudoscalar interaction Lagrangian terms:

$$\mathcal{L}_{scalars} = -\sqrt{2}\gamma_{\sigma\pi\pi}\left(\sigma\partial_\mu\pi^+\partial_\mu\pi^- + ...\right) - \frac{\gamma_{\sigma K\bar{K}}}{\sqrt{2}}\left(\sigma\partial_\mu K^+\partial_\mu K^- + ...\right)$$
$$- \sqrt{2}\gamma_{f_0\pi\pi}\left(f_0\partial_\mu\pi^+\partial_\mu\pi^- + ...\right) - \frac{\gamma_{f_0 K\bar{K}}}{\sqrt{2}}\left(f_0\partial_\mu K^+\partial_\mu K^- + ...\right)$$
$$- \gamma_{\kappa K\pi}\left(\kappa^0\partial_\mu K^-\partial_\mu\pi^+ + ...\right) . \qquad (6)$$

For generality we are not assuming any model to relate these couplings to each other. As discussed in [1], the derivative coupling is the one which would follow from a chiral invariant model. Also, the terms shown here are the particular ones needed to compute the $\pi^+ K^+$ scattering amplitude. The coupling constants $\gamma_{\sigma\pi\pi}$, $\gamma_{f_0\pi\pi}$ and $\gamma_{f_0 K\bar{K}}$ were estimated phenomenologically in [2], yielding the values $|\gamma_{\sigma\pi\pi}| = 7.81$ GeV^{-1}, $|\gamma_{f_0\pi\pi}| = 2.43$ GeV^{-1}, $|\gamma_{f_0 K\bar{K}}| = 10$ GeV^{-1}.

For now we take $\gamma_{\sigma K\bar{K}} = \gamma_{\sigma\pi\pi}$ which would be the case if the scalar σ was a member of an $SU(3)$ multiplet (implying in turn some particular quark substructure for the scalars). In our final analysis we will however be more general and consider the effect of varying the magnitude and sign of $\gamma_{\sigma K\bar{K}}$.

The σ contribution to the invariant amplitude is

$$A_\sigma^{1/2}(s,t,u) = \frac{\gamma_{\sigma\pi\pi}\gamma_{\sigma K\bar{K}}}{4} \frac{(t-2m_\pi^2)(t-2m_K^2)}{m_\sigma^2 - t}. \tag{7}$$

The $f_0(980)$ amplitude has an identical structure with $\sigma \to f_0$ everywhere. We shall take $m_\sigma = 0.55$ GeV and $m_{f_0} = 0.98$ GeV. These t-channel contributions are both positive and actually make the unitarity violation slightly worse, with $R_0^{1/2} > \frac{1}{2}$ at about 900MeV.

Now let us consider the strange scalar κ contribution. Its regularized $I = \frac{1}{2}$ invariant amplitude is:

$$A_\kappa^{1/2}(s,t,u) = \frac{\gamma_{\kappa K\pi}^2}{8}\left[\frac{3(s - m_\pi^2 - m_K^2)^2}{m_\kappa^2 - s - im_\kappa G'_\kappa \theta(s - s_{th})} - \frac{(u - m_\pi^2 - m_K^2)^2}{m_\kappa^2 - u - im_\kappa G'_\kappa \theta(u - s_{th})}\right]. \tag{8}$$

This regularization is formally crossing symmetric (the u-channel regularization term will vanish in the physical region). We will treat m_κ, $\gamma_{\kappa K\pi}$ and G'_κ as independent parameters. Analogously to the treatment of the light broad $\sigma(560)$, we have introduced a possible deviation from the pure Breit-Wigner form by allowing G'_κ to be a free parameter. The first term in (8) is a direct channel pole and should be extremely important at energies around m_κ. Thus, as in the $\pi\pi$ case it may be used to cure the unitarity violation of the $J = 0$ partial wave amplitude. Since the real part of a direct channel resonance contribution turns sharply negative just above the resonance energy and the graph in Fig. 2 rises above the positive unitarity bound at around 900 MeV we are led to choose m_κ to lie roughly around this energy. We see from Fig. 2, which is a plot of $R_0^{1/2}$ including the contribution of the scalar mesons that it is now easy to ensure that the amplitude lies roughly within the unitarity bound.

Comparison with Experiment for the $I = \frac{1}{2}$ Channel

The magnitude and phase of the experimental $I = \frac{1}{2}$ s-wave amplitude are given in Fig. 15 of Aston et al [7]. We have translated these to the real part $R_0^{1/2}(s)$, which is required for our approach, and show the results in Fig. 3 where we see an interesting dip at around 1400 MeV. In order to extend our description of πK scattering to the region above 1GeV, we first include only resonances which will contribute s-channel pole terms. This seems reasonable in light of [2] where it was found that, for the analogous $\pi\pi$ partial wave, the effects of the resonances in the equivalent energy regime tended to cancel amongst themselves, and could, along with those of inelasticity, be absorbed in minor adjustments of the parameters describing the light scalar. Thus we include in addition only the relatively narrow strange scalar resonance $K_0^*(1430)$. From our point of view the most interesting question is whether our model including the κ meson provides the correct background structure to explain the overall shape of $R_0^{1/2}$ in this 1400GeV region.

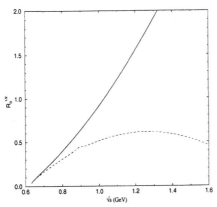 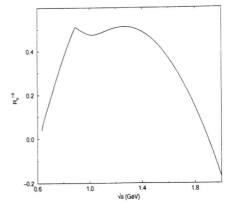

FIGURE 2. LEFT: Contribution of current algebra (solid line), and current algebra + vectors (dashed-line) to $R_0^{1/2}$. RIGHT: Contribution of current algebra + vectors + σ + $f_0(980)$ + κ to $R_0^{1/2}$ for κ parameters chosen as $m_\kappa = 900$MeV, $\gamma_{\kappa K\pi} = 4.8$GeV^{-1} and $G_\kappa' = 280$MeV.

In the case of $\pi\pi$ scattering, the interplay between the narrow resonance $f_0(980)$ and its background was introduced as a regularization of the direct channel resonance pole which is $\propto \dfrac{1}{s - m_*^2}$. In the vicinity of the resonance, upon projection into the appropriate partial wave, one sets the amplitude equal to

$$\frac{e^{2i\delta} m_* \Gamma_*}{m_*^2 - s - im_*\Gamma_*} + e^{i\delta}\sin\delta, \qquad (9)$$

where m_* and Γ_* are the resonance mass and width, while δ is the background phase which is assumed to be constant in the neighborhood of the resonance. This form automatically makes the amplitude unitary in this region.

Clearly the behaviour of this amplitude in the neighbourhood of the resonance depends upon the value of δ. As an example, in Fig. 3 we plot the real part of this amplitude for $\delta \approx \dfrac{\pi}{2}$ and $\delta \approx \dfrac{\pi}{4}$. The former furnished the correct behavior at the $f_0(980)$ resonance in the case of $\pi\pi$ scattering [2], and we can see qualitatively that one might expect the latter to enable good agreement with the experimental $K\pi$ channel picture shown also in Fig. 3.

Now let us consider the detailed application of this mechanism to $K\pi$ scattering. The contribution of the $K_0^*(1430)$ to the $I = \tfrac{1}{2}$ channel is structurally similar to that of the κ in (8). The real part of this contribution to the regularized invariant amplitude is

$$\mathrm{Re}\left[A_*^{1/2}(s,t,u)\right] = \frac{\gamma_*^2}{8}\mathrm{Re}\left[e^{2i\delta\theta(s-s_{th})}\frac{3(s - m_\pi^2 - m_K^2)^2}{m_*^2 - s - im_*G_*'\theta(s-s_{th})}\right] - \frac{1}{3}s \to u. \quad (10)$$

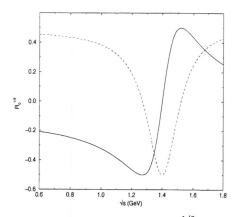 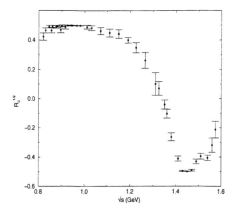

FIGURE 3. LEFT: Shape of $R_0^{1/2}$ derived from Eq. (9) for resonance ($m = 1.4$ GeV and $\Gamma = 0.25$ GeV) in the presence of a background. Plot shows two choices for the background phase - $\delta_{BG} = \frac{\pi}{2}$ (solid line) and $\delta_{BG} = \frac{\pi}{4}$ (dashed line). RIGHT: Experimental Data for $R_0^{1/2}$.

Here we have denoted quantities associated with the $K_0^*(1430)$ by a star subscript. The quantity γ_* is defined in terms of the $K_0^*(1430)$ partial width into $K\pi$ by:

$$\Gamma\left(K_0^*(1430) \to K\pi\right) = \frac{3\gamma_*^2 q(m_*^2)(m_*^2 - m_\pi^2 - m_K^2)}{64\pi m_*^2}, \qquad (11)$$

where $q(s)$ is the momentum transfer. The background phase δ will not be considered an arbitrary parameter but shall be the constant quantity defined from

$$\frac{1}{2}\sin 2\delta = \widetilde{R}_0^{1/2}(s = m_*^2), \qquad (12)$$

where $\widetilde{R}_0^{1/2}(s)$ is the real part of the partial wave amplitude previously computed as the sum of the crossing symmetric current algebra, vector, σ, $f_0(980)$ and κ pieces. With these arrangements the total invariant amplitude is formally crossing symmetric. Finally, for the sake of generality, we shall consider G'_* to be a fitting parameter, not necessarily equal to $\Gamma\left(K_0^*(1430) \to K\pi\right)$. This allows for the possibility of some inelasticity.

We fit the theoretical amplitude, which consists of the real part of the partial wave projection of (10) added to $\widetilde{R}_0^{1/2}(s)$, defined above, to the experimental data displayed in Fig. 3. The fitted parameters are shown in Table 1. We obtain the three $K_0^*(1430)$ parameters self-consistently from our model rather than taking them from [7]. The graphical comparison between experiment and the fitted amplitude, is shown in Fig. 4, corresponding to a background phase given by $\sin 2\delta = 0.937$.

TABLE 1. Comparison of different fits in the $J = 0$, $I = \frac{1}{2}$ channel, corresponding to different choices of $\gamma_{\sigma K \bar{K}}$.

Fitted Parameter	$\gamma_{\sigma K \bar{K}} = \gamma_{\sigma \pi \pi}$	$\gamma_{\sigma K \bar{K}} = 0$	$\gamma_{\sigma K \bar{K}} = -\gamma_{\sigma \pi \pi}$
m_κ	897 ± 2.1 MeV	951 ± 0.7 MeV	998 ± 1.1 MeV
G'_κ	322 ± 6.0 MeV	277 ± 10.6 MeV	195 ± 5.3 MeV
$\gamma_{\kappa K \pi}$	5.0 ± 0.07 GeV^{-1}	4.32 ± 0.16 GeV^{-1}	4.04 ± 0.08 GeV^{-1}
m_*	1385 ± 3.3 MeV	1365 ± 2.5 MeV	1349 ± 2.1 MeV
G'_*	266 ± 9.5 MeV	201 ± 9.8 MeV	148 ± 5.6 MeV
γ_*	4.3 ± 2.1 GeV^{-1}	$3.7 \pm .1$ GeV^{-1}	3.1 ± 0.05 GeV^{-1}
χ^2	4.0	9.0	25.7

We find from (11) that $\Gamma(K_0^*(1430) \to \pi K) = 238$ MeV and as a result (identifying G'_* as the total width) our estimate of the branching ratio of $K_0^*(1430)$ to decay to πK is B$[K_0^*(1430) \to \pi K] = \Gamma_{K_0^*(1430)}/G'_* = 0.895$. This quantity is comparable to the 0.93 obtained in [7]. Similarly, the (first column of Table 1) mass and width we obtain - 1385 MeV and 266 MeV - are in reasonable agreement with their [7] respective values - 1429 MeV and 287 MeV.

We obtain the deviation of our κ parameterization from a pure Breit-Wigner shape by noting that near the resonance the $J = 0$ partial wave projection of (8) is:

$$\frac{m_\kappa G_\kappa}{m_\kappa^2 - s - i m_\kappa G'_\kappa}, \tag{13}$$

where the perturbative width G_κ is given by the form analogous to (11). $\frac{G_\kappa}{G'_\kappa} = 1$ is the pure Breit-Wigner situation. We obtain $\frac{G_\kappa}{G'_\kappa} = 0.13$ which is similar to $\frac{G_\sigma}{G'_\sigma} = 0.29$, previously obtained [2] for the σ. It seems that such deviations for the low mass scalars are a characteristic feature of our model. Ordinarily, when the resonance is a dominant feature by itself, the Breit-Wigner form may be regarded as equivalent to unitarity near the resonance. However, in our model, there are several different interfering contributions in the low mass region and all work together to keep the partial wave amplitude within the unitarity bound.

Comparison with Experiment for the $I = \frac{3}{2}$ Channel

Since there are no $I = \frac{3}{2}$ resonances in our model, there are no s-channel poles in this isospin channel, and hence this analysis depends little on the details of the regularizations. As in the $I = \frac{1}{2}$ case, cancellations of individual contributions to the partial wave amplitude act to preserve the unitarity bound. The experimental points for the real part $R_0^{3/2}$ were translated from Fig. 12 of [8] and are displayed in our Fig. 4.

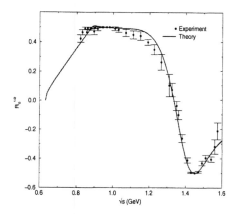 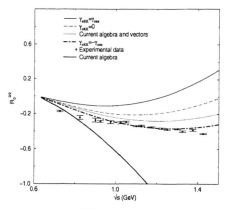

FIGURE 4. LEFT: Comparison of the theoretical prediction of $R_0^{1/2}$ with its experimental data (for choice $\gamma_{\sigma K \bar{K}} = \gamma_{\sigma \pi \pi}$). RIGHT: Comparison of various predictions for $R_0^{3/2}$ with experiment.

Fig. 4 also shows various predictions from our model. Firstly, we see that the current algebra prediction alone quite soon departs from the data points and begins to violate the unitarity bound at around 900 MeV. Inclusion of the vector mesons restores unitarityy and can be seen to give a much better fit to the data.

It turns out that the only important scalar contribution to this channel comes from σ meson exchange, which depends of course on $\gamma_{\sigma K \bar{K}}$. Fig. 4 shows the results for the three choices of $\gamma_{\sigma K \bar{K}}$ given in Table 1. The best choice for the $I = \frac{3}{2}$ amplitude is the case $\gamma_{\sigma K \bar{K}} = -\gamma_{\sigma \pi \pi}$ which unfortunately yields the fit with the highest χ^2 for the $I = \frac{1}{2}$ analysis. Actually the general trend of the data is reproduced for all values of $\gamma_{\sigma K \bar{K}}$ shown.

Since there are no large direct channel resonance contributions, the $I = \frac{3}{2}$ amplitude may be especially sensitive to exchanged resonances in the range above 1 GeV which we are currently neglecting. This is in contrast to the $I = \frac{1}{2}$ amplitude which contains fitting parameters that can absorb the effects of higher resonance exchanges. It would be interesting to examine the effects of these higher mass resonances.

DISCUSSION

We have found that a large N_c motivated approximate treatment of πK scattering can give a crossing symmetric and unitary amplitude as a fit to the existing experimental data. A novel feature of this approach, which is analogous to that employed for $\pi \pi$ scattering in [2], is to start with the invariant perturbative amplitude which is manifestly crossing symmetric. This results in individual contributions dramatically violating the partial wave unitarity bounds. We rely on cancellations among these competing contributions to rescue unitarity. In our framework this suggests the existence of a light strange scalar resonance κ which has parameters mass $m_\kappa = 897$ MeV, width $G_\kappa = 322$ MeV and coupling $\gamma_{\kappa K \pi} = 5\text{GeV}^{-1}$.

If one accepts the existence of the $\kappa(900)$ and $\sigma(560)$, in addition to the $f_0(980)$ and $a_0(980)$, then there is a full set of candidates for a possibly unconventional (i.e. not of pure $q\bar{q}$ type) low mass (less than 1GeV) scalar nonet (see for example [10]). The nature of such a nonet is of great interest - see [9] for a recent discussion. A useful clue may arise from knowledge of the pattern of $0^+0^-0^-$ coupling constants defined in Eq. (6). The numerical values obtained in our approach are given below Eq. (6) and in Table 1.

ACKNOWLEDGMENTS

With thanks to my advisor Joseph Schechter and to Amir H. Fariborz and Francesco Sannino. The work at Syracuse has been supported in part by the US DOE under contract DE-FG-02-85ER 40231 and that of F. Sannino in part under contract DE-FG-02-92ER-40704.

REFERENCES

1. D. Black, A.H. Fariborz, F. Sannino, and J. Schechter, hep-ph/9804273 (to appear in Phys. Rev. **D56** (1998)).
2. F. Sannino and J. Schechter, Phys. Rev. **D52**, 96 (1995). and M. Harada, F. Sannino and J. Schechter, Phys. Rev. **D54**, 54 (1996).
3. F.E. Close, *An Introduction to Quarks and Partons*, Academic Press (1979).
4. Review of Particle Properties, Phys. Rev. **D54** (1996).
5. S. Ishida, M. Ishida, T. Ishida, K. Takamatsu and T. Tsuru, Prog. Theor. Phys. **98**, 621 (1997). See also M. Ishida and S. Ishida, Talk given at 7th International Conference on Hadron Spectroscopy (Hadron 97), Upton, NY, 25-30 Aug. 1997, hep-ph/9712231.
6. E. Witten, Nucl. Phys. **B160**, 57 (1979). See also S. Coleman, *Aspects of Symmetry*, Cambridge University Press (1985). The original suggestion is given in G. 't Hooft, Nucl. Phys. **B72**, 461 (1974).
7. D. Aston *et al*, Nucl. Phys **B296**, 493 (1988).
8. P. Estabrooks *et al*, Nucl Phys. **B133**, 490 (1978).
9. V. Elias, A.H. Fariborz, Fang Shi and T.G. Steele, Nucl. Phys. **A633**, 279 (1998).
10. R. L. Jaffe, Phys. Rev. **D15**, 281 (1977).

The Influence of Gauge Boson Mass on the Most Attractive Channel Hypothesis

F. S. Roux[1]

Department of Physics, University of Toronto, 60 St. George Strest, Toronto, Ontario, Canada M5S 1A7

Abstract. It is well known that leading order analyses indicate that the singlet channel, when present, is the Most Attractive Channel (MAC) for the formation of fermion condensates. Here we report that, in the presence of nonzero gauge boson masses, next-to-leading order corrections to the kernel of the gap equation can be large and of opposite sign to that of the leading order. This would indicate that the gap equation may allow self-consistent solutions where the condensate appears in a non-singlet channel which would spontaneously break the gauge symmetry.

I MOTIVATION AND HISTORY OF MAC

One of the remaining puzzles of nature is the masses of the elementary particles. The standard model of elementary particles contains the masses as free parameters in terms of Yukawa couplings. Although this parameterizes the masses well, it does not present an explanation for their origin. It is unsatisfying that there can be so many fundamental mass scales in nature. Therefore one would expect that some mechanism is responsible for the generation of these mass values. Such a mechanism must be able to generate mass scales 19 orders of magnitude below the Planck scale. In addition it must generate different mass scales ranging over more than 5 orders of magnitude.

These are difficult requirements to meet. However, there are some aspects in quantum field theory that may be useful in understanding such a mechanism. One is that the spontaneous dynamical breaking of a chiral symmetry would produce fermion masses that are of the order of the scale where the breaking occurs. Another is that the breaking of a flavor or family symmetry may be responsible for the diversity in the mass scales of the various fermions. Dynamical symmetry breaking requires strong dynamics and is therefore an inherently non-perturbative effect. Unfortunately, non-perturbative calculations are difficult.

---

[1] This work was done with B. Holdom and is reported in[1].

In view of this Cornwall [2] and Raby, Dimopolous and Susskind [3] made a dynamical assumption in the late 70's to aid the formulation of models. Here is the statement of this dynamical assumption as quoted from Ref. [3]:[2]

> When the gauge forces between fermions in a given channel are attractive, bound states will form when the running coupling becomes large enough. If the coupling becomes even larger, these composites become massless. Further increase in the coupling will cause the vacuum to rearrange and a spontaneous symmetry breaking condensate will form [4]. The nature of the condensate, its transformation properties under symmetry groups are determined by the quantum numbers of the *most attractive channel* (MAC).

A leading order analysis (ladder approximation) showed that the most attractive channel is determined by a group theory factor (the channel factor),

$$CF = \frac{1}{2}[C_2(r_1) + C_2(r_2) - C_2(R)] \qquad (1)$$

where $C_2(r_n)$ denotes the second Casimir constant of the irreducible representation, r_n. Here r_1 and r_2 denote the respective irreducible representations of the two fermions and R denotes the irreducible representations of the channel. From this expression one can see that if a singlet channel can form, it will be the most attractive channel.

The dynamical assumption together with the channel factor became known as the Most Attractive Channel or MAC hypothesis. However, more than a decade later we still do not have a working dynamical theory of fermion mass that is based on the MAC hypothesis!

The reason for this may be that the MAC hypothesis is too restrictive. It dictates the channel of the condensate and therefore to a large extent what the breaking pattern looks like. The results of these restrictions were extremely complicated models to which the moose diagrams bear witness.

On the other hand, if one can lift these restrictions the breaking patterns in chiral theories would be more open. Channels that are repulsive according to MAC may in fact be attractive. Thus, with additional effects such as 4-fermion interactions the most attractive channel may turn out to be a non-singlet channel, even in vector theories. One may, for example, be able to explain the large top mass in terms of a new gauge symmetry that breaks near 1 TeV by forming condensates of fourth family fermions in a non-singlet channel. [5]

It may appear strange that anyone would pay any attention to a leading order analysis when the phenomenon under investigation is clearly non-perturbative. The justification for the MAC hypothesis is based on two aspects. One is that a leading order stability analysis indicates that the critical value for the coupling constant

---

[2] The term *channel* is used to distinguish the different possible bound states that can form. It denotes the irreducible representation under which the bound state transforms.

where the condensate forms is still small enough that one can do a perturbative analysis. The critical value is found to be

$$\alpha_{crit} = \frac{\pi}{3CF} \tag{2}$$

where CF denotes the channel factor as defined in (1). This value is still small enough compared to $\alpha = 4\pi$, which can be regarded as the value where the interaction becomes truly strong. One may however ask to what extent the critical value of α may change beyond a leading order analysis.

The other justification came from the next-to-leading order analysis that was performed by Appelquist, Lane and Mahanta in 1988 [6]. They considered a gauge coupling that does not run (*i.e.* a walking theory) by choosing the number of flavors such that the one-loop beta function vanishes: $11C_A - 4N_f T(R) = 0$. To alleviate the complicated calculations they further made the assumption that the next-to-leading order kernel is proportional to leading order kernel up to logarithms. They also implicitly assumed that the gauge symmetry does not break by leaving the gauge bosons massless. Their calculations of corrections for various channels gave the largest corrections to be on the order of 20%.

Here we present a next-to-leading order analysis that differs in the following aspects from that of Ref. [6]. We only include diagrams that are leading order in $1/N_c$ or $1/N_f$ where N_c is the number of colors and N_f is the number of flavors, however, we do not restrict ourselves only to walking theories. We allow the possibility of gauge symmetry breaking by including a possible non-zero gauge boson mass. No assumptions are made about the shape of the next-to-leading order kernel. Instead we consider the kernel at the dominant momentum region.

II THE EFFECTIVE ACTION

The calculation is performed in the conventional way starting with the CJT effective action [7]

$$\Gamma[S] = -\text{Tr}\{\ln(S^{-1})\} + \text{Tr}\{(S^{-1} - \not{\partial})S\} - \sum(2\text{PI}), \tag{3}$$

where the traces include integration over the 4-momentum. S denotes the full propagator, which contains the dynamical mass function, $\Sigma(p)$. At second order in the dynamical mass function one finds

$$\Gamma \sim \int \Sigma(p)^2 p\, dp - \frac{1}{2} \int \int \Sigma(p)\Sigma(k) F(p,k) dp\, dk. \tag{4}$$

Here $F(p,k)$ is the kernel function, which determines whether the massless solution becomes unstable.[3]

---

[3] The massless solution will make $\Gamma = 0$. If the kernel function is large enough it will allow massive solutions that will make Γ negative and that are for this reason more stable than the

The kernel consist of the following diagrams, where the cross denotes a mass insertion. At leading order in the coupling constant there is only one diagram, which is shown in Figure 1:

FIGURE 1. Leading order diagram.

At next-to-leading order the diagrams that are of leading order in $1/N_c$ are shown in Figure 2:

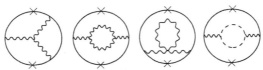

FIGURE 2. Next-to-leading order diagrams that are leading order in $1/N_c$.

and the single diagram that is of leading order in $1/N_f$ is shown in Figure 3:

FIGURE 3. Next-to-leading order diagram that is leading order in $1/N_f$.

All above diagrams are proportional to the channel factor of (1). Apart from the influence of the gauge boson mass, the channel factor gives the only influence of the condensing channel. One can therefore parameterize the kernel function as follows

$$F(p,k) = \frac{2}{\pi} \left[C_2(r_1) + C_2(r_2) - C_2(R) \right] \left[\alpha F_1(p,k) + \alpha^2 \mathcal{R} F_2(p,k) \right] + O(\alpha^3) \quad (5)$$

Where F_1 and F_2 respectively denote the leading and next-to-leading order terms, up to normalization, in the kernel function. The normalization is done by choosing the constant $\mathcal{R}$ such that $F_1(p,k) = F_2(p,k)$ at the dominant momentum region. In this way $\mathcal{R}$ gives the size of the next-to-leading order term relative to the leading order term.

The leading order term is

---

massless solution. The aim of this analysis is therefore not to find the shape of the mass function (*i.e.* to solve a gap equation) but instead to determine whether there exist solutions that are more stable than the massless solution. Although a gap analysis can determine the massive solutions it does not reveal whether nature would prefer such solutions.

$$F_1(p,k) = \frac{1}{pk}\left[p^2 + k^2 + M^2 - \sqrt{(p^2 + k^2 + M^2)^2 - 4p^2k^2}\right] \qquad (6)$$

where M is the gauge boson mass and p and k are the magnitudes of the momenta that flow through the two mass insertions respectively. When the gauge boson is massless, the expression in (6) becomes

$$F_1 = \min\left(\frac{k}{p}, \frac{p}{k}\right). \qquad (7)$$

Note that F_1 is dominant where $p \approx k$ and it decreases as $\min(k/p, p/k)$ away from this region. F_2 is also dominant for $p \approx k$ but does not decrease as $\min(k/p, p/k)$ away from $p = k$ as assumed in Ref. [6].

For the massless case, $M = 0$, the dominant momentum region is determined by the shape of the dynamical mass function, $\Sigma(p)$. In the massive case, $M \neq 0$, the dominant momentum region is approximately equal to the gauge boson mass M. It cannot be much smaller because the gauge boson mass would behave like an IR cutoff and it cannot be much larger because then it would revert back to the massless case.

III RESULTS

The value of $\mathcal{R}$ is calculated for the dominant momentum regions in both cases: $M = 0$ and $M \neq 0$. In the massless case the dominant momentum region implies that $p = k = \mu$ where μ is the renormalization scale. For this case we find that

$$\mathcal{R} = 0.33 C_A - 0.21 N_f T(R) + 0.15 C_F. \qquad (8)$$

Here C_A denotes the second Casimir constant of the adjoint representation, C_F denotes the second Casimir constant of the fundamental representation and T(R) is 1/2.

For the massive case the dominant momentum region gives $M = p = k = \mu$. In this case we find

$$\mathcal{R} = -0.26 C_A - 0.045 N_f T(R) - 0.12 C_F. \qquad (9)$$

The expression in (8) and (9) where evaluated for the specific cases where $N_f = 3$ and $N_f = 16$ and where in both cases $N_c = 3$. The results are summarized in Table 1.

The second case, where the gauge boson is massless and the number of fermions is large, is closest to the case considered in Ref. [6]. We found a larger contribution than they did, which may be due to the fact that we did not make the assumption that F_2 decreases as $\min(k/p, p/k)$ away from $p = k$.

Note that the value of $\mathcal{R}$ is fairly large in the massless case but it becomes smaller and eventually negative as the number of fermions is increased. The value of $\mathcal{R}$

TABLE 1. Numerical values for $\mathcal{R}$

	N_c	N_f	$\mathcal{R}$
$M = 0$	3	3	0.9
$M = 0$	3	16	-0.4
$M \neq 0$	3	3	-1.0
$M \neq 0$	3	16	-1.3

starts off negative in the massive case and becomes even more negative when the number of fermions is increased. Furthermore the magnitude of $\mathcal{R}$ in the massive case indicates that the second order dominates over the first order. If any credence could be given to this observation by itself, it would indicate that what would be the most attractive channel at leading order becomes the most repulsive channel at the second order and vice versa. In actual fact this signals the breakdown of perturbation theory and therefore indicates that this type of analysis cannot make any statement about which channel will be the most attractive.

IV SUMMARY AND CONCLUSIONS

We performed a stability analysis to next-to-leading order in α, neglecting all diagrams that are subleading in $1/N_c$ and $1/N_f$. We included a possible non-zero gauge boson mass and did not make any assumptions about the shape of the next-to-leading order kernel function.

The results indicate that in the presence of a non-zero gauge boson mass the next-to-leading order term becomes as large as the leading order term with the opposite sign. This implies that perturbation theory breaks down. Hence identifying the most attractive channel on the basis of the leading order contribution to the kernel function (*i.e.* using the value of the channel factor) is misleading.

REFERENCES

1. B. Holdom and F.S. Roux, hep-ph/9804311.
2. J.M. Cronwall, Phys. Rev. **D10**, 500 (1974).
3. S. Raby, S. Dimopolous and L. Susskind, Nucl. Phys. **B169**, 373 (1980).
4. Y. Nambu and G. Jona-Lasinio, Phys. Rev. **122**, 345 (1961).
5. B. Holdom, Phys. Rev. **D54**, 1068 (1996), hep-ph/9512298; Phys. Rev. **D57**, 357 (1998), hep-ph/9705231.
6. T. Appelquist, K. Lane and U. Mahanta, Phys. Rev. Lett. **61**, 1553 (1988).
7. J.M. Cronwall, R. Jackiw and E. Tomboulis, Phys. Rev. **D10**, 2428 (1974).

The $<\bar{f}f>$ Condensate Component of the Anomalous Magnetic Moment of a Fundamental Dirac Fermion

Victor Elias and Kevin B. Sprague[†]

[†]*Department of Applied Mathematics*
The University of Western Ontario
London, Ontario N6A 5B7 CANADA

Abstract. Fermion-antifermion condensate contributions to the anomalous magnetic moment of a condensing fermion are considered. The real part of such contributions is shown to be zero, and a nonzero imaginary part is found *below* the threshold for producing fermion-antifermion pairs. The calculation is shown to be gauge-parameter independent provided that the mass used to characterize the fermion propagator is the same as that appearing in the operator product expansion, suggestive of a dynamical mass.

I INTRODUCTION

Anomalous magnetic moments (AMMs) of fermions have long played a central role in the development and testing of 'new physics'. The determination, to high accuracy, of electron and muon AMMs stands as an important confirmation of the veracity of quantum electrodynamics (QED), while the static quark model does well to predict the AMMs of the proton and neutron. Here we investigate a nonperturbative contribution to the AMM of a condensing fermion.

In quantum chromodynamics (QCD), the chiral noninvariance of the vacuum may be described by order-parameters referred to as condensates. These condensates may be characterized, as in QCD sum-rule applications [1], by nonvanishing vacuum expectation values (VEVs) of normal ordered products of field operators. Such quantities enter into Feynman amplitudes via normal ordered pieces in the Wick-Dyson expansion of time ordered products of fields. In QED these pieces vanish, as the normal ordering prescription necessarily annihilates a purely perturbative vacuum. However, in a vacuum with nonperturbative content (*ie.* QCD), they do not vanish.

The quark-antiquark condensate of QCD characterizes the following VEV, understood as a solution to the free Dirac equation with the condensate entering through an appropriately chosen initial condition [2,3]:

$$<0|:\psi_i^a(y)\bar{\psi}_j^b(z):|0> = -\frac{\delta^{ab}}{3}<\bar{q}q>\sum_{r=0}^{\infty}C_r\left[-im\gamma^\mu(y_\mu-z_\mu)\right]_{ij}^r, \quad (1.1a)$$

$$C_r = \left\{ \begin{array}{ll} \frac{1}{(r/2)![(r+2)/2]!2^{r+2}} \, , & r \text{ even,} \\ \frac{1}{[(r-1)/2]![(r+3)/2]!2^{r+2}} \, , & r \text{ odd.} \end{array} \right\} \quad (1.1b)$$

[i,j/a,b are Dirac/colour indices.] In principle, every Feynman amplitude whose Wick-Dyson expansion of time ordered fields contains a fermion propagator term,

$$<0|T\psi_i^a(y)\bar{\psi}_j^b(z)|0> = \int \frac{d^4p}{(2\pi)^4} e^{-ip\cdot(y-z)} [S_F(p)]_{ij}^{ab}, \quad (1.2)$$

must also contain a term such as (1.1). This provides a mechanism through which vacuum condensates can enter into the field-theoretical side of QCD sum-rule calculations [4]. It is interesting to note that the *only* signature of nonabelian physics in the derivation of (1.1) is an overall colour-summation factor of $3(=\delta^{aa})$ in the denominator, a factor that can easily be absorbed into a redefinition of the fermion-antifermion condensate $<\bar{f}f>$ for the abelian case (henceforth define $\delta^{ab}<\bar{q}q> \to <\bar{f}f>$ in (1.1a)).

It is our intention to determine whether the fermion-antifermion condensate contribution to the anomalous QED magnetic moment of a fermion field is calculable through the manipulation of techniques [5,6] adapted from QCD sum-rule applications.

II EXTRACTING THE ANOMALOUS MAGNETIC MOMENT FROM THE ELECTROMAGNETIC VERTEX CORRECTION

Fig.1a: Fig.1b:

We can obtain a value for $\mathcal{K}$, the AMM of a fundamental fermion, simply by comparing different expressions for the electromagnetic vertex correction. One can define the unrenormalized vertex correction (Fig.1b) to the electromagnetic vertex (Fig.1a), with incoming and outgoing fermion momenta p_1^μ and p_2^μ respectively, to be $\bar{u}(p_2)\Lambda^\mu(p_2,p_1)u(p_1)$, where $[q^\mu \equiv p_2^\mu - p_1^\mu]$

$$\Lambda^\mu(p_2, p_1) \equiv e^2 Q^2 \left[R(q^2)\gamma^\mu + \frac{2S(q^2)}{m}(p_1^\mu + p_2^\mu) \right], \qquad (2.1)$$

such that the unrenormalized vertex, to second order, is $-ieQ\Gamma^\mu \equiv -ieQ(\gamma^\mu + \Lambda^\mu(p_2, p_1))$ with eQ the electromagnetic fermion charge. This unrenormalized vertex can be expressed as follows in terms of the renormalized vertex form factors $F_1(q^2)$, $\mathcal{K}F_2(q^2)$:

$$\begin{aligned}
&\bar{u}(p_2)[\gamma^\mu + \Lambda^\mu(p_2, p_1)] u(p_1) \\
&= \bar{u}(p_2) \left[(1 + e^2 Q^2 [R(q^2) + 4S(q^2)])\gamma^\mu - 2e^2 Q^2 S(q^2) i\sigma^{\mu\nu} q_\nu/m \right] u(p_1) \\
&\equiv Z\, \bar{u}(p_2) \left[F_1(q^2)\gamma^\mu + i\sigma^{\mu\nu} q_\nu \mathcal{K} F_2(q^2)/2m \right] u(p_1),
\end{aligned} \qquad (2.2)$$

where Z is a (divergent) rescaling factor, the value of which can be found using the renormalization condition $F_1(0) = 1$, in which case, to order-e^2, $Z = 1 + e^2 Q^2 (R(0) + 4S(0))$. To leading order in e^2, one then finds that

$$F_1'(q^2) = 1 + e^2 Q^2 \left[(R'(0) + 4S'(0))q^2 + \mathcal{O}(q^4) \right] + \mathcal{O}(e^4), \qquad (2.3)$$

$$\mathcal{K} F_2(q^2) = -4e^2 Q^2 S(q^2) + \mathcal{O}(e^4). \qquad (2.4)$$

The $q^2 \to 0^+$ limit of Eq. (2.4) gives the $\mathcal{O}(e^2)$ anomalous magnetic moment of QED, which (in contrast to F_1) devolves solely from the vertex correction and is insensitive to additional (vacuum-polarization, self-energy, and bremsstrahlung) diagrams.

The purely perturbative contribution to the unrenormalized vertex correction (2.1) is found to be

$$\begin{aligned}
[\Lambda^\mu(p_2, p_1)]^{pert} &= -i(Qe)^2 \int \frac{d^4 k}{(2\pi)^4} \frac{\gamma_\tau (\slashed{p}_2 - \slashed{k} + m)\gamma^\mu(\slashed{p}_1 - \slashed{k} + m)\gamma^\tau}{k^2[(p_2-k)^2 - m^2][(p_1-k)^2 - m^2]} \\
&= -i(Qe)^2 \left\{ \left[-2 \slashed{p}_1 \gamma^\mu \slashed{p}_2 + 4m(p_1^\mu + p_2^\mu) - 2m^2 \gamma^\mu \right] I(p_2, p_1) \right. \\
&\quad + \left[2\gamma^\rho \gamma^\mu \slashed{p}_2 + 2 \slashed{p}_1 \gamma^\mu \gamma^\rho - 8mg^{\mu\rho} \right] I_\rho(p_2, p_1) \\
&\quad \left. - 2\gamma^\rho \gamma^\mu \gamma^\sigma I_{\rho\sigma}(p_2, p_1) \right\}.
\end{aligned} \qquad (2.5)$$

The integrals I, I_ρ and $I_{\rho\sigma}$ are respectively defined by

$$\begin{aligned}
&[I(p_2, p_1);\, I_\rho(p_2, p_1);\, I_{\rho\sigma}(p_2, p_1)] \\
&\equiv \int \frac{d^n k}{(2\pi)^n} \frac{[1;\, k_\rho;\, k_\rho k_\sigma]}{(k^2 - \epsilon^2)\left[(k-p_2)^2 - m^2\right]\left[(k-p_1)^2 - m^2\right]}.
\end{aligned} \qquad (2.6)$$

In (2.6), infrared divergences are regulated by the "photon-mass" ϵ and we have analytically continued the integrals to n dimensions. These integrals are easily evaluated by standard methods, and expressions for $R(q^2)$ and $S(q^2)$ (for small

q^2) can be obtained by comparison of (2.1) and (2.5). The anomalous magnetic moment is then found to be finite:

$$\mathcal{K} \lim_{q^2 \to 0+} F_2(q^2) = -4e^2 Q^2 \lim_{q^2 \to 0+} S(q^2) = e^2 Q^2 / 8\pi^2, \qquad (2.7)$$

the famous result first obtained by Schwinger and Feynman [7,8]. Henceforth, all insertions of $q^2 = 0$ are to be understood as limits from the right. The F_1 form-factor slope in the vertex correction (2.3) is also a classical result of perturbative QED,

$$F_1'(0) = e^2 Q^2 \left[R'(0) + 4S'(0) \right] = -\frac{e^2 Q^2}{24\pi^2 m^2} \left[\frac{3}{4} + \ln\frac{\epsilon^2}{\mu^2} \right]. \qquad (2.8)$$

III $< \bar{f} f >$ CONTRIBUTION TO $\mathcal{K} F_2(q^2)$

Fig.2a: Fig.2b:

Methodologically, we can tune the perturbative calculation in the previous section to suit our needs in the nonperturbative regime. Recall the Wick expansion for a time ordered product of fermion field operators:

$$T\psi(x)\bar{\psi}(y) = < 0|T\psi(x)\bar{\psi}(y)|0 > + :\psi(x)\bar{\psi}(y):$$

and consider taking the VEV of both sides. In perturbative calculations, the normal ordered piece is routinely discarded. However, in nonperturbative calculations, it becomes an order parameter describing chiral symmetry breaking, so that our new Feynman rules must take the VEV of this piece into account [4]. Effectively, we are allowing the fermion to condense. This introduces two new nonvanishing graphs (Fig.2) to be added to the amplitude (2.5) (a third graph, wherein both internal fermion lines are allowed to condense, happens to vanish!).
Configuration-space nonperturbative propagators in the new graphs replace their fermion propagator counterparts with the following expression [6,2,3]:

$$< 0| :\psi(x)\bar{\psi}(y): |0> = \int d^4k \, e^{-ik\cdot(x-y)} (\gamma \cdot k + m)\mathcal{F}(k), \qquad (3.1a)$$

where

$$\int d^4k \mathcal{F}(k) e^{-ik\cdot x} \equiv - <\bar{f}f> J_1(m\sqrt{x^2})/(6m^2\sqrt{x^2}), \qquad (3.1b)$$

with $<\bar{f}f>$ identified as the (appropriately normalized) fermion-antifermion condensate of (1.1). Expressions for the nonperturbative propagator contributions to Fig.2a and Fig.2b are found by making the following substitutions [6] in the Feynman-rule version of the vertex correction (first line of 2.5):

$$Fig(2a): \quad \frac{1}{(k_2 - p_1)^2 - m^2} \to -i(2\pi)^4 \mathcal{F}(k_2 - p_1); \tag{3.2}$$

$$Fig(2b): \quad \frac{1}{(k_2 - p_2)^2 - m^2} \to -i(2\pi)^4 \mathcal{F}(k_2 - p_2). \tag{3.3}$$

The net effect of these changes is to reproduce the Feynman amplitude of (2.5), but with the Feynman integrals (2.6) altered as follows:

$$\int \frac{d^n k}{(2\pi)^n} \frac{[1;\ k_\rho;\ k_\rho k_\sigma]}{(k^2 - \epsilon^2)[(k-p_2)^2 - m^2][(k-p_1)^2 - m^2]}$$
$$\to -i \int \frac{d^4 k [1;\ k_\rho;\ k_\rho k_\sigma] \mathcal{F}(k-p_1)}{(k^2 - \epsilon^2)[(k-p_2)^2 - m^2]} - i \int \frac{d^4 k [1;\ k_\rho;\ k_\rho k_\sigma] \mathcal{F}(k-p_2)}{(k^2 - \epsilon^2)[(k-p_1)^2 - m^2]} \tag{3.4}$$

which are UV finite. Thus we retain the form of the amplitude (2.5), but with the integrals $I, I_\rho, I_{\rho\sigma}$ now being given (after a trivial shift of integration variable) by

$$[I, I_\rho, I_{\rho\sigma}] = -i \int \frac{d^4 k [1;\ k_\rho + p_{1\rho};\ (k_\rho + p_{1\rho})(k_\sigma + p_{1\sigma})] \mathcal{F}(k)}{[(k+p_1)^2 - \epsilon^2][(k+p_1-p_2)^2 - m^2]}$$
$$-i \int \frac{d^4 k [1;\ k_\rho + p_{2\rho};\ (k_\rho + p_{2\rho})(k_\sigma + p_{2\rho})] \mathcal{F}(k)}{[(k+p_2)^2 - \epsilon^2][(k+p_2-p_1)^2 - m^2]}. \tag{3.5}$$

To proceed further, we work on-shell with a Feynman parameter z, making use of the mass-shell relation [6] $k^2 \mathcal{F}(k) = m^2 \mathcal{F}(k)$ and the following integral relations [9] ($p^2 > 0$):

$$R_1 \equiv \int d^4 k\ \mathcal{F}(k) = -<\bar{f}f>/12m \tag{3.6}$$

$$R_2(p,\ \mu) \equiv \int \frac{d^4 k\ \mathcal{F}(k)}{(p-k)^2 - \mu^2 + i|\epsilon|} = -\frac{<\bar{f}f>}{24m^3 p^2}$$
$$\times \left[p^2 + m^2 - \mu^2 - \sqrt{[p^2 - (m-\mu)^2][p^2 - (m+\mu)^2]}\right], \tag{3.7}$$

$$R_3(p,\ \mu) \equiv \int \frac{d^4 k\ \mathcal{F}(k)}{[(p-k)^2 - \mu^2 + i|\epsilon|]^2}$$
$$= \frac{<\bar{f}f>}{24m^3 p^2} \left[1 - \frac{p^2 + m^2 - \mu^2}{[p^2 - (m-\mu)^2][p^2 - (m+\mu)^2]}\right]. \tag{3.8}$$

Noting that $\mu^2 = m^2 z^2$, $p^2 = m^2(1-z)^2 + q^2 z$, we then find that

$$\Delta S(q^2) = 2m^2 \int_0^1 dz(1-z) \left[-\frac{<\bar{f}f>(1-z)}{36m[m^2(1-z)^2 + q^2 z]^2} \right.$$
$$- \frac{[5m^2(1-z)^2 + q^2 z(1-4z)]}{6[m^2(1-z)^2 + q^2 z]^2} R_2[z, q^2]$$
$$\left. - \frac{[m^2 q^2 z(1-z)(3+5z) + q^4 z^2(1+2z)]}{6[m^2(1-z)^2 + q^2 z]^2} R_3[z, q^2] \right]. \tag{3.9}$$

Notice that R_2 and R_3 develop imaginary parts only when $0 < q^2 < 4m^2$. Such a branch cut between $q^2 = 0$ and $q^2 = 4m^2$ is also seen to occur in $<q\bar{q}>$ contributions to two-point current-correlation functions [10,11].

Explicit evaluation of (3.9) for $0 < q^2 < 4m^2$ yields the following results:

$$Re[\Delta S(q^2)] = 0, \tag{3.10}$$

$$Im[\Delta S(q^2)] = -\frac{<\bar{f}f>}{12m\sqrt{4m^2 q^2 - q^4}} \tag{3.11}$$

where $\Delta \mathcal{K} F_2(q^2) = -4e^2 Q^2 \Delta S(q^2)$.

IV DISCUSSION

A Gauge Invariance

It is easy to show that the calculation is gauge invariant provided that a single fermion mass characterizes the perturbative and nonperturbative fermion propagators. In an arbitrary covariant gauge, the contribution of Fig.2a is proportional to

$$\int d^4k \, \mathcal{D}^{\tau\sigma}(k, \xi) \frac{\mathcal{F}(k-p_1)\gamma_\tau(\not{k}-\not{p}_2-m)\gamma_\mu(\not{k}-\not{p}_1-m)\gamma_\sigma}{[(k-p_2)^2 - m^2]}$$
$$\equiv \Lambda^{(a)}(p_2, p_1) - (1-\xi)\Lambda_\xi^{(a)}(p_2, p_1), \tag{4.1}$$

where ξ is the photon-propagator gauge parameter in $\mathcal{D}^{\tau\sigma}(k) = \frac{g^{\tau\sigma}}{k^2} - (1-\xi)\frac{k^\tau k^\sigma}{k^4}$. In (4.1), $\Lambda^{(a)}$ is the (Feynman-gauge) contribution we have already considered, and $\Lambda_\xi^{(a)}$ is the contribution arising from the extra propagator term. On-shell gauge-parameter independence translates into the requirement that $\bar{u}(p_2)\Lambda_\xi^{(a)}(p_2, p_1)u(p_1)$ vanish, or eqivalently, that the $k^\tau k^\sigma/k^4$ term in $\mathcal{D}^{\tau\sigma}$ does not contribute to the on-shell vertex correction. To demonstrate this, we consider (after a trivial shift of variables)

$$\Lambda_\xi^a(p_2,\ p_1)u(p_1) \qquad (4.2)$$
$$= \int \frac{d^4k\ (\not{k}+\not{p}_1)(\not{k}+\not{p}_1-\not{p}_2-m)\gamma_\mu \mathcal{F}(k)(\not{k}-m)(\not{k}+\not{p}_1)u(p_1)}{((k+p_1)^2)^2\,[(k+p_1-p_2)^2-m^2]}.$$

The function $\mathcal{F}(k)$ is a Dirac scalar that can be moved past gamma-matrices. We note that $k^2\mathcal{F}(k) = m^2\mathcal{F}(k)$, a consequence of (3.1a) being a solution [3] of the free Dirac equation $(\gamma\cdot\partial_x + im) < 0| : \psi(x)\bar\psi(y) : |0> = 0$ [this is explicit in the series expansion (1.1a)]. We then find that

$$\mathcal{F}(k)(\not{k}-m)(\not{k}+\not{p}_1)u(p_1) = \mathcal{F}(k)(k^2-m^2)u(p_1) = 0,$$

which, when substituted into (4.2), necessarily implies that $\Lambda_\xi^a(p_2,\ p_1)u(p_1) = 0$. Similarly, the corresponding contribution from Fig.2b vanishes. Hence the vertex correction $\bar u(p_2)\Lambda^\mu(p_2,\ p_1)u(p_1)$ is manifestly gauge-parameter independent on shell.

It is important to realize that any attempt to distinguish between the perturbative and nonperturbative propagator masses will destroy the gauge-parameter independence of the result (e.g., [12]). The gauge-parameter independence of electroweak two-point functions has similarly been shown [13] to be contingent, for a given flavour, upon such an identification. Since the mass appearing in the VEV (1.1) is necessarily dynamical, gauge-parameter independence suggests that the fermion mass m be understood as dynamical rather than Lagrangian in origin [15,16], a reflection of the chiral noninvariance of the vacuum.

B $Im(\Delta S(q^2))$

The structure function $\Delta S(q^2)$ is purely imaginary (3.11) when the momentum transfer q^2 is between 0 and $4m^2$. Imaginary parts of Feynman amplitudes correspond to physical states. Although $\Delta S(q^2)$ is imaginary below the $q^2 = 4m^2$ kinematic threshold for the production of a physical fermion-antifermion ($\bar f f$) pair, this $0 < q^2 < 4m^2$ branch cut, when augmented by the purely-perturbative $\bar f f$-production branch cut beginning at $q^2 = 4m^2$, may be associated with the $q^2 = 0$ production threshold for the Goldstone bosons required by assuming explicit chiral symmetry breaking; i.e., by assuming a condensate exists. Thus m must be regarded as a dynamical mass rather than an explicit Lagrangian mass; e.g. the pion is massless only in the limit of Lagrangian chiral symmetry (zero current-quark mass).

These results suggest that the nonzero imaginary part of $\Delta S(q^2)$ is in fact a kinematic manifestation of the Goldstone theorem, with the kinematic production threshold at $q^2 = 0$ corresponding to the massless meson anticipated from a *dynamical* breakdown ($< \bar f f > \neq 0$) of Lagrangian chiral symmetry.

C Speculations Concerning Quarks

QED and QCD quark-interactions cannot be treated in isolation. Hence the contribution of the $<\bar{q}q>$ condensate to quark electromagnetic properties cannot be overlooked. This requires that we consider the exchange of intermediate photons *and* gluons in Fig.2, suggesting that all factors of $e^2 Q^2$ in section III be replaced by $[e^2 Q^2 + g_s^2 T(R)]$. Although this change becomes problematical in the $q^2 \to 0$ limit where QCD becomes nonperturbative, there exists some evidence for a freezing out of the effective QCD coupling g_s at small momentum transfers [18–22].

Having acknowledged our concerns, we may proceed to speculate on the applicability of (3.10) and (3.11) to quarks. The former equation appears to indicate an absence of any $<\bar{q}q>$ contributions to the quark magnetic moment, provided m is understood to be the $\mathcal{O}(300 MeV)$ dynamical mass characterizing the static quark model. Moreover, the shift in the kinematic production threshold from $q^2 = 4m^2$ to $q^2 = 0$ may be indicative of QCD's transition from a gauge theory of quarks and gluons to a genuine low-energy theory of hadrons; *i.e.* pions.

ACKNOWLEDGEMENTS: We are grateful for discussions with D.G.C. McKeon, V. A. Miransky and T. G. Steele, and for support from the Natural Sciences and Engineering Research Council of Canada.

REFERENCES

1. M.A. Shifman, A.I. Vainshtein, and V.I. Zakharov, *Nucl. Phys.* **B147**, 385 and 448(1979).
2. V. Elias, T.G. Steele, and M.D. Scadron, *Phys. Rev.* **D38**, 1584 (1988).
3. F.J. Yndurain, *Z. Phys. C* **42**, 643 (1989).
4. P. Pascual and R. Tarrach (1984) *QCD: Renormalization for the Practitioner*, Lecture Notes in Physics 194, H. Araki, J. Ehlers, K. Hepp, R. Kippenhahn, H.A. Weidenmüller, and J. Zittartz, eds., Springer-Verlag, Berlin, pp. 168-191 (1984).
5. E. Bagan, M.R. Ahmady, V. Elias, and T.G Steele, T. G. *Phys. Lett. B* **305**, 151 (1993).
6. E. Bagan, M.R. Ahmady, V. Elias, and T.G. Steele, *Z. Phys. C* **61**, 157 (1994).
7. J. Schwinger, *Phys. Rev.* **73**, 416 (1948).
8. R.P. Feynman, *Phys. Rev.* **76**, 749 and 769 (1949).
9. V. Elias, K.B. Sprague, *to be published in Int. J. Th. Phys.* (1998).
10. E. Bagan, J.I. LaTorre, and P. Pascual *Z. Phys. C* **32**, 43 (1986).
11. V. Elias, J.L. Murison, M.D. Scadron, and T.G. Steele, *Z. Phys. C* **60**, 235 (1993).
12. H. He, *Z. Phys. C* **69**, 287 (1996).
13. M.R. Ahmady, V. Elias, R.R. Mendel, M.D. Scadron, T.G. Steele, *Phys. Rev. D* **39**, 2764 (1989).
14. P. Pascual and E. de Rafael, *Z. Phys. C* **12**, 127 (1982).
15. V. Elias and M.D. Scadron, *Phys. Rev. D* **30**, 647 (1984).
16. L.J. Reinders and K. Stam, *Phys. Lett. B* **180**, 125 (1986).

17. H.D. Politzer, *Nucl. Phys. B* **117**, 397 (1976).
18. A.C. Mattingly and P.M. Stevenson *Phys. Rev. Lett.* **69**, 1320 (1992).
19. P.M. Stevenson, *Phys. Lett. B* **331**, 187 (1994).
20. J. Ellis, M. Karliner and M.A. Samuel, *Phys. Lett. B* **400**, 176 (1997).
21. L.R. Baboukhadia, V. Elias, and M.D. Scadron, *J. Phys. G* **23**, 1065 (1997).
22. E. Gardi and M. Karliner, Preprint [hep-ph/9802218] (1998).
23. S.J. Brodsky and G.P. Lepage, Exclusive processes in quantum chromodynamics, in *Perturbative Quantum Chromodynamics*, A. H. Mueller, ed., World Scientific, Singapore, pp. 153-156 (1989).
24. K. Higashijima, *Phys. Rev. D* **29**, 1228 (1984).

Deriving N=2 S-dualities from Scaling

Alex Buchel

Newman Laboratory, Cornell University, Ithaca NY 14853

Abstract. S-dualities in scale invariant $N = 2$ supersymmetric field theories are derived by embedding those theories in asymptotically free theories with higher rank gauge groups. S-duality transformations on the couplings of the scale invariant theory follow from the geometry of the embedding of the scale invariant theory in the Coulomb branch of the asymptotically free theory.

INTRODUCTION

S-duality is the quantum equivalence of classically inequivalent field theories. The paradigmatic example is the strong-weak coupling duality of $N = 4$ supersymmetric Yang-Mills under which theories with couplings τ and $-1/\tau$ are identified.

In this talk I discuss S-dualities for scale invariant $N = 2$ supersymmetric field theories with matter in the fundamental representation of gauge groups [1-5]. The main evidence for S-dualities in $N = 2$ theories has come from the spectrum of BPS saturated states [6] and from low-energy effective actions [7]. By relating S-duality transformations in scale invariant $N = 2$ gauge theories to global symmetries in asymptotically free theories, it was shown in [8] that S-dualities of gauge theories with a single gauge group factor are, in fact, *exact* equivalences of their quantum field theories. S-dualities of the theories with product gauge groups are expected to have a more complicated structure of identifications on the classical coupling space [1-5]. Nonetheless, it is still possible to relate S-dualities of these theories to global symmetries of higher rank asymptotically free theories [9].

The basic idea is to regard the marginal couplings of the scale invariant theory as the lowest components of $N = 2$ vector multiplets—complex scalar "Higgs" fields—in an enlarged theory. Then the coupling space $\mathcal{M}$ of the scale invariant theory is realized as a submanifold of the Coulomb branch $\mathcal{C}$ of the enlarged theory. Any S-duality identifications of different points of $\mathcal{M}$ are interpreted as equivalences on $\mathcal{C}$. By choosing the enlarged theory appropriately, these equivalences on $\mathcal{C}$ can be made manifest as (spontaneously broken) global symmetries.

S-DUALITY FOR SIMPLE GAUGE GROUPS

The scale invariant $N = 2$ theories with a single SU, SO, or Sp gauge group and quarks in the fundamental (defining) representation all have low-energy effective theories that are invariant under identifications of τ under a discrete group isomorphic to[1] $\Gamma^0(2) \subset SL(2, \mathbf{Z})$ [10,11]. $\Gamma^0(2)$ is the subset of $SL(2, \mathbf{Z})$ matrices with even upper off-diagonal entry, or equivalently, is the subgroup of $SL(2, \mathbf{Z})$ generated by $\widetilde{T} : \tau \to \tau + 2$ and $S : \tau \to -1/\tau$, acting on the classical coupling space $\mathcal{M}_{\rm cl} = \{{\rm Im}\tau > 0\}$. (We have taken the gauge coupling to be $\tau = \frac{\vartheta}{\pi} + i\frac{8\pi}{g^2}$, differing by a factor of two from the usual definition.) $\Gamma^0(2)$ is characterized more abstractly as the group freely generated by two generators $\widetilde{T}$ and S subject to one relation $S^2 = 1$. Dividing $\mathcal{M}_{\rm cl}$ by the S-duality group gives the the quantum coupling space $\mathcal{M} = \mathcal{M}_{\rm cl}/\Gamma^0(2)$. Generally, the relations satisfied by the generators $\{\widetilde{T}, S\}$ encode the holonomies in $\mathcal{M}$ around the fixed points of the $\Gamma^0(2)$ action. For example, for the action given above, $\mathcal{M}$ has three such points: the weak coupling point $\tau = +i\infty$ fixed by $\widetilde{T}$, an "ultra-strong" coupling point $\tau = 1$ fixed by $\widetilde{T}S$, and a $\mathbf{Z}_2$ point $\tau = i$ fixed by S. In fact, this data plus the fact that the topology of $\mathcal{M}$ is that of a two sphere summarizes the physically meaningful information about the space of couplings. Its particular realization as a fundamental domain of $\Gamma^0(2)$ acting on the τ upper half plane is dependent on which coordinate τ we use; without an independent non-perturbative definition of τ, the only conditions we can physically impose are on its behavior at arbitrarily weak coupling. Thus the physical content of a statement of S-duality is nothing more than a characterization of the topology of the space of couplings and the holonomies around its singular points.

Consider the scale invariant $SU(r)$ theory with $2r$ fundamental quarks. The Coulomb branch of the theory is described by the curve [10]

$$y^2 = \prod_{a=1}^{r}(x - \phi_a)^2 + (h^2 - 1)\prod_{j=1}^{2r}(x - \mu_j - h\mu), \tag{1}$$

parameterized by a gauge coupling h, quark masses μ_j and μ, and Higgs vevs ϕ_a. h is a function of the coupling such that $h^2 \sim 1 + 64e^{i\pi\tau}$ at weak coupling, μ_j (satisfying $\sum_j \mu_j = 0$) are the eigenvalues of the mass matrix transforming in the adjoint of the $SU(2r)$ flavor group, the singlet mass μ is charged under the $U(1)$ "baryon number" global symmetry, and the ϕ_a (satisfying $\sum_a \phi_a = 0$) are the eigenvalues of the adjoint Higgs field; only flavor and gauge invariant combinations of the μ_j and ϕ_a appear as coefficients in (1).

In the scale invariant theory (setting the masses to zero) in this parameterization of the curve, the coupling parameter space is the h plane. There are two weak coupling points $h = \pm 1$ and a point $h = \infty$ where the low energy effective theory on the Coulomb branch is singular. The low energy effective action is invariant under the $\mathbf{Z}_2$ identification $h \to -h$ (with $\mu \to -\mu$). Identifying the h-plane under

---

[1] For the $SU(2)$ scale invariant theory the S-duality group is enlarged to $SL(2, \mathbf{Z})$ [7].

the action of this $\mathbf{Z}_2$ gives a quantum coupling space $\mathcal{M}$ with one weak coupling singularity ($h=1$), a $\mathbf{Z}_2$ fixed point ($h=0$), and a singular point ($h=\infty$). This matches the description given above of the fundamental domain of $\Gamma^0(2)$ and is the low energy evidence for this S-duality.

We now review the scaling argument of [8] which derives this S-duality by embedding the scale invariant theory in the asymptotically free $SU(r+1)$ theory with $2r$ quarks. We can flow to the scale invariant $SU(r)$ theory by Higgsing the $SU(r+1)$ theory so that $\phi_a = M$, $1 \le a \le r$ and $\phi_{r+1} = -rM$ and assigning the singlet mass $\mu = M$ to keep the $2r$ quarks massless. These tunings are also valid in the quantum theory, since upon applying them to the asymptotically free $SU(r+1)$ curve

$$y^2 = \prod_{a=1}^{r+1}(x-\phi_a)^2 - \Lambda^2(x-\mu)^{2r} \qquad (2)$$

and shifting $x \to x + M$, the curve factorizes as

$$y^2 = x^{2r}\left[(x+(r+1)M)^2 - \Lambda^2\right]. \qquad (3)$$

We recognize the x^{2r} factor of (3) as the singularity corresponding to the scale invariant vacuum of the $SU(r)$ theory with $2r$ massless quarks. We identify the dimensionless parameter M/Λ which varies along the scale invariant singularity with (some holomorphic function of) the gauge coupling of the scale invariant theory. Denote by $\mathcal{G}$ the submanifold (parameterized by M) of scale invariant vacua of the Coulomb branch. From the degenerations of (3) $\mathcal{G}$ has a weak coupling point at $M = \infty$ and two ultra-strong coupling points (singularities) at $M = \pm\Lambda/(r+1)$. Furthermore there is a non-anomalous $\mathbf{Z}_2 \subset U(1)_R$ which acts on the Higgs fields as $\phi \to -\phi$ (and when appropriately combined with a global flavor rotation takes $\mu \to -\mu$), so that the M plane is identified under $M \to -M$,[2] giving a single ultra-strong coupling point and a $\mathbf{Z}_2$ orbifold point at $M = 0$, thus deriving the content of the conjectured S-duality for the SU series.

Although this argument used the low energy effective action of the asymptotically free theory, it provides more than just low energy evidence for the S-duality. By tuning vevs on the Coulomb branch $\mathcal{C}$ of the asymptotically free theory to approach the scale invariant fixed line $\mathcal{G} \subset \mathcal{C}$, we can deduce exact information about the scale invariant theories.[3] Since the manifold of fixed points $\mathcal{G}$ is related to the coupling space $\mathcal{M}$ of the scale invariant theory by a holomorphic map (by $N=2$ supersymmetry), $\mathcal{G}$ must be a multiple cover of $\mathcal{M}$, and hence the topology of $\mathcal{G}$ will constrain the topology of $\mathcal{M}$, giving exact S-duality relations. These identifications

---

[2]) Although one should not quotient the space of vacua by the action of a spontaneously broken global symmetry, because we are interpreting this submanifold of the Coulomb branch as the *coupling* space of the scale invariant theory, it is legitimate (indeed necessary) to do so.
[3]) This same type reasoning is used in [12] to deduce exact equivalences between conformal field theories and supergravity theories.

are exact in the the scale invariant theory because by approaching $\mathcal{G}$ arbitrarily closely on the Coulomb branch we can make the effect of any irrelevant operators from the asymptotically free theory as small as we like. In essence, this argument assumes only that the scale invariant theory has a gap in its spectrum of dimensions of irrelevant operators.[4]

The explicit factorization of (3) into the conformal factor of the scale invariant $SU(r)$ theory and the factor responsible for additional singularities (which we interpreted as singularities on the space of couplings) is possible because of the special hyperelliptic representation of this curve. Solutions of $N = 2$ gauge theories with product gauge groups are encoded in curves which are not hyperelliptic. To generalize the preceding argument to product gauge groups it is useful to reformulate it in an M-theory/IIA brane language.[5]

An *ad hoc* object from the field theory perspective—a Riemann surface whose complex structure encodes the low energy couplings on the Coulomb branch—is given a physical interpretation in the M-theory formulation of $N = 2$ gauge theories [14,1]. Consider the type IIA configuration of intersecting branes depicted in Fig. 1. Two (NS) fivebranes extend in the directions $x^0, x^1, \ldots, x^5$ and are located at $x^7 = x^8 = x^9 = 0$ and at some specific values of x^6. $r+1$ (D) fourbranes stretch between the fivebranes, extend over $x^0, \ldots, x^3$ and are located at $x^7 = x^8 = x^9 = 0$ and at a point in $v = x^4 + ix^5$. Note that these fourbranes are finite in the x^6 direction. This configuration represents the Coulomb branch vacua of an $SU(r+1)$ gauge theory in the four dimensions $x^0, \ldots, x^3$, with the position of the fourbranes in v corresponding to the vevs of the adjoint scalar in the $N = 2$ vector multiplet. For the model in hand we choose $v = M$ for r of the fourbranes and $v = -rM$ for the remaining one. In Fig. 1 we have also shown $2r$ semi-infinite fourbranes which also extend over $x^0, \ldots, x^3$ and are located at $x^7 = x^8 = x^9 = 0$. r of these fourbranes end on the left fivebrane and the other r end on the right one and correspond to the $2r$ hypermultiplets (quarks) of the $SU(r+1)$ gauge theory. Their (fixed) v coordinates encode the masses of the hypermultiplets. With the choice of mass parameters as in (2), one gets the configuration of Fig. 1.

This brane configuration is singular where the ends of the fourbranes end on the fivebrane world volume. These singularities are resolved by lifting the construction to M theory [1] where the IIA fourbranes become M theory fivebranes wrapped around the eleventh dimension x^{10}. In fact, the whole configuration of branes in Fig. 1 appears as a single M theory fivebrane with world volume $\mathbf{R}^4 \times \Sigma$ where $\mathbf{R}^4$ is the four-dimensional low energy space-time and Σ is the Riemann surface describing the Coulomb branch of the $SU(r+1)$ gauge theory with $2r$ quarks. Let $t = e^{-s}$ with $s = (x^6 + ix^{10})/R$, R being the radius of the eleventh dimension. Then Σ is holomorphically embedded in $\mathbf{C}^2 = \{t, v\}$ as the curve [1]

---

[4]) By the arguments of [13] this is plausibly also the condition for the scale invariant theory to be conformally invariant.
[5]) In [4] S-dualities for a broader class of theories are derived by considering type II strings on Calabi-Yau three-folds.

$$(v - M)^r t^2 + \tilde{f} (v - M)^r (v + rM) t + (v - M)^r = 0. \tag{4}$$

From the Type IIA perspective (4) describes bending fivebranes as $t \to \infty$ or 0. Asymptotically, the separation of the fivebranes reads

$$s_2 - s_1 = \ln \frac{t_1}{t_2} \simeq 2 \ln \tilde{f} v \tag{5}$$

which describes the running of the β-function of the asymptotically free $SU(r+1)$ gauge theory with $2r$ quarks, with $1/\tilde{f}$ identified with the strong coupling scale Λ of the $SU(r+1)$ theory, and v interpreted as the energy scale at which we measure the coupling.

The scaling limit, implicit in (3), involves looking at the scale invariant $SU(r)$ theory at energy scales much smaller than either M or Λ. Only far below these scales can one sensibly talk about the scale invariance of the $SU(r)$ gauge theory which appears as an effective low energy description after integrating out modes charged under the decoupled $U(1)$ gauge factor. The decoupling of the $U(1)$ gauge factor from the brane point of view means that we consider a small region of the brane construction near $v = M$, which locally looks like a finite $SU(r)$ gauge

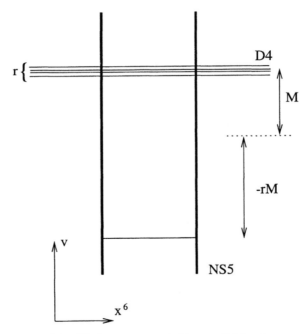

FIGURE 1. A configuration of fivebranes connected by parallel fourbranes realizing the embedding of the scale invariant $SU(r)$ gauge theory into an asymptotically free $SU(r+1)$ theory. The horizontal dotted line marks the position of the origin in the v-plane.

theory. To describe the scale invariant geometry near $v = M$ we thus introduce a local coordinate $x = v - M$ and consider $x \ll \{\Lambda, M\}$ which is the geometrical realization of the scaling to the fixed point theories. From (4) we find the description of this region to be

$$x^r \left(t^2 + \frac{M(r+1)}{\Lambda} t + 1 \right) = 0, \qquad (6)$$

which indeed describes the scale invariant $SU(r)$ theory (at the origin of its Coulomb branch) with coupling parameter

$$f = M(r+1)/\Lambda. \qquad (7)$$

Thus, we have tuned to a one complex dimensional submanifold of the Coulomb branch $\{M\} \equiv \mathcal{G} \subset \mathcal{C}$ of the $SU(r+1)$ theory, which realizes some multiple cover $\widetilde{\mathcal{M}}$ of the coupling space of the scale invariant $SU(r)$ theory. At certain points in $\mathcal{G}$, specifically $M = \pm 2\Lambda/(r+1)$ and $M \to \infty$, (6) develops additional singularities that we interpret through (7) as singularities ("punctures") of $\widetilde{\mathcal{M}}$. $\widetilde{\mathcal{M}}$ is not quite the coupling manifold $\mathcal{M}$ of the $SU(r)$ theory since, as in the field theoretical arguments above, there is a $\mathbf{Z}_2$ identification $M \to -M$ coming from the unbroken subgroup of the $U(1)_R$ symmetry:

$$\mathcal{M} \simeq \widetilde{\mathcal{M}}/\mathbf{Z}_2. \qquad (8)$$

This once again gives the expected S-duality of the $SU(r)$ theory.

S-dualities can be derived for the scale invariant SO and Sp theories along the same lines. One can also provide a similar interpretation of the derivation in brane language.

Before proceeding to the generalization of this construction to the product gauge group theories, we comment on the definition of S-duality *groups*. It is natural to define the S-duality group Γ by $\mathcal{M} = \mathcal{M}_{\text{cl}}/\Gamma$. Using the fact that $\pi_1(\mathcal{M}_{\text{cl}}) = 1$ one might be tempted to identify $\Gamma = \pi_1(\mathcal{M})$, the fundamental group of $\mathcal{M}$. However, as we have just seen, Γ does not act freely on $\mathcal{M}_{\text{cl}}$, so $\pi_1(\mathcal{M})$ does not capture all of Γ. In particular, we must enlarge π_1 to include the holonomies around the orbifold singularities of $\mathcal{M}$ [15]. For example, we can consider fixed points of discrete identifications as being punctures, and identify a closed loop around a $\mathbf{Z}_2$ orbifold singularity $M = 0$ of $\mathcal{M}$ with the order two element of the S-duality group. Adding to this the other nontrivial element of the fundamental group of $\mathcal{M}$ represented by a closed loop around the $M^2 = 4\Lambda^2/(r+1)^2$ puncture, we recover the full S-duality group $\Gamma^0(2)$ of the scale invariant $SU(r)$ theory.

S-DUALITY IN PRODUCT GAUGE GROUPS

$N = 2$ gauge theories with products of SU, SO and Sp groups were solved in [1–3]. The derivation of S-dualities for these models is a straightforward extension of the scaling arguments of the previous section. We present here only the results, the details can be found in [9].

The quantum coupling space $\mathcal{M}^{(n)}$ of scale invariant $\otimes_i SU(k_i)$ and $\otimes_i SO(k_{2i-1}) \otimes Sp(k_{2i})$ gauge theories with n gauge group factors and matter in the fundamental and bifundamental representations is identified with

$$\mathcal{M}^{(n)} = \mathcal{M}_{0,n+3;2} \tag{9}$$

where $\mathcal{M}_{0,n+3;2}$ is the moduli space of a genus zero Riemann surface with $n+3$ marked points, two of which are distinguished and ordered while the other $n+1$ are indistinguishable. $\mathcal{M}^{(n)}$ is isomorphic to $\widetilde{\mathcal{M}}^{(n)}/\mathbf{Z}_{n+1}$, where $\widetilde{\mathcal{M}}^{(n)} = \{f_i\}$, $f_i \in \mathbf{C}$, $i = 1, \ldots, n$, with punctures determined by the vanishing locus of the discriminant of

$$y^{n+1} + f_1 y^n + \ldots + f_i y^{n+1-i} + \ldots + 1 = 0 \tag{10}$$

along with values of f_i at which one or more roots of (10) go to zero or infinity. $\mathbf{Z}_{n+1}$ identifies the moduli f_i under $f_i \to \omega_{n+1}^{p \cdot i} f_i$, for $p = 1, \ldots n$ with $\omega_{n+1} = e^{2\pi i/(n+1)}$. The S-duality group Γ_n for these theories is identified with $\pi_1(\mathcal{M}_{0,n+3;2})$, extended to include holonomies around the orbifold points as discussed in the previous section.

That all the scale invariant "cylindrical" models studied in [1,2] have the same quantum coupling space $\mathcal{M}_{0,n+3,2}$, determined only by the number of gauge group factors is intuitively clear: $N = 2$ gauge theories with n factors in the gauge groups were derived in [1,2] from a configuration with $n+1$ straight NS fivebranes which can be associated [1] with $n+1$ unordered points on a cylinder. The moduli space of these points have "punctures" (where the low energy effective description of the theory is singular) whenever two or more NS fivebranes coincide or move to infinity. This can be cast in the form of the vanishing of the discriminant of an $n+1$ order polynomial as in (10). Although models with symplectic and orthogonal groups require the introduction of an orientifold $O4$ plane [2], orientifold projection do not restrict the NS fivebranes moduli and the quantum coupling space appears the same as for the unitary groups.

Unlike the previous example, the $\otimes_i SU(k_i) \otimes SO(N)$ and $\otimes_i SU(k_i) \otimes Sp(N)$ series of [5] have a different quantum coupling space (and therefore a different S-duality group). In these cases the coupling space is $\mathcal{M}^{(n)} = \mathcal{Q}_{0,n,2}$ (n is the total number of gauge group factors), where $\mathcal{Q}_{0,n,2}$ is the moduli space of genus zero Riemann surfaces with two distinguished and marked points and n unordered pairs of points with the further condition that, in terms of any complex coordinate z on the sphere, the product of the coordinates within each pair is the same for all pairs. $\mathcal{M}^{(n)} \simeq \widetilde{\mathcal{M}}^{(n)}/\mathbf{Z}_2$ where $\widetilde{\mathcal{M}}^{(n)} = \{f_i\}$, $f_i \in \mathbf{C}$, $i = 1, \ldots, n$ and $\mathbf{Z}_2$ identifies $f_i \to (-1)^i f_i$. Punctures on $\widetilde{\mathcal{M}}^{(n)}$ are at values of f_i for which either the discriminant of the polynomial

$$y^{2n} + f_1 y^{2n-1} + \ldots + f_{n-1} y^{n+1} + f_n y^n + f_{n-1} y^{n-1} + \ldots + f_1 y + 1 = 0 \tag{11}$$

vanishes or some or its roots go to zero or infinity. It is always possible to factorize a polynomial of the form (11) into one of the form

$$\prod_{i=1}^{n}(y^2 + z_i y + 1) = 0, \qquad (12)$$

thus giving an isomorphism between $\widetilde{\mathcal{M}}^{(n)} = \{f_i\}$ and $\widetilde{\mathcal{Q}}^{(n)} = \{z_i$ modulo permutations$\}$. The $\mathbf{Z}_2$ identification $f_i \to (-1)^i f_i$ becomes $z_i \to -z_i$. With this presentation it is clear that $\mathcal{M}^{(n)} = \widetilde{\mathcal{Q}}^{(n)}/\mathbf{Z}_2 \simeq \mathcal{Q}_{0,n,2}$.

CONCLUSION

The uniform way in which the complicated quantum coupling spaces of the $N = 2$ scale invariant theories were derived by this scaling procedure suggests that it should be effective more generally. In particular it would be interesting to extend this argument to study S-dualities in scale invariant $N = 1$ theories, some classes of which have been proposed and studied in [16,17].

Another open question involves the physics of the "ultra-strong" coupling points in the coupling spaces of $N = 2$ theories. Unlike the $N = 4$ theories where the $SL(2, \mathbf{Z})$ S-duality identifies all the ultra-strong (Im$\tau = 0$) points with the weak coupling limit of the theory, we have seen above that the S-dualities of the $N = 2$ scale invariant theories are generically smaller, and leave ultra-strong points (or manifolds) in their coupling spaces. One counterexample is the scale invariant $SU(2)$ theory with four fundamental quark hypermultiplets, studied in [7]. There it was found that the theory in fact has the whole $SL(2, \mathbf{Z})$ duality and no ultra-strong points. This could also be derived through scaling arguments [8] by comparing the embeddings of $SU(2)$ into $SU(3)$ and $Sp(4)$ asymptotically free theories. One should note that the scaling arguments presented here do not claim to capture all possible S-dualities—there may be further identifications of the coupling space which are missed since our arguments only show that $\mathcal{M}^{(n)}$ is some multiple cover of the true coupling space.

This leaves open the possibility that there are further identifications of the coupling spaces of the $N = 2$ theories, perhaps relating the ultra-strong coupling points to some other weakly coupled physics. One place where we know such further identifications must exist are in theories with $SU(2)$ gauge group factors: for in the limit that the other factors decouple, the $SU(2)$ factors must have the full $SL(2, \mathbf{Z})$ duality of [7].

ACKNOWLEDGMENTS

Author thanks his collaborator P.C. Argyres. This work is supported in part by NSF grant PHY-9513717.

REFERENCES

1. E. Witten, *Nucl. Phys.* **B500**, 3 (1997).
2. K. Landsteiner, E. Lopez, and D.A. Lowe, *Nucl. Phys.* **B507**, 197 (1997).
3. A. Brandhuber, J. Sonnenschein, S. Theisen, and S. Yankielowicz, *Nucl. Phys.* **B504**, 175 (1997).
4. S. Katz, P.Mayr, and C. Vafa, *Adv. Theor. Math. Phys.* **1**, 53 (1998).
5. K. Landsteiner, and E. Lopez, *Nucl. Phys.* **B516**, 273 (1998).
6. C. Montonen, and D. Olive, *Phys. Lett.* **B72**, 117 (1977); E. Witten, and D. Olive, *Phys. Lett.* **B78**, 97 (1978); H. Osborn, *Phys. Lett.* **B83**, 321 (1979); A. Sen, *Phys. Lett.* **B329** 217 (1994).
7. N. Seiberg, and E. Witten, *Nucl. Phys.* **B431**, 484 (1994).
8. P.C. Argyres *Adv. Theor. Math. Phys.* **2**, 61 (1998).
9. P.C. Argyres, and A. Buchel, hep-th/9804007, to appear in *Phys. Lett.* **B**.
10. P.C. Argyres, M.R. Plesser, and A.D. Shapere, *Coulomb phase of N=2 supersymmetric QCD, Phys. Rev. Lett.* **75**, 1699 (1995).
11. P.C. Argyres and A.D. Shapere, *Nucl. Phys.* **B461**, 437 (1996).
12. J. Maldacena, hep-th/9711200.
13. J. Polchinski, *Nucl. Phys.* **B303**, 226 (1988).
14. A. Klemm, W. Lerche, P. Mayr, C. Vafa, and N. Warner, *Nucl. Phys.* **B477**, 746 (1996).
15. P.C. Argyres, *Nucl. Phys.* B (Proc. Suppl.) **61A**, 149 (1998).
16. R. Leigh and M. Strassler, *Nucl. Phys.* **B447**, 95 (1995).
17. S. Kachru and E. Silverstein, *Phys. Rev. Lett.* **80**, 4855 (1998); A. Lawrence, N.Nekrasov, and C. Vafa, hep-th/9803015; M. Bershadsky, Z. Kakushadze, and C. Vafa, hep-th/9803076; Z. Kakushadze, hep-th/9803214.

Symmetry of Quantum Matrix Models

C.-W. H. Lee[1] and S. G. Rajeev

Department of Physics and Astronomy, University of Rochester, Rochester, New York 14627.

Abstract. We discuss some general aspects of quantum matrix models, and introduce an algebraic approach to them by expressing their symmetry in terms of a Lie algebra.

Let us introduce the notion of a quantum matrix model first [1]. A quantum matrix model is a model in which quantum fields consist of $N \times N$ *matrices* of annihilation and creation operators, and physical states and observables are built out of *traces* of these matrices. A typical annihilation operator is of the form $a^{\mu_1}_{\mu_2}(k)$ where μ_1 and μ_2 can be regarded as column and row indices, and k tells us to which matrix this annihilation operator belongs in a multi-matrix model with matrices 1, 2, ..., and Λ where Λ is a positive integer. The corresponding creation operator is $a^{\dagger \mu_2}_{\mu_1}(k)$. These operators satisfy the following (anti-)commutation relation:

$$\left[a^{\mu_1}_{\mu_2}(k_1), a^{\dagger \mu_3}_{\mu_4}(k_2)\right]_{\pm} = \delta_{k_1 k_2} \delta^{\mu_3}_{\mu_2} \delta^{\mu_1}_{\mu_4}. \tag{1}$$

Here we take the anti-commutator if both k_1 and k_2 are labels for fermions, and the commutator otherwise.

Quantum matrix models arise in many areas of physics. The one with the longest history is perhaps quantum chromodynamics (QCD) [2], [3], [4], [5]. In the light-cone formalism of QCD, a and $a^\dagger$ are annihilation and creation operators for gluons, N is the number of colors, μ's are color quantum numbers, and k is the linear momentum of the gluon concerned. (The range of values k can take can be made finite by, e.g., putting the spacetime in a torus.)

Supersymmetric Yang-Mills (SYM) theory is another example in which quantum matrix models show up [6], [7]. Here both bosons and fermions are in the adjoint matter field representation, and thus it is very natural for us to regard them as matrix fields. Depending on the actual physical system which an SYM theory is used to describe, these matrix fields can have a variety of physical meanings. For example, if the spacetime is essentially flat, in the low energy regime the dynamics of Dp-branes can be approximated very well by an $\mathcal{N} = 1$ SYM theory dimensionally reduced from 10 down to $p+1$ [8]. Here we have a $(p+1)$-dimensional gauge field

---

[1] speaker.

theory with $9 - p$ adjoint matter fields of bosons and a number of adjoint matter fields of fermions. N, the dimension of the matrices, is the number of D-branes in the system, μ is a label for a particular D-brane, and k, ranging from $p + 1$ to 9, labels the transverse dimensions. In the classical theory, diagonal elements of the matrices give the coordinates of the D-branes in the k-th dimension, and off-diagonal elements tell us the distances between the corresponding D-branes in that transverse dimension.

Closely associated with D-branes is the M-theory conjecture [9], namely that in the infinite momentum frame, M-theory is exactly described by the $N \to \infty$ limit of 0-brane quantum mechanics. A natural corollary of this conjecture is that light-front type-IIA superstring theory can be described by an $\mathcal{N} = 1$ SYM theory dimensionally reduced from 10 to 2 [10] in the large-N limit. The Hamiltonian thus obtained is essntially the Green-Schwarz light-front string Hamiltonian, except that the fields are matrices. This is reminiscent of an earlier attempt to formulate string theory by string-bits [11], another quantum matrix model. (A comprehensive review on SYM theory, D-branes and M-theory can be found in Ref. [12].)

The above examples show that the large-N limit of quantum matrix models is of interest, and it should be worthwhile to explore ways to tackle these models.

Effort has been made to solve matrix models in the large-N limit, and there has been some progress in solving some matrix models *classically* in the large-N limit [13], and in solving some matrix models *numerically* [5], [7]. Neverthelss, it will be nice if we have other perspectives, as these may shed light on some properties of matrix models which are hard to understand in the classical and numerical approaches.

One important aspect of matrix models is the underlying symmetry. History teaches us that discovering the symmetry of a system can greatly improve our understanding of it. For example, the underlying symmetry of a hydrogenic atom is $SO(4)$; the Hamiltonian can be written as a rational function of the generators of the associated Lie algebra. The representation theory of $so(4)$ then yields the spectrum. As another example, the XXX model (Heisenberg model) and the XXZ model possess the Lie group $SU(2)$ and the quantum group $U_q(sl(2))$ as their symmetries respectively [14]. Exploiting the Lie algebra or the quantum group help us a lot in understanding the physical properties of these integrable models also. The $SO(4)$ symmetry in the one-dimensional Hubbard model [15] also shapes the spectrum of it. In the case of conformal field theory, it is well known that its symmetry is governed by the Virasoro algebra. The highest weight representation of the Virasoro algebra then help us find out the spectrum of conformal field theory. A related example is the Wess-Zumino-Witten model which provides an excellent example for an application of the Kac-Moody algebra. (A collection of reprint articles on Kac-Moody algebras, Virasoro algebras and the Wess-Zumino-Witten model can be found in Ref. [16].)

The lesson is that if we know how to characterize the symmetry of a physical system by a Lie algebra or its generalization like a quantum group, we have a deeper understanding of the properties of this physical system like its spectrum. Is there

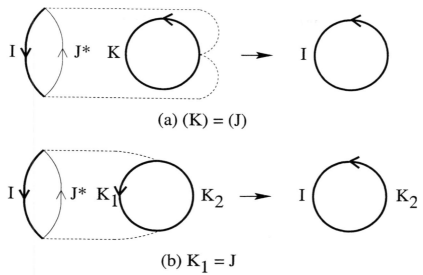

FIGURE 1. The action of a color-invariant observable on a color-invariant state in the large-N limit. Fig. (a) shows the first term on the R.H.S. of Eq.(3) whereas Fig. (b) shows the second term. The dotted lines join the segments to be 'contracted' together. Note J carries an asterisk to indicate that this sequence is put in reverse.

an underlying symmetry for quantum matrix models? The answer is affirmative.

To see what this Lie algebra is, let us go back to the physical system with physical states and observables built out of the annihilation and creation operators satisfying Eq.(1). For the sake of simplicity, let us confine ourselves to bosonic operators (to see how the following formalism can be generalized to include fermions also, the reader may go to Ref. [17]). In the language of gauge theory, a color-invariant state is a linear combination of states of the form

$$\Psi^{(K)} \equiv N^{-c/2} a^{\dagger \nu_2}_{\nu_1}(k_1) a^{\dagger \nu_3}_{\nu_2}(k_2) \cdots a^{\dagger \nu_1}_{\nu_c}(k_c)|0\rangle. \qquad (2)$$

Here we sum over all possible values of the row and column indices. The superscript K is the integer sequence $k_1, k_2, \ldots, k_c$. We put K inside a pair of parentheses to indicate that a cyclic permutation of these integers will produce the same state. The factor of N in this equation serves as a normalization factor in the large-N limit. A color-invariant observable is a linear combination of operators either of the form

$$g^I_J \equiv N^{-(a+b-2)/2} a^{\dagger \mu_2}_{\mu_1}(i_1) a^{\dagger \mu_3}_{\mu_2}(i_2) \cdots a^{\dagger \nu_b}_{\mu_a}(i_a) \cdot \\ a^{\nu_{b-1}}_{\nu_b}(j_b) a^{\nu_{b-2}}_{\nu_{b-1}}(j_{b-1}) \cdots a^{\mu_1}_{\nu_1}(j_1)$$

or of the form

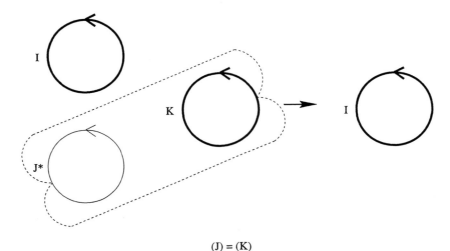

(J) = (K)

FIGURE 2. The action of another color-invariant observable on a color-invariant state in the large-N limit.

$$\tilde{f}^{(I)}_{(J)} \equiv N^{-(a+b)/2} a^{\dagger \mu_2}_{\mu_1}(i_1) a^{\dagger \mu_3}_{\mu_2}(i_2) \cdots a^{\dagger \mu_1}_{\mu_a}(i_a) \cdot$$
$$a^{\nu_{b-1}}_{\nu_b}(j_b) a^{\nu_{b-2}}_{\nu_{b-1}}(j_{b-1}) \cdots a^{\nu_b}_{\nu_1}(j_1).$$

Again we sum over all possible values of the row and column indices. I is the integer sequence i_1, i_2, ..., i_a, and J is the sequence j_1, j_2, ..., j_b. The factors of N are chosen to ensure that once we act these operators on states of the form given by Eq.(2), we will get other normalizable states.

It is known that in the large-N limit, if we operate g^I_J on $\Psi^{(K)}$, any segment of any cyclic permutation of K identical to J will be replaced with I [3], [18]:

$$g^I_J \Psi^{(K)} = \delta^K_{(J)} \Psi^{(I)} + \sum_{K_1 K_2 = (K)} \delta^{K_1}_J \Psi^{(IK_2)}. \tag{3}$$

In Eq.(3), a summation like $\sum_{K_1 K_2 = (K)}$ means that we sum over all possible cyclic permutation of K, and all possible pairs of K_1 and K_2 such that the concatenated sequence is exactly this cyclically permuted sequence. $\delta^{K_1}_J$ gives 1 if K_1 is exactly equal to J, and 0 otherwise; $\delta^K_{(J)}$ gives the number of cyclic permutations of J such that the permuted sequence is exactly K. Fig. 1 gives us a graphical representation of Eq.(3).

Likewise, if we operate $\tilde{f}^{(I)}_{(J)}$ on $\Psi^{(K)}$, we will get a multiple of $\Psi^{(I)}$ if (J) and (K) are the same equivalence class:

$$\tilde{f}^{(I)}_{(J)} \Psi^{(K)} = \delta^K_{(J)} \Psi^{(I)}. \tag{4}$$

Eq.(4) is depicted in Fig. 2.

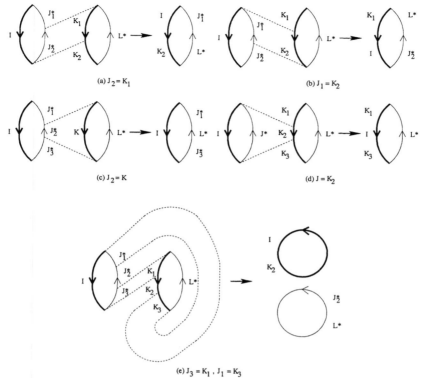

FIGURE 3. Diagrammatic representations of some terms on the R.H.S. of Eq.(6).

Note that if we take the vector space of all finite linear combinations of color-invariant states of the form in Eq.(2) as the defining representation space for the color-invariant observables g_J^I and $\tilde{f}_{(J)}^{(I)}$, then these color-invariant observables are *not* linearly independent.

Now comes the crux of the whole formalism. Even though it is not possible to define the product $g_J^I g_L^K$ by its action of an arbitrary state $\Psi^{(M)}$ [19], it is possible to construct a *commutator* between g_J^I and g_L^K in such a way that

$$\left[g_J^I, g_L^K\right]\Psi^{(M)} \equiv g_J^I(g_L^K \Psi^{(M)}) - g_L^K(g_J^I \Psi^{(M)}) \tag{5}$$

is satisfied [19]. In fact, it turns out that

$$\left[g_J^I, g_L^K\right] = \delta_J^K g_L^I + \sum_{J_1 J_2 = J} \delta_{J_2}^K g_{J_1 L}^I + \sum_{K_1 K_2 = K} \delta_J^{K_1} g_L^{I K_2}$$
$$+ \sum_{\substack{J_1 J_2 = J \\ K_1 K_2 = K}} \delta_{J_2}^{K_1} g_{J_1 L}^{I K_2} + \sum_{J_1 J_2 = J} \delta_{J_1}^K g_{L J_2}^I + \sum_{K_1 K_2 = J} \delta_J^{K_2} g_L^{K_1 I}$$

$$+ \sum_{\substack{J_1 J_2 = J \\ K_1 K_2 = K}} \delta_{J_1}^{K_2} g_{LJ_2}^{K_1 I} + \sum_{J_1 J_2 J_3 = J} \delta_{J_2}^{K} g_{J_1 L J_3}^{I} + \sum_{K_1 K_2 K_3 = K} \delta_{J}^{K_2} g_{L}^{K_1 I K_3}$$

$$+ \sum_{\substack{J_1 J_2 = J \\ K_1 K_2 = K}} \delta_{J_2}^{K_1} \delta_{J_1}^{K_2} \tilde{f}_{(L)}^{(I)} + \sum_{\substack{J_1 J_2 J_3 = J \\ K_1 K_2 = K}} \delta_{J_3}^{K_1} \delta_{J_1}^{K_2} \tilde{f}_{(J_2 L)}^{(I)}$$

$$+ \sum_{\substack{J_1 J_2 = J \\ K_1 K_2 K_3 = K}} \delta_{J_2}^{K_1} \delta_{J_1}^{K_3} \tilde{f}_{(L)}^{(IK_2)}$$

$$+ \sum_{\substack{J_1 J_2 J_3 = J \\ K_1 K_2 K_3 = K}} \delta_{J_3}^{K_1} \delta_{J_1}^{K_3} \tilde{f}_{(J_2 L)}^{(IK_2)} - (I \leftrightarrow K, J \leftrightarrow L). \quad (6)$$

Likewise, the commutators

$$\left[g_J^I, \tilde{f}_{(L)}^{(K)}\right] = \delta_{(J)}^{K} \tilde{f}_{(L)}^{(I)} + \sum_{K_1 K_2 = (K)} \delta_J^{K_1} \tilde{f}_{(L)}^{(IK_2)} - \delta_{(L)}^{I} \tilde{f}_{(J)}^{(K)} - \sum_{L_1 L_2 = (L)} \delta_{L_2}^{I} \tilde{f}_{(L_1 J)}^{(K)} \quad (7)$$

and

$$\left[\tilde{f}_{(J)}^{(I)}, \tilde{f}_{(L)}^{(K)}\right] = \delta_{(J)}^{K} \tilde{f}_{(L)}^{(I)} - \delta_{(L)}^{I} \tilde{f}_{(J)}^{(K)} \quad (8)$$

satisfy relations similar to Eq.(5)

The proof of the above three equations can be found in Ref. [19]. Due to the lack of space, we can only show a number of generic diagrams for some of the terms in the above equations. (A more complete collection of diagrams can be found in Ref. [19].) Fig. 3 shows some terms in Eq.(6). Fig. 4 illustrates some representative terms in Eqs.(7) and (8).

To understand the nature of this Lie algebra, let us see if it can be reduced to some other familiar Lie algebras in some special cases. If we set $\Lambda = 1$, then it turns out that this Lie algebra is an extension of the Virasoro algebra without any central extension by $gl_{+\infty}$ [20], the algebra of infinite-dimensional matrices with finite numbers of non-zero entries [19]. As the Virasoro algebra is intimately related to conformal symmetry, some properties of quantum matrix models may be related to conformal symmetry, too. Moreover, the Virasoro algebra can also be viewed as the algebra of vector fields on a circle. Thus it may be possible to realize the Lie algebra introduced in this paper on a non-commutative manifold.

Finally, we remark that we have found a symmetry algebra for the operators acting on *open strings* [21], and a superalgebra acting on strings with both fermionic and bosonic degrees of freedom [17]. Also discussed in our papers mentioned in this proceedings are some examples of exactly integrable matrix models of which we have some supersymmetric generalizations recently [22]. We conjecture that the Cartan

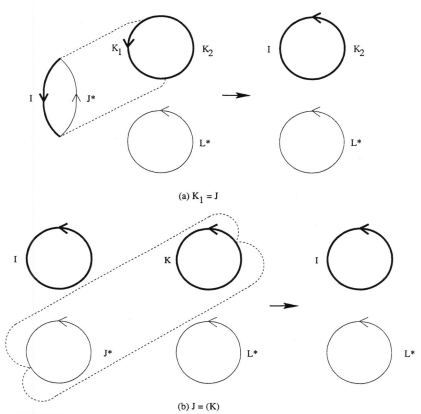

FIGURE 4. Diagrammatic representations of the R.H.S. of Eqs.(7) and (8).

subalgebras of these Lie (super)-algebras should have some intimate relationship with the conserved charges in these models. Moreover, there may be a relationship between the Yang-Baxter equation and these Lie (super)-algebras.

We acknowledge the support in part provided by the U.S. Department of Energy under grant DE-FG02-91ER40685.

REFERENCES

1. C.-W. H. Lee and S. G. Rajeev, Phys. Rev. Lett. **80**, 2285 (1998).
2. G. 't Hooft, Nucl. Phys. **B75**, 461 (1974); E. Brezin, C. Itzykson, G. Parisi, and J. B. Zuber, Commun. Math. Phys. **59**, 35 (1978); E. Witten, Nucl. Phys. **B160**, 57 (1979); B. Sakita, Quantum Theory of Many-Variable Systems and Fields, Singapore, World Scientific Lecture Notes in Physics, 1, World Scientific, Singapore, 1985.
3. C. B. Thorn, Phys. Rev. **D20**, 1435 (1979).

4. S. G. Rajeev, Phys. Lett. **B209**, 53 (1988); in *Proceedings of the 11th Annual Montreal-Rochester-Syracuse-Toronto Meeting, Syracuse, 1989* Syracuse University, Syracuse, NY, 1989, p.78; Phys. Rev. **D42**, 2779 (1990); Phys. Rev. **D44**, 1836 (1991).
5. K. Demeterfi, I. R. Klebanov, and G. Bhanot, Nucl. Phys. **B418**, 15 (1994); F. Antonuccio and S. Dalley, Nucl. Phys. **B461**, 275 (1996).
6. D. Kutasov, Nucl. Phys. **B414**, 33 (1994).
7. Y. Matsumura, M. Sakai, and T. Sakai, Phys. Rev. **D52**, 2446 (1995); D. Gross, A. Hashimoto, and I. R. Klebanov, Phys. Rev. **D57**, 6420 (1998).
8. R. G. Leigh, Mod. Phys. Lett. **A4**, 2767 (1989); E. Witten, Nucl. Phys. **B460**, 335 (1996); J. Polchinski, e-print hep-th/9611050.
9. T. Banks, W. Fischler, S. H. Shenkar, and L. Susskind, Phys. Rev. **D55**, 5112 (1997); T. Banks, e-print hep-th/9710231; D. Bigatti and L. Susskind, e-print hep-th/9712072.
10. R. Dijkgraaf, E. Verlinde, and H. Verlinde, Nucl. Phys. **B500**, 43 (1997); Nucl. Phys. Proc. Suppl. **62**, 348 (1998).
11. O. Bergman and C. B. Thorn, Phys. Rev. **D52**, 5980 and 5997 (1995).
12. W. Taylor IV, e-print hep-th/9801182.
13. M. R. Douglas, Phys. Lett. **B238**, 176 (1990); P. Di Francesco, P. Ginsparg and J. Zinn-Justin, Phys. Rep. **254**, 1 (1995); G. W. Semenoff and R. J Szabo, Int. J. Mod. Phys. **A12**, 2135 (1997).
14. C. Gómez, M. Ruiz-Altaba, G. Sierra, *Quantum Groups in Two-Dimensional Physics*, Cambridge University Press, 1996.
15. C. N. Yang and S. C. Zhang, Mod. Phys. Lett. **B4**, 759 (1990).
16. P. Goddard and D. Olive (editors), *Kac-Moody and Virasoro Algebras: A Reprint Volume for Physicists*, Advanced Series in Mathematical Physics Vol. 3, World Scientific, 1988.
17. C.-W. H. Lee and S. G. Rajeev, to be published.
18. C.-W. H. Lee, Ph. D. thesis, in preparation.
19. C.-W. H. Lee and S. G. Rajeev, e-print hep-th/9806002, to be published in J. Math. Phys..
20. V. G. Kac, *Infinite Dimensional Le Algebras* (Third Edition) Cambridge University Press, 1990, p.112.
21. C.-W. H. Lee and S. G. Rajeev, e-print hep-th/9712090, to be published in Nucl. Phys. B.
22. C.-W. H. Lee and S. G. Rajeev, e-print hep-th/9806019, to be published in Phys. Lett. B.

Perturbed Conformal Field Theory: A Tool for Investigating Integrable Models

Marco Ameduri

Newman Laboratory of Nuclear Studies
Cornell University
Ithaca, NY 14853—USA

Abstract. Zamolodchikov's formalism of perturbed conformal field theory is reviewed, and its applications to the analysis of new families of integrable models are discussed. We examine the integrability properties of a class of generalized sine-Gordon models.

INTRODUCTION

Until the relatevily recent developments in the study of supersymmetric gauge theories [1], which have made it possible to obtain exact information about four-dimensional quantum field theories (QFTs)[1], the study of QFTs in two dimensions was the only way to gain an exact understanding of their structure. For many of these models it has been possible to compute their full S-matrix, thereby determining completely the spectrum of the theory. Progress in the computation of the exact correlation functions has also been made[2].

In recent years many two-dimensional integrable models have found an important application in the description of condensed matter systems (see [4] and references therein.)

The field of integrable models also displays a wealth of sofisticated mathematical structures—*e.g.* infinite dimensional Lie algebras and quantum groups.

For all these reasons the open problem of a classification of two-dimensional integrable models is very important, both for its theoretical and practical aspects. In this talk I wish to review an important approach to the problem, due to A.B. Zamolodchikov [5], where massive models are regarded as perturbations of a conformal field theory (CFT). I will also very briefly mention current applications of the method.

---

[1] For a review, see [2].
[2] For a review, see [3].

CONFORMAL FIELD THEORY

A physical system at a second order phase transition can be described in terms of a conformally invariant quantum field theory, which can be regarded as a fixed point of the renormalization group transformations (for a review and a rich set of references see for instance [6].)

In dimension higher than two the group of conformal transformations is finite dimensional, so that the presence of conformal symmetry is not enough for a complete characterization of a QFT. In two dimensions the conformal group is infinite dimensional (any analytic function generates a local conformal transformation); this fact powerfully constrains the structure of a two-dimensional CFT, to allow for a complete classification based on the representation theory of the conformal algebra.

In this section we recall some basic notions of CFT in two dimensions, mainly with the purpose of establishing our notation[3]. We will work with complex coordinates

$$z = x + iy, \quad \bar{z} = x - iy. \tag{1}$$

As a consequence of translation and scale invariance, the complex components of the energy-momentum tensor

$$T \equiv T_{zz}, \quad \bar{T} \equiv T_{\bar{z}\bar{z}}, \quad \Theta \equiv 4T_{z\bar{z}}, \tag{2}$$

will satisfy the following relations:

$$\Theta = 0, \quad \partial_{\bar{z}} T = 0, \quad \partial_z \bar{T} = 0. \tag{3}$$

Eq. (3) implies that T is an analytic function of the coordinates, while $\bar{T}$ is antianalytic. One can therefore define a Laurent mode expansion for the energy-momentum tensor:

$$T(z) = \sum_{n=-\infty}^{\infty} \frac{L_n}{z^{n+2}}. \tag{4}$$

From the operator product expansion (OPE) of T with itself

$$T(z)T(w) = \frac{c/2}{(z-w)^4} + \frac{2T(w)}{(z-w)^2} + \frac{\partial_w T(w)}{z-w}, \tag{5}$$

one can easily show that the Laurent modes L_n in (4) satisfy the following algebra, known as the Virasoro algebra:

$$[L_n, L_m] = (n-m)L_{n+m} + \frac{c}{12}n(n-1)\delta_{n+m,0}. \tag{6}$$

---

[3] For a comprehensive set of reprints see [7].

The constant c in (5) and (6) is known as the central charge, and can be considered the defining parameter of the CFT. Relations similar to (4)–(6) hold for the antianalytic component $\overline{T}$.

A field $\Phi(z,\bar{z})$ that under the conformal transformation $z \to f(z)$, $\bar{z} \to \bar{f}(\bar{z})$ changes according to the rule

$$\Phi(z,\bar{z}) \to \left(\frac{df}{dz}\right)^{\Delta} \left(\frac{d\bar{f}}{d\bar{z}}\right)^{\overline{\Delta}} \Phi\left(f(z), \bar{f}(\bar{z})\right) \tag{7}$$

is called a primary field of conformal weight $(\Delta, \overline{\Delta})$; the spin of the field and its scaling dimension are therefore $(\Delta - \overline{\Delta})$ and $(\Delta + \overline{\Delta})$.

The OPE of a primary field Φ of weight $(\Delta, \overline{\Delta})$ with the energy-momentum tensor is

$$T(z)\Phi(w) = \frac{\Delta}{(z-w)^2}\Phi(w) + \frac{\partial_w \Phi}{z-w}. \tag{8}$$

As already mentioned, it is possible to classify the operator content of a conformal field theory according to the irreducible representations of the Virasoro algebra, with the primary fields playing the role of the highest weight vectors. An important class of conformal fields theories is formed by the so-called minimal models, for which the operator algebra (the set of all OPEs) closes on a finite number of primary fields. The central charge of a given unitary minimal model is given by

$$c = 1 - \frac{6}{p(p+1)}, \quad p = 1, 2, \ldots . \tag{9}$$

INTEGRALS OF MOTION IN CFT

One can easily prove the existence of an infinite number of conserved currents in a given CFT. The action of the Virasoro modes L_n on the space of local fields can be naturally defined:

$$(L_n \Phi)(z, \bar{z}) = \oint_z d\zeta (\zeta - z)^{n+1} T(\zeta) \Phi(z, \bar{z}) \tag{10}$$

If we act with L_n on the identity operator $\mathbf{I}$ we generate an infinite tower of holomorphic fields: using (10) one finds

$$(L_{-2}\mathbf{I}) = T(z), \tag{11}$$

$$\vdots \tag{12}$$

$$(L_{-n}\mathbf{I}) = \frac{1}{(n-2)!}\partial_z^{n-2} T(z), \tag{13}$$

$$(L_{-2}L_{-2}\mathbf{I}) =: T^2(z): . \tag{14}$$

We see that all the currents so defined are holomorphic, and therefore satisfy a conservation equation of the type

$$\partial_{\bar{z}}\left(L_{-m}\ldots L_{-n}\mathbf{I}\right)(z) = 0. \tag{15}$$

We conclude that in a CFT there exists an infinite number of conserved currents. In the rest of the talk we will generically label them by T_s, where s refers to the spin of the current.

We are now ready to address the central question: If a perturbation that spoils the conformal invariance is added to the theory, will these conservation laws remain valid? The question is of crucial importance; as discussed in the talk by B. Gerganov at this conference, the existence of an infinite number of conservation laws is enough to establish the integrability of the model.

PERTURBATIONS OF A CFT

Let us consider a theory defined by the action

$$S_{\text{pert}} = S_{\text{CFT}} + \lambda_r \int \Phi_r(x, y) d^2x, \tag{16}$$

where S_{CFT} is the (formal) action of a conformal field theory, and Φ_r is a relevant field of conformal dimension (Δ_r, Δ_r); the dimension of λ_r therefore is $(1 - \Delta_r, 1 - \Delta_r)$.

Let us consider one of the conserved currents T_s of spin s of the given CFT. In the perturbed theory T_s will no longer be holomorphic; rather it will satisfy an equation of the type

$$\partial_{\bar{z}} T_s(z, \bar{z}) = \lambda_r R^{(1)}_{s-1} + \ldots + \lambda_r^n R^{(n)}_{s-1} + \ldots. \tag{17}$$

For dimensional consistency the dimension of each of the operators $R^{(n)}_{s-1}$ must be $(s - n\epsilon, 1 - n\epsilon)$, where $\epsilon \equiv 1 - \Delta_r$. Assuming that the dimension of the fields in the perturbed theory (16) remains identical to the dimension of the corresponding fields in the CFT[4], we can immediately conclude that, since there are no fields of negative dimension, the series (17) is actually finite. Moreover, fore some specific choice of the perturbing operator Φ_r, only the first term can arise:

$$\partial_{\bar{z}} T_s = \lambda_r R^{(1)}_{s-1}. \tag{18}$$

The operator $R^{(1)}_{s-1}$ appearing in (18) can be calculated from the OPE in the conformal theory between T_s and the perturbation Φ_r; indeed one has [5]

$$\partial_{\bar{z}} T_s = \lambda_r \oint_z \frac{d\zeta}{2\pi i} \Phi_r(\zeta, \bar{z}) T_s(z). \tag{19}$$

One can prove, for the sake of illustration, the following facts:

---

[4] For a discussion of this subtle point, see the original paper [5].

1. The energy-momentum tensor is conserved for any choice of the perturbation. One has

$$\partial_{\bar{z}} T = \partial_z \Theta, \tag{20}$$

where $\Theta = \lambda_r (\Delta_r - 1) \Phi_r$.

2. For the current T_4, one in general has

$$\partial_{\bar{z}} T_4 = \lambda_r (\Delta_r - 1) \left(2 L_{-2} L_{-1} + \frac{\Delta_r - 3}{6} L_{-1}^3 \right) \Phi_r. \tag{21}$$

Therefore, as we see from (21), in general the conservation law is spoiled, since the r.h.s. cannot be written as a total ∂_z derivative.

For some choice of the perturbing operator though, it is possible to write the r.h.s. of (18) as a ∂_z derivative:

$$\partial_{\bar{z}} T_s = \partial_z \Theta_{s-2}, \tag{22}$$

which implies that the conservation law, albeit modified, survives the addition of a perturbation, so that the resulting massive field theory remains integrable.

Prototypical cases are the minimal models S_p perturbed by any one of the operators[5] $\Phi_{1,3}$, $\Phi_{1,2}$, $\Phi_{2,1}$:

$$S_p^{(1,3)} = S_p + \lambda \int \Phi_{1,3} d^2 x, \tag{23}$$

$$S_p^{(1,2)} = S_p + \lambda \int \Phi_{1,2} d^2 x, \tag{24}$$

$$S_p^{(2,1)} = S_p + \lambda \int \Phi_{2,1} d^2 x, \tag{25}$$

for which an infinite number of conserved currents exist, with spins given by

$$s = 2n - 1, \quad n = 1, 2, \ldots \quad \text{for} \quad S_p^{(1,3)}, \tag{26}$$

and

$$s = 1, 6n \pm 1, \quad n = 1, 2, \ldots \quad \text{for} \quad S_p^{(1,2)}, S_p^{(2,1)}. \tag{27}$$

---

[5] The indices refer to the standard labeling of the Kac table (see [6]).

FURTHER APPLICATIONS

The importance of the formalism of perturbed CFT lies on the fact that the simple formula (19) can be applied to test for the presence of conserved currents in more general cases. Recently it has allowed for the construction of conservation laws (both local and non-local) in multi-field generalizations of the sine-Gordon model [4], [8]:

$$\mathcal{L} = \sum_{i=1}^{2} \frac{1}{2} (\partial_\mu \phi_i)^2 + \lambda \cos(\beta_1 \phi_1) \cos(\beta_2 \phi_2), \qquad (28)$$

and

$$\mathcal{L} = \sum_{i=1}^{3} \frac{1}{2} (\partial_\mu \phi_i)^2 + \lambda \left[\cos(\alpha_1 \phi_1 + \alpha_2 \phi_2) e^{\beta \phi_3} + \cos(\alpha_1 \phi_1 - \alpha_2 \phi_2) e^{-\beta \phi_3} \right], \qquad (29)$$

provided that the couplings are constrained as follows:

$$\beta_1^2 + \beta_2^2 = 4\pi, \quad \text{or} \quad 8\pi, \qquad (30)$$
$$\alpha_1^2 + \alpha_2^2 - \beta^2 = 4\pi, \qquad (31)$$

for the models (28), (29) respectively.

An extensive investigation of further generalizations of this class of models is the goal of our current investigations [9].

REFERENCES

1. Seiberg N., and Witten E., *Nucl. Phys.* **B426**, 19 (1994) and *Nucl. Phys.* **B431**, 484 (1994).
2. Harvey J.A., in *Fields, Strings and Duality—TASI 96*, Greene B., and Efthimiou C.J. eds., World Scientific, 157 (1997).
3. Mussardo G., *Phys. Rep.* **218**, 215 (1992).
4. Lesage F., Saleur H., and Simonetti P., *Tunneling in quantum wires I: Exact solution of the spin isotropic case*, cond-mat/9703220 and *Tunneling in quantum wires II: A new line of IR fixed points*, cond-mat/9707131.
5. Zamolodchikov A.B., in *Advanced Studies in Pure Mathematics*, vol. 19, Jimbo M., Miwa T., and Tsuchiya A. eds., 641 (1989).
6. Di Francesco P., Mathieu P., and Sénéchal D., *Conformal Field Theory*, New York, Springer Verlag, 1997.
7. *Conformal Invariance and Applications to Statistical Mechanics*, Itzykson C., Saleur H., and Zuber J.B. eds., World Scientific, 1988.
8. Fateev V.A., *Nucl. Phys.* **B473**, 509 (1996).
9. Ameduri M., Efthimiou C.J., and Gerganov B., work in progress.

Investigation by Monte Carlo Renormalization of 2-D Simplicial Quantum Gravity Coupled to Gaussian Matter

Eric B. Gregory*, Simon M. Catterall* and Gudmar Thorleifsson[†]

*Physics Department, Syracuse University, Syracuse, NY 13244, USA
[†]Facultät für Physik, Universität Bielefeld, D-33615, Bielefeld, Germany

Abstract. We extend a recently proposed real space renormalization group scheme for dynamical triangulations to situations where the lattice is coupled to continuous scalar fields. Using Monte Carlo simulation in combination with a linear, stochastic blocking scheme for the scalar fields we are able to determine the leading eigenvalues of the stability matrix with good accuracy both for $c_M = 1$ and $c_M = 10$ theories. We show how this method provides an estimate of the anomalous dimension for $c_M = 10$ matter coupled to two dimensional quantum gravity.

INTRODUCTION

Dynamical triangulations are a powerful tool for the study of systems involving manifolds with irregular and fluctuating geometry. One of the most notable successes of this method is the field of two dimensional quantum gravity. The central idea is to take a smooth manifold with irregular geometry and replace it with a *simplicial* manifold, that is, one built out of small bits of flat space called *simplices*. In our two dimensional studies, the simplicial manifold is a collection of equilateral triangles glued together. This regularization scheme has provided analytical understanding, through matrix model techniques, of 2-D quantum gravity and bosonic string theories, which are modeled as a dynamical triangulation coupled to scalar fields. Additionally, this approach makes these problems well suited for investigation with the numerical simulation techniques often employed in statistical mechanics. This is important because, to date, meaningful analytic results are not available for systems with matter central charge c_M greater than one. Renormalization group techniques have proved very useful in studying the critical behavior of statistical mechanics systems. Specifically the Monte Carlo Renormalization Group (MCRG) is well suited to many problems for which we lack analytical understanding. Recently, Thorleifsson and Catterall [2] developed a successful MCRG procedure for dynamical triangulations. Our goal has been to extend this method to the case of dynamical triangulations coupled to scalar fields, allowing us to investigate the sectors inaccessible to analytic solutions.

MODEL AND NUMERICAL APPROACH

Upon discretising the 2-D quantum gravity path integral, we obtain the partition function,

$$Z = \int \mathcal{D}[g_{\mu\nu}] \mathcal{D}[\text{matter}] e^{-S} \longrightarrow Z = \sum_\tau \sum_{\text{matter}} e^{-S}. \quad (1)$$

The integrations over all metrics becomes a sum over all possible triangulations, τ. The action, S is a sum of the Euclidean Einstein-Hilbert action, S_{E-H}, and a matter-gravity piece, S_{m-g}. In this work we confine ourselves to studying two dimensional systems with fixed volume N and fixed topology χ. In this case S_{E-H} is a constant. In particular we concern ourselves with the case of

$$S_{m-g} = \int d^D x \sqrt{g} (\partial \phi)^2 \longrightarrow S_{m-g} = \sum_{\mu=1}^{D} \sum_{<ij>} \left(\phi_i^\mu - \phi_j^\mu \right)^2. \quad (2)$$

The second sum is over all nearest neighbor lattice points. The *matter central charge*, $c_M = D$, counts the number of matter fields per node. Meaningful analytic solutions are available only for $0 \leq c_M \leq 1$ and $25 \leq c_M$.

Once we have generated the lattice, we update it with standard link flip moves. We remove the common link of two adjacent triangles and replace it with a link connecting the two nodes that previously did not have a link connecting them. We accept or reject such a move on the basis of the outcome of a Metropolis test.

Updates of the Gaussian fields proceed in two ways. The first is a Metropolis update. We propose a small change in the value of a field at a node and then use a Metropolis test to determine whether the change is accepted.

MONTE CARLO RENORMALIZATION

A powerful tool for investigating the critical exponents of statistical systems is the renormalization group. The usual approach is to use some "course graining" procedure whereby the system is blocked or replaced by one with fewer degrees of freedom. Doing so effectively integrates out the short distance fluctuations, while preserving the long-distance physics of the system.

In general, upon application of the renormalization group the action $S(\phi)$, expanded on a suitable basis of operators, changes

$$S = \sum_\alpha K_\alpha \mathcal{O}_\alpha \xrightarrow{P} S' = \sum_\alpha K'_\alpha \mathcal{O}'_\alpha. \quad (3)$$

Here the K's are coupling constants and the $\mathcal{O}$'s are operators defined on the bare lattice while their primed counterparts denote the corresponding quantities on the blocked lattice. Repeated iteration of the renormalization group operation on a critical system causes the coupling constants to flow to fixed points:

$$\{K\}^{(0)} \xrightarrow{P} \{K\}^{(1)} \xrightarrow{P} \ldots \xrightarrow{P} \{K\}^* \tag{4}$$

Near a fixed point one can write can write a linear approximation for the change in the coupling constants under further blocking. Explicitly, the couplings, after $k+1$ iterations of the renormalization group, are given by

$$K_\alpha^{(k+1)} - K_\alpha^* \simeq \sum_\beta \frac{\partial K_\alpha^{(k+1)}}{\partial K_\beta^{(k)}}|_{\mathbf{K}=\mathbf{K}^*} \left(K_\beta^{(k)} - K_\beta^* \right) \equiv \sum_\beta T_{\alpha\beta} \delta K_\beta^{(k)}. \tag{5}$$

The object $T_{\alpha\beta}$ is referred to as the stability matrix. Solving

$$\sum_\beta T_{\alpha\beta} u_\beta^i = \lambda_i u_\alpha^i \tag{6}$$

for the eigenvalues, λ, allows one to determine the critical exponents, y, which are given by the expression

$$\lambda = b^y. \tag{7}$$

Here $b = N^{(k)}/N^{(k+1)}$ is the volume blocking factor, the ratio of the number of degrees of freedom in the bare system to the number of degrees of freedom in the system after being renormalized once.

Some systems lend themselves well to analytic renormalization. In these cases explicit expressions for the stability matrix $T_{\alpha\beta}$ can be found. Such is not the case for dynamical triangulations. One cannot write down an exact map from a given triangulation to a coarse one carrying the same long distance geometrical information. Hence we use Monte Carlo Renormalization. It is possible individually renormalize an ensemble of triangulations, measure operator expectation values at the different blocking levels on each, and then determine the elements of the stability matrix using correlators between the different operators.

In dealing with fields on a dynamical triangulation one must find a renormalization group prescription that integrates out the small scale features of both the geometry and of the field configuration, while preserving the large scale physics. In this investigation we will first block the geometry independently from the fields, then assign new blocked scalar fields to the renormalized lattice. We use the node decimation method developed by Thorleifsson and Catterall [2] to block the geometry and take their method of blocking Ising spins on a random triangular lattice and extend it to the case of Gaussian scalar fields on the lattice.

BLOCKING THE GEOMETRY

The general idea for blocking the lattice geometry utilizes a scheme called node decimation. This proceeds by randomly picking nodes and removing them from the triangulation. Removing a node with coordination number q will leave a q-sided hole that must be randomly triangulated. In practice this is done by first picking a node to be removed, then flipping links around the selected node until it has a coordination number of three. Now when the node is removed the three-sided hole in the triangulation can be replaced by a triangle. The link flips in the intermediate step are performed without any Metropolis test and are hence independent of the Gaussian fields. As long as a proposed link flip is geometrically possible, that is, flipping the link will result in a good triangulation, the move is performed. Very occasionally there is a node for which it's not possible to reduce its coordination number to three. In these cases, after a large number of flip attempts, the node is abandoned and we choose a new node to remove.

BLOCKING SCALAR FIELDS

We have investigated using a *linear* RG transformation to assign the block lattice fields. For the surviving nodes new scalar fields, ϕ' are constructed, which are a function of the original fields ϕ from the unblocked lattice.

$$\phi'_i = \xi \left[\alpha \phi_i + (1-\alpha) \frac{1}{q_i} \sum_{j=1}^{q_i} \phi_j \right] + \frac{1}{\sqrt{a_w}} \chi. \qquad (8)$$

Notice that blocking is *stochastic* – χ is a Gaussian random variable with $\langle \chi^2 \rangle = 1$. We will argue that the range of the fixed point action is dependent on the auxiliary parameter a_w.

The sum is over all the bare lattice nodes j that were neighbors of the blocked node i. We have introduced a relative weight, α between the contribution of the bare lattice field at node i and the average of the fields of the bare lattice neighbors of node i. In principle, one must choose α carefully so that it is appropriate for the blocking factor $b = N'/N$ being used. Clearly as $b \to 1$, α should approach one, and as b grows α should decrease. However, we have found that that in practice this scheme is fairly robust under reasonable choices of α.

In general we must also rescale the blocked fields by an overall factor ξ – the *field renormalization constant*. Choosing the appropriate value of ξ turns out to be one of the most challenging aspects of this method. We follow the approach of Bell and Wilson in their study of real space renormalization of scalar fields on a regular lattice. One must first understand the roles of a_w and ξ. In general, with any choice of a_w, renormalization of an infinite lattice will lead to a local fixed point. These fixed points have the same continuum limit and the same critical exponents, but differ in the range of the interaction of their fixed point actions. In practice, of course, we must renormalize a finite system, so we must pick a_w to bring us as close to a fixed point as possible, in as few blockings as possible. This is evidenced by a plateau in the eigenvalues, expectation values or in actual coupling constant values if they are measurable. Additionally we want the fixed point action to be as local as possible so that all the significant interactions are still expressed, even when the system has been blocked down to a small lattice. These requirements may conflict so a compromise may be necessary in choosing a_w. In contrast, with an infinite lattice, only a single choice of ξ will generate a flow from a local action to a non-trivial fixed point. We denote this special value as ξ^*. There may be flows to other fixed points, but none start at the local Hamiltonian. On a finite lattice, however other (non-local) fixed points may be reached with $\xi \neq \xi^*$, even when starting from the local, critical action. Thus, to accurately investigate the critical behavior of the local fixed point with a finite lattice, one must accurately tune ξ to ξ^*. This is possible because, near the local fixed point there is some remnant of the rescaling freedom that existed on the infinite lattice. So, only very near this local fixed point should we see a redundant marginal eigenvalue corresponding to perturbations along the line of equivalent local fixed points that were accessible through rescalings of the local action on an infinite lattice.

Thus, having chosen an $a_w \sim a_{w_{\text{local}}}$ we can determine the correct renormalization constant ξ^* by requiring that there be a (approximately) marginal operator in the spectrum of the stability matrix. We have found this approach yields accurate and stable estimates for ξ^* at $c_M = 1$ both for scalar fields in flat space and coupled to gravity (where KPZ predicts $\xi^* = 1$).

Now the procedure is as follows. We first generate a collection of statistically independent lattice-field configurations using the Monte Carlo methods described above. We renormalize each, scanning a range of a_w, and storing the value of operators measured at different blocking levels. Once we identify the most optimal value of a_w, we repeat the process, this time holding a_w constant and scanning in ξ until we find a marginal subleading eigenvalue.

We have utilized the following basis of operators in our MCRG analysis.

$$\sum_i \phi_i \Box^m \phi_i \qquad m = 0, 1, 2.. \ . \tag{9}$$

We have defined $\Box \phi_i$ as

$$\Box \phi_i = \sum_{\langle ij \rangle} \phi_j - q_i \phi_i. \tag{10}$$

By recursively applying this operator as defined above, we see that $\phi_i \Box^m \phi_i$ involves nodes m links away from node i. Hypothetically, a lattice of volume N could require a basis with operators up to order $m \sim N^{1/D_h}/2$. Using the stochastic blocking allows the interaction to fall off fast enough that good results can be obtained with a basis containing operators up to $m = 4$ for a typical run.

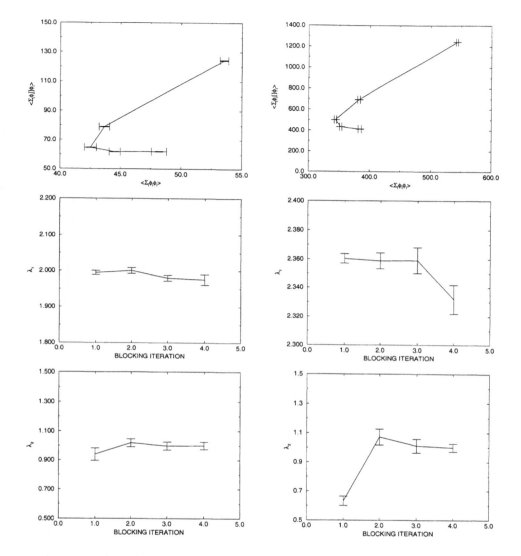

FIGURE 1. (upper) $\langle \sum_i \phi_i \Box \phi_i \rangle$ plotted against $\langle \sum_i \phi_i \phi_i \rangle$ for different blocking iterations, but the same target volume, $N = 250$. Top lattice data is in the upper right. Leading (middle) and sub-leading (lower) eigenvalue of the stability matrix for $N_0 = 4000$ lattice. All plots are for $c_M = 1$ blocked using $a_w = 100$ and $\xi = 1.00$.

FIGURE 2. (upper) $\langle \sum_i \phi_i \Box \phi_i \rangle$ plotted against $\langle \sum_i \phi_i \phi_i \rangle$ for different blocking iterations, but the same target volume, $N = 250$. Leading (middle) and sub-leading (lower) eigenvalues of the stability matrix for $N_0 = 4000$ lattice. All plots are for $c_M = 10$ blocked using $a_w = 100$ and $\xi = 0.92$.

RESULTS

The $c_M = 1$ simulation served as somewhat of a test case, as analytical results are available. We found that the stability of the leading and sub-leading eigenvalues was remarkably insensitive to the choice of a_w, as long as $a_w > 50$. We judged $a_w = 100$ to be optimal for $c_M = 1$.

Figure 1 (upper graph) shows how the expectation values of two operators flow under blocking of a $c_M = 1$ system. Here finite size effects are removed by using measurements from lattices of the same size. By starting with bare lattices of different sizes we produce blocked lattices of the same target volume through a different number of RG iterations. This plot shows evidence of fixed point behavior in the flows of operator expectation values, and suggests that the closest approach to the fixed point is between the second and third blocked lattices. For this reason, we will use the eigenvalues from the third blocking as the best estimate of the fixed point eigenvalues.

One can see in Fig. 1 (lower graph) that the field renormalization constant, $\xi = 1.00(3)$ gives a stable sub-leading eigenvalue consistent with unity. This combination of $\xi = 1$ and $a_w = 100$ yields a leading eigenvalue $\lambda_1 = 1.980(90)$ (Fig. 1, middle) The error on ξ is determined by measuring how far ξ can be changed in either direction before the mean of the sub-leading eigenvalue, λ_2 is one standard deviation from unity. We quote the dominant error for λ_2 which is due to our uncertainty in determining ξ. The statistical error for λ_2 is 0.008 and the uncertainty in choosing the optimal a_w contributes an error of 0.01.

We studied $c_M = 10$ dynamical triangulations using the same procedure. We use the value $a_w = 100$ as the most optimal, though again, values of $a_w \geq 100$ all seem to produce eigenvalues that are stable under blocking. For $c_M = 10$ many of the choices of a_w give a complex sub-leading eigenvalue in the first iteration of the blocking procedure. We took this to mean that the system starts out farther from the fixed point. This is not terribly troubling, as one can choose a_w such that in the successive blockings all of the eigenvalues are real.

Using the same analysis as we did for $c_M = 1$, we find that $\xi = 0.92(4)$ gives a sub-leading eigenvalue that is nearly one for $c_M = 10$. Although for $c_M = 10$ statistical errors made picking out ξ^* a delicate procedure, $\xi = 0.92$ seemed to produce the most stable eigenvalues as well as being nearest to one. This gives $\lambda_1 = 2.359(200)$. Here we just quote the dominant error due to the uncertainty in determining ξ^*. Figure 2 summarizes the $c_M = 10$ renormalization results.

We completed this analysis using a basis of the first four operators of the type described in (9), that is $m = 0, 1, 2, 3$. We find that using larger bases is troublesome, causing the appearance of complex eigenvalues. This may be due to lingering problems with the locality of the action.

DISCUSSION

Although there exists no analytic RG calculation for dynamical lattices, using the results of matrix method calculations one can nevertheless make strong predictions of the appropriate value of ξ, the field renormalization constant, for the case of $c_M = 1$. Additionally, using simple dimensional arguments, one can develop expectations about the value of some of the eigenvalues of the stability matrix.

One can argue that on a lattice with fractal dimension d_H, a volume rescaling of $b = s^{d_H}$ requires the lattice fields be rescaled like

$$\phi_{\text{lat}} \to \phi_{\text{lat}} b^{-\frac{\beta}{d_H}}, \tag{11}$$

implying,

$$\xi = b^{-\frac{\beta}{d_H}}, \tag{12}$$

where β is the the length dimension of the fields. This is a generalization of the field renormalization constant given in [7]. If we assume the action to be dimensionless, then we can count dimensions to show that the undressed length dimension for a Gaussian field in flat two-dimensional space is zero:

$$S = \int d^2x (\partial \phi)^2 \implies \beta_0 = 0 \tag{13}$$

In the presence of quantum gravity however, the field will in general develop an anomalous length dimension. For $c_M \leq 1$ we can use the well known KPZ [1] expressions,

$$\beta - \frac{\beta(1-\beta)}{1-\gamma_s} = \beta_0 \quad \text{and} \quad \gamma_s = \frac{1}{12}\left(c_M - 1 - \sqrt{(25-c_M)(1-c_M)}\right). \tag{14}$$

to find, β, the dressed field dimension So for $c_M = 1$,

$$\beta = \beta_0 = 0, \tag{15}$$

implying $\xi = 1$. For $c_M = 1$ our MCRG result of $\xi^* = 1.00(3)$ can be taken as an independent numerical determination of β consistent with (15). For $c_M = 10$, our result of $\xi^* = 0.92(4)$ would suggest that $\frac{\beta}{d_H} = 0.12(6)$.

Although meaningful analytical results are not available for $c_M > 1$, we can check that the critical exponents extracted from the eigenvalues of the stability matrix are consistent with those obtained through finite size scaling analysis. Simple arguments can be made to show that if an operator O at criticality scales like

$$\langle O \rangle \sim N^\mu \tag{16}$$

on lattices with volume N, then there should be a leading eigenvalue

$$\lambda_1 = b^\mu. \tag{17}$$

Consider the leading operator O_1 again for $c_M = 10$. Its finite size scaling is shown in Fig. 3 and yields a finite size scaling exponent of $\mu = 1.25(1)$. This is consistent with the leading eigenvalue $\lambda_1 = 2.36(20)$ extracted from the stability matrix (Fig. 2).

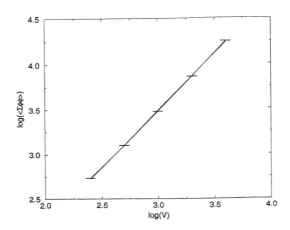

FIGURE 3. A log-log plot of the scaling of $\langle \sum_i \phi_i \phi_i \rangle$ vs. volume for the bare triangulation with $c_M = 10$. The data is fit to $\langle \sum_i \phi_i \phi_i \rangle \sim V^{1.25(1)}$.

CONCLUSIONS

We have successfully extended the renormalization group scheme for dynamical triangulations based on node decimation to the case of continuous scalar fields. Results for the wave function renormalization constant ξ and hence $\frac{\beta}{d_H}$ together with the leading eigenvalue of the stability matrix are obtained for $c_M = 1$ and $c_M = 10$ which are consistent with theoretical expectations.

We gratefully acknowledge the support of the United States Department of Energy in the form of grant DE-FG-02-85ER40231.

REFERENCES

1. F. David, Mod. Phys. Lett. **B186**, 379 (1987); J. Distler and H. Kawai, Nucl. Phys. **B321**, 509 (1989); V.G. Knizhnik, A.M. Polyakov and A. Zamolodchikov, Mod. Phys. Lett. **A3**, 819 (1988).
2. G. Thorleifsson, S. Catterall, Nucl. Phys. **B461**, 350 (1996).
3. C.B. Lang, Nucl. Phys. **B265**, 630 (1986).
4. T.L. Bell and K.G. Wilson, Phys. Rev. **B11**, 3431 (1975).
5. F. David, *Simplicial Quantum Gravity and Random Lattices*, (hep-th/9303127), Lectures given at Les Houches Summer School on Gravitation and Quantization, Session LVII, Les Houches, France, 1992; J. Ambjørn, *Quantization of Geometry*, (hep-th/9411179), Lectures given at Les Houches Summer School on Fluctuating Geometries and Statistical Mechanics, Session LXII, Les Houches, France 1994; P. Di Francesco, P. Ginzparg and J. Zinn-Justin, Phys. Rep. **254**, 1 (1995).
6. A. Bennett, Nucl. Phys. **B300**, 253 (1988).

7. T.L. Bell and K.G. Wilson, Phys. Rev. **B10**, 3935 (1974).
8. S. Catterall, G. Thorleifsson, M. Bowick, and V. John, Phys. Lett. **B354**, 58 (1995).
9. J. Ambjørn, J. Jurkiewicz and Y. Watabiki, Nucl. Phys. **B454**, 313 (1995).
10. S. Jain and S.D. Mathur, Phys. Lett. **B286**, 239 (1992).
11. J. Ambjørn, S. Jain, and G. Thorleifsson, Phys. Lett. **B307**, 34 (1993); J. Ambjørn, and G. Thorleifsson, Phys. Lett. **B323**, 7 (1994).

Phase structure of 3D dynamical triangulations with a boundary

Simeon Warner*, Simon Catterall*, Ray Renken[†]

* Department of Physics, Syracuse University, Syracuse, NY 13210, USA
[†] Department of Physics, Unviersity of Central Florida, Orlando, FL 32816, USA

Abstract. We present the results of Monte Carlo simulation of 3D dynamical triangulations with a boundary. Three phases are indentified and characterized. In addition to two phases similar to those of compact 3D dynamical triangulations, one of these phases is a new, boundary dominated phase; a simple argument is presented to explain its existence. First order transitions are shown to occur along the critical lines separating phases.

INTRODUCTION

Dynamical triangulation models arise from simplicial discretizations of continuous Riemannnian manifolds. A manifold is approximated by glueing together a set of equilateral simplices with fixed edgelengths. This glueing ensures that each face is shared by exactly two distinct simplices – the resultant simplicial lattice is called a triangulation. In the context of Euclidean quantum gravity it is natural to consider a weighted sum of all possible triangulations as a candidate for a regularized path integral over metrics. Physically distinct metrics correspond to inequivalent simplicial triangulations. This prescription has been shown to be very successful in two-dimensions (see, for example, [1]).

Most analytic studies and almost all numerical work done so far has been restricted to compact manifolds like the sphere. In this paper we develop techniques that allow us to extend numerical studies to simplicial manifolds with boundaries. Specifically, we study the 3-disk created by inserting an S^2 boundary into a triangulation of the sphere S^3. This allows us to compute an object which is the simplicial equivalent of the 'wavefunction of the Universe' [2]:

$$\psi[h] = \int Dg e^{-S(g)} \tag{1}$$

The functional integral over 3-metrics g is restricted to those with 2-metric h on the boundary. The simplicial analog is simply

$$\psi(T_2) = \sum_{T_3} e^{-S_L(T_3)} \tag{2}$$

Thus the probability amplitude for finding a particular 2-triangulation T_2 is obtained by counting (with some weight) all 3-triangulations T_3 which contain T_2 as their boundary. A natural lattice action S_L can be derived from the continuum action by straightforward techniques [3]. It contains both the usual Regge curvature piece familiar from compact triangulations together with a boundary cosmological constant term. In three-dimensions the curvature is localized on links. If L_M denotes the set of links in the bulk of the 3-triangulation (excluding the boundary) and $L_{\partial M}$ those in the boundary the action can be written

$$S_L = \kappa_1 \left(\sum_{h \in L_M} (2\pi - \alpha n_h) + \sum_{h \in L_{\partial M}} (\pi - \alpha n_h) \right) \tag{3}$$

The quantity $\alpha = \arccos(1/3)$ and n_h is the number of simplices sharing the link (hinge) h. Typically S_L will also contain a bulk cosmological constant that can be used to tune the simulation volume. The resultant action can be rewritten in the form

$$S_b = -\kappa_0 N_0 + \kappa_3 N_3 + \kappa_b N_2^b \tag{4}$$

where N_2^b is the area of the boundary. Here, κ_3 is used to tune the volume of the system. We are thus left with a two-dimensional phase space parameterized by κ_0 and κ_b conjugate to the number of nodes and the number of boundary triangles. It is trivial to generalize this to, for example, four dimensions. The partition function for the system is then

$$Z = \sum_T e^{-S_b} \tag{5}$$

where the sum is over triangulations, T.

Various other extended phase diagrams have been studied for 3-dimensional dynamical triangulations including adding spin matter [4,5], adding gauge matter [6,7], and adding a measure term [8]. Much of this work was motivated by the desire to find a continuous phase transition. No such transitions have been found.

SIMULATION

Our simulation algorithm is an extension of the algorithm for compact manifolds in arbitrary dimension described by Catterall [9]. Consider the environment of any vertex in a 3-triangulation — it is composed of simplices making up a trivial 3-ball. The boundary of this 3-ball is just the sphere, S^2. A boundary with the topology of S^2 can thus be created in the original triangulation by removing these simplices.

If the original triangulation corresponded to the sphere S^2 the topology of the new triangulation is that of a 2-disk.

In practice we simulate a compact manifold with one 'special' vertex. This vertex and all the simplices sharing it are ignored during any measurement. In this way every triangulation of our marked sphere S^2 is in one-to-one correspondence with a triangulation of the 2-disk. Notice that the usual compact manifold moves applied to all simplices (including those sharing the special vertex) will in general change the boundary of the D-disk. Indeed these moves are ergodic with respect to the boundary. Furthermore, the proof that these moves satisfy a detailed balance relation goes through just as for the compact case. The one extra restriction is simple — one must never delete the special vertex. With this trick we can trivially extend our compact codes to the situation in which a $S^{(D-1)}$ boundary has been added - we are merely simulating a compact lattice with an action which singles out a special vertex and its neighbour simplices. This contrasts with the set of additional boundary moves used by Adi et al [10] for simulations in 2-dimensions.

Measurement routines do not include the special vertex or any simplices connected to it. For example, the number of 3-simplices in the system with boundary is the number of 3-simplices in the whole simulation minus the number of 3-simplices sharing the special vertex. The size of the boundary is simply the number of 3-simplices sharing the special vertex.

We have used the Metropolis Monte Carlo [11] scheme with usual update rule:

$$p(\text{accept move}) = min\{e^{-\Delta S_b}, 1\} \tag{6}$$

and in this way we explore the space of unlabeled triangulations with the action S_b (equation 4).

In 3-dimensions there are just 4 types of move: vertex insertion, vertex deletion and exchange of a link with a face (two moves: link to face or face to link). Where these moves take place on sections of the triangulation involving the special vertex we take care to count changes in the numbers of simplices inside and outside of the boundary but otherwise the moves are the same as for the bulk.

PHASE DIAGRAM

Most of the simulation for this work has been performed with nominal volume, $N_3 = 2000$. In all runs the κ_3 was used to tune the nominal system volume for each given κ_0 and κ_b. Series of runs varying either κ_0 or κ_b were made and the vertex susceptibility used to search for phase transitions. We define the vertex susceptibility, χ, to be normalized with respect to the number of 3-simplices:

$$\chi = \frac{1}{N_3}(\langle N_0^2 \rangle - \langle N_0 \rangle^2) \tag{7}$$

The points shown in figure 1 are taken from the positions of peaks in the vertex susceptibility. There are three phases which we characterize as: phase 1 - crumpled,

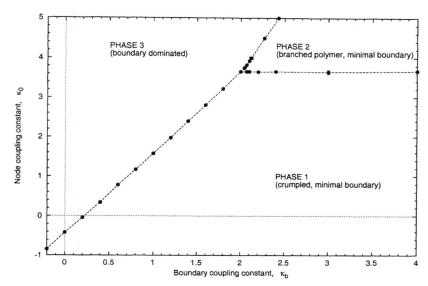

FIGURE 1. Phase diagram for 3-dimensional dynamical triangulation with a boundary. All points have error bars in either κ_b or κ_0, most cannot be seen because they are smaller than the symbols. Nominal simulation volume, $N_3 = 2000$.

minimal boundary; phase 2 - branched polymer, minimal boundary; and phase 3 - boundary dominated.

In phases 1 and 2 the boundary is simply 4 triangles (2-simplices) connected to form a tetrahedral hole. The system is essentially like a compact manifold with one marked 3-simplex — the tetrahedral hole. In phase 3 the boundary is large — typically a substantial fraction of the bulk volume.

Simple argument for boundary dominated phase

Here we argue that the boundary dominated phase can be explained by considering an effective action written in terms of the boundary size. We show that in certain circumstances a large boundary will decrease this action. Otherwise one of the minimal boundary phases will be favored.

Consider the action:

$$S_b = -k_0 N_0 + k_b N_2^b \tag{8}$$

We ignore the volume term as this is kept fixed during the simulation. If we note that the boundary is itself a 2-sphere then we know that:

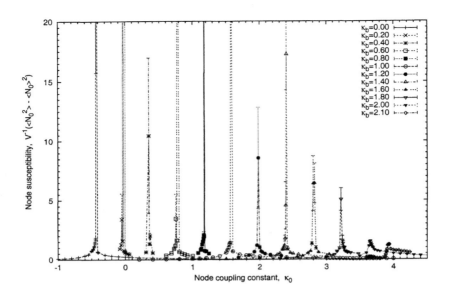

FIGURE 2. Sample of node susceptibility data for different values of the boundary coupling constant, κ_b.

$$N_2^b = 2(N_0^b - 2) \tag{9}$$

and

$$N_0 = N_0^b + N_0^i \tag{10}$$

where N_0 is the number of vertices, N_0^b is the number of vertices on the boundary, N_0^i is the number of internal vertices, and N_2^b is the number of 2-simplices (triangles) on the boundary, So we may rewrite the action:

$$S_b = -\kappa_0 N_0^i + (2\kappa_b - \kappa_0)N_0^b \tag{11}$$

If we now consider N_0^i fixed and note that the number of manifolds with boundary size N_2^b is governed by an exponential factor $e^{\kappa_b^c N_2^b} = e^{2\kappa_b^c N_0^b}$, where κ_b^c is a new constant, we may then write an effective action for the number of boundary vertices:

$$S_{eff} \approx (-\kappa_0 + 2(\kappa_b - \kappa_b^c))N_0^b \tag{12}$$

The presence of small or large boundaries is then determined by the sign of this action. We thus expect the phase transition at $\kappa_0 = 2(\kappa_b - \kappa_b^c)$ which is in good agreement with what we see.

a)

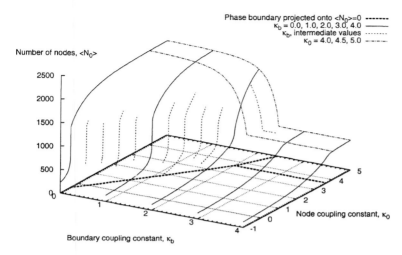

b)

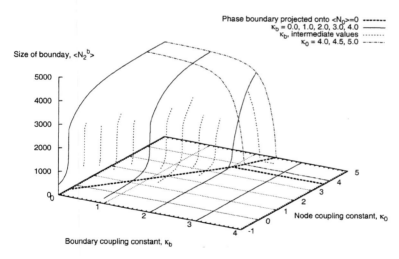

FIGURE 3. a) Number of nodes, $\langle N_0 \rangle$, as a function of κ_0 and κ_b. Note that we see three distinct areas with different values of $\langle N_0 \rangle$: the boundary dominated phase (small κ_b, large κ_0) with large $\langle N_0 \rangle$, the crumpled phase (small κ_0) with small $\langle N_0 \rangle$, and the branched polymer phase (large κ_b and κ_0) with intermediate $\langle N_0 \rangle$. b) Boundary size, $\langle N_2^b \rangle$, as a function of κ_0 and κ_b. Note that we see two distinct areas with different values of $\langle N_2^b \rangle$: the boundary dominated phase (small κ_b, large κ_0) with large boundary, and the crumpled and branched-polymer phases with small boundary. Nominal simulation volume, $N_3 = 2000$.

Order of transitions

Simulations of compact manifolds in 3 and 4-dimensions are known to have a first order phase transition between crumpled and branched polymer phases (3d [12], 4d [13,14]).

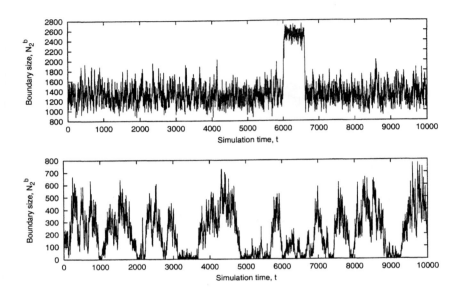

FIGURE 4. Time series showing the boundary size (N_2^b) during simulation. The upper plot is at the transition between the crumpled and boundary dominated phases ($k_0 = -0.423$, $k_b = 0$). The lower plot is at the transition between the branched polymer and boundary dominated phases ($k_0 = 5$, $k_b = 2.43$). Nominal simulation volume, $N_3 = 2000$, and time is in units of $100 N_3$ attempted updates.

Our Monte Carlo time series show strong bistability on all three phase boundaries (see figure 4). We take this to indicate that all three phase transitions are first order.

CONCLUDING REMARKS

We have demonstrated an effective algorithm for simulating 3D dynamical triangulations with a boundary.

We have identified three phases in 3-dimensional dynamical triangulations with a boundary and mapped the boundaries within the range of couplings $-1 < \kappa_0 < 5$ and $-0.5 < \kappa_b < 4$. The observed phases include the crumpled and branched-polymer phases seen in triangulations of compact manifolds, and also a new, boundary dominated phase. The existence of this phase, and the shape of the phase boundary on the κ_0-κ_b phase diagram, is predicted by a simple argument.

Strong bistability in the time series at the phase transitions indicates that all transitions within the range of couplings studied are first order. Further results will be reported in [15].

ACKNOWLEDGMENTS

Simon Catterall was supported in part by DOE grant DE-FG02-85ER40237. Ray Renken was supported in part by NSF grant PHY-9503371.

REFERENCES

1. J Ambjorn. Quantization of geometry. hep-th/9411179 (1994).
2. J B Hartle and S W Hawking. Phys. Rev. D **12** 2960–2975 (1983).
3. J B Hartle and R Sorkin. Gen. Rel. and Grav., **13** 541–549 (1981).
4. Ray L Renken, Simon M Catterall and John B Kogut. Nucl. Phys., **B 422** 677–689 (1994).
5. J Ambjorn, C Kristjansen, Z Burda and J Jurkiewicz. Nucl. Phys. **B** Proc. Suppl., 771– (1993).
6. Ray L Renken, Simon M Catterall and John B Kogut. Nucl. Phys. **B 389** 601–610 (1993).
7. J Ambjorn, J Jurkiewicz, S Bilke, Z Burda and B Petersson. Mod. Phys. Lett. **A 9** 2527 (1994).
8. Ray L Renken, Simon M Catterall and John B Kogut. hep-lat/9712011 (1997).
9. S Catterall. Comp. Phys. Comm. **87** 409–415 (1995).
10. E Adi, M Hasenbusch, M Marcu, E Pazy, K Pinn and S Solomon. hep-lat/9310016 (1993).
11. Nicholas Metropolis, Arianna W Rosenbluth, Marshall N Rosenbluth, Augusta H Teller and Edward Teller. J. Chem. Phys. **21** (1953).
12. J Ambjorn, D V Boulatov, A Krzywicki and S Varsted. Phys. Lett. **B 276** 432–436 (1992).
13. P Bialas, Z Burda, A Krzywicki and B Petersson. Nucl. Phys. **B 472** 293 (1996).
14. B de Bakker. Phys. Lett. **B 389** 238 (1996).
15. Simeon Warner, Simon Catterall and Ray Renken. hep-lat/9808006 (1998).

The Experimental Status of the Deficit in Charmed Semileptonic B Decays

René Janicek

*Department of Physics, McGill University,
Montréal, Québec, H3A 2T8 Canada*

for the CLEO Collaboration

Abstract. We present the status of the deficit in semileptonic decays of B mesons. All the results presented are based on measurements of the inclusive and exclusive semileptonic decays at the CLEO II detector at the Cornell CESR collider facility. In particular, detailed results on recent measurement of the $B \to D^{**}l\nu$ are given, and an outline of the search for nonresonant decays is presented.

I INTRODUCTION

CLEO II [1] is an experiment that studies e^+e^- physics at the $\Upsilon(4S)$ resonance. Semileptonic B decays (figure 1) proceed mainly via $b \to c$ transitions accompanied by the emission of a lepton and an anti-neutrino ($\ell\bar{\nu}_\ell$). We distinguish inclusive decays ($b \to c\ell\bar{\nu}_\ell$) which include the whole semileptonic spectrum, from exclusive decays ($B \to D\ell\bar{\nu}_\ell$, $B \to D^*\ell\bar{\nu}_\ell$, $B \to D^{**}\ell\bar{\nu}_\ell$, etc.) which have well-defined final states. Currently the sum of the different exclusive channels branching fractions doesn't saturate the total inclusive branching fraction.

Here we attempt to resolve this deficit by measuring two new exclusive decay channels, $B \to D^{**}\ell\bar{\nu}_\ell$ and $B \to D\pi\ell\bar{\nu}_\ell$.

The world average branching ratio for the inclusive ($b \to c\ell\bar{\nu}_\ell$) transition is 10.18% based on measurements taken at the $\Upsilon(4S)$ resonance. The two well-established exclusive measurements are $B \to D\ell\bar{\nu}_\ell$ and $B \to D^*\ell\bar{\nu}_\ell$ with respective world averages of 1.94% and 5.05%, as summarized in table 1. $3.19 \pm 0.54\%$ of the inclusive branching ratio remains unaccounted for by exclusive channels, this discrepancy is known as the semileptonic deficit in B decays. A significant deficit clearly justifies the need to look for higher order resonances or nonresonant contributions to saturate the inclusive rate.

In the following section we discuss the new exclusive measurement by CLEO of $B^- \to D_1^0 \ell^- \bar{\nu}_\ell$ and $B^- \to D_2^{*0} \ell^- \bar{\nu}_\ell$. In the second section, we describe progress in the search for $B^- \to D^+\pi^- e^- \bar{\nu}_e$ transitions.

CP452, *Toward the Theory of Everything*: MRST'98
edited by J. M. Cline, M. Knutt, G. D. Mahlon, and G. D. Moore
© 1998 The American Institute of Physics   1-56396-845-2/98/$15.00

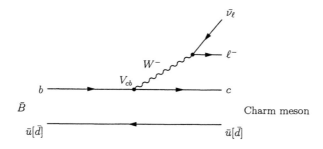

FIGURE 1. Feynman diagram for the semileptonic decay of the $\bar{B}$ meson.

TABLE 1. Charmed semileptonic B decays

Decay Mode	Branching Fraction
$B^- \to X_c e^- \bar{\nu}_e$	$(10.18 \pm 0.40)\%$
$B^- \to D^0 e^- \bar{\nu}_e$	$(1.94 \pm 0.26)\%$
$B^- \to D^{*0} e^- \bar{\nu}_e$	$(5.05 \pm 0.25)\%$
Inclusive-Exclusive	$(3.19 \pm 0.54)\%$

II ANALYSIS DETAILS AND RESULTS: $B \to D^{**} \ell \bar{\nu}_\ell$

In this section, we report on the recent measurement of D_1 and D_2^* by Cleo [2], the so called narrow D^{**}. Respectively these resonances correspond to the $^{2S+1}L_J = {}^1P_1$ and $^{2S+1}L_J = {}^3P_2$ states of the $c\bar{u}$ system, with masses 2.42 GeV ($\Gamma \simeq 19$ MeV) and 2.46 GeV ($\Gamma \simeq 23$ MeV).

The data used represents an integrated luminosity of 3.10 fb^{-1} on the $\Upsilon(4S)$ resonance, corresponding to 3.29×10^6 $B\bar{B}$ events, and 1.61 fb^{-1} at a center-of-mass energy ~ 55 MeV below the $\Upsilon(4S)$ resonance. Details of the detector behavior can be found in [1].

We reconstruct the two D_J^0 in the decay channel $D_J^0 \to D^{*+}\pi^-$, where $D^{*+} \to D^0\pi^+$, and $D^0 \to K^-\pi^+$ or $D^0 \to K^-\pi^+\pi^0$.

We require that the $K^-\pi^+$ and $K^-\pi^+\pi^0$ be within 16 MeV/c^2 and 25 MeV/c^2 of the PDG [3] value of the D^0 mass respectively ($\sim 2\sigma$). The $D^0\pi^+$ candidates must have a $\delta m = M(D^0\pi^+) - M(D^0)$ within 2 MeV/c^2 of the known $D^{*+} - D^0$ mass difference [3]. The D^{*+} candidates are then combined with additional π^-s to form D_J^0 candidates. Finally the D_J^0s are combined with the leptons in the event, either electrons or muons, to form candidates for $B^- \to D_J^0 \ell^- \bar{\nu}_\ell$ decays.

To reduce background we select $D^{*+}\pi^-\ell^-$ combinations consistent with $B^- \to D_J^0 \ell \bar{\nu}_\ell$ and reject $D^{*+}\pi^-\ell^-$ combinations consistent with $\bar{B}^0 \to D^{*+}\ell\bar{\nu}_\ell$. This is accomplished by requiring that $|\cos\theta_{B-D_J\ell}| \leq 1$ and $\cos\theta_{B-D^*\ell} < -1$ where

TABLE 2. Extracted yields from figure 2 for the two signal channels. Corresponding backgrounds to the signal are included.

	D_1^0	D_2^{*0}
ON Resonance Yield (N_{D_J})	56.6 ± 11.9	10.3 ± 9.4
Non-$B\bar{B}$ bkg	2.3 ± 2.7	1.5 ± 2.8
Fake Lepton bkg	0.8 ± 0.6	0.0 ± 0.3
Final Yield (n_{D_J})	53.5 ± 12.2	8.8 ± 9.8

$$\cos\theta_{B-D_J\ell} = \frac{|\mathbf{p}_{D_J\ell}|^2 + |\mathbf{p}_B|^2 - |\mathbf{p}_{\bar{\nu}_\ell}|^2}{2|\mathbf{p}_B||\mathbf{p}_{D_J\ell}|} \quad (1)$$

and

$$\cos\theta_{B-D^*\ell} = \frac{|\mathbf{p}_{D^*\ell}|^2 + |\mathbf{p}_B|^2 - |\mathbf{p}_{\bar{\nu}_\ell}|^2}{2|\mathbf{p}_B||\mathbf{p}_{D^*\ell}|} \quad (2)$$

The signal is extracted using the mass difference $\delta M_J = M(D^{*+}\pi^-) - M(D^{*+})$. Multiple counting is avoided by selecting the best candidate per event based on $M(\pi^0)$, $M(D^0)$, δm, and $M^2(\nu_\ell) \simeq M_B^2 + M^2(D_J\ell) - 2E_B E(D_J\ell)$.

Both decay modes of the D^0 are combined and the final plot of the δM_J is shown in figure 2. The fit to the data in this figure is an unbinned likelihood fit. The function used for this fit is a threshold background function plus Breit-Wigner resonance functions with the masses and widths of the two narrow D_J^0 resonances fixed. Each Breit-Wigner function is convoluted with a Gaussian that describes the detector resolution. Results of the fit are summarized in table 2.

The non-$B\bar{B}$ background is obtained by performing the same analysis on off-resonance data. The fake lepton background is generated by repeating the analysis on tracks that are not leptons and scaling appropriately. Both backgrounds are then subtracted from the final yields.

The final branching fractions are obtained from

$$\mathcal{B}(B^- \to D_J^0 \ell^- \bar{\nu}_\ell) = \frac{n_{D_J}/\epsilon_{D_J}}{4\,N_{\Upsilon(4S)}\,f_{+-}\,\mathcal{B}(D_J^0 \to D^{*+}\pi^-)\mathcal{B}(D^{*+} \to D^0\pi^+)\,\mathcal{B}(D^0 \to K^-\pi^+(\pi^0))} \quad (3)$$

where n_{D_J} are the final yields, ϵ_{D_J} the different reconstruction efficiencies, all obtained from Monte Carlo generated events based on the ISGW2 model [4]. We assume that the branching fractions of $\Upsilon(4S)$ to charged and neutral $B\bar{B}$ pairs are 50% each. We also assume that $\mathcal{B}(D_1^0 \to D^{*+}\pi^-) = 67\%$ and $\mathcal{B}(D_2^{*0} \to D^{*+}\pi^-) = 20\%$. This is based on isospin conservation and CLEO measurements [5]. Branching fractions for the D^* and D^0 are obtained from [3]. We find that

$$\mathcal{B}(B^- \to D_1^0 \ell^- \bar{\nu}_\ell) = (0.56 \pm 0.13 \pm 0.08 \pm 0.04)\,\%, \quad (4)$$
$$\mathcal{B}(B^- \to D_2^{*0} \ell^- \bar{\nu}_\ell) < 0.8\,\% \quad (90\%\ \text{C.L.}). \quad (5)$$

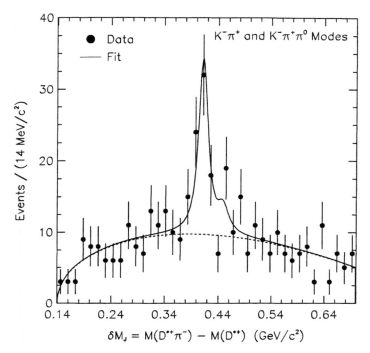

FIGURE 2. The δM_J distribution for the $B^- \to D_1^0 \ell^- \bar{\nu}_\ell$ and $B^- \to D_2^{*0} \ell^- \bar{\nu}_\ell$ ($\ell = e$ and μ) candidates. The two D^0 modes are combined. The solid line is the signal and background function fitted to the data. The dashed line is only the background function.

III ANALYSIS DETAILS AND RESULTS: $B \to D\pi\ell\bar{\nu}_\ell$

Adding the result outlined in the previous section for D_1^0 to the exclusive decays still doesn't saturate the inclusive branching fraction. A deficit of $2.6 \pm 0.6\%$ remains. Obviously there will be some contribution from the D_2^{*0} on which we set an upper limit and the broad D^{**} resonances, but one should also expect some part of this deficit to be covered by the nonresonant decays. This reasoning provides the motivation behind the analysis discussed below. If they are strong enough, the nonresonant channels could also be an important source of background to the other exclusive decays: $B \to Dl\bar{\nu}_\ell$, $B \to D^*l\bar{\nu}_\ell$ and $B \to D^{**}l\bar{\nu}_\ell$, affecting the determination of quantities like V_{cb}.

There are four $B \to D\pi l\bar{\nu}_\ell$ nonresonant decays:

- $B^- \to D^+\pi^-\ell^-\bar{\nu}_\ell$
- $B^- \to D^0\pi^0\ell^-\bar{\nu}_\ell$
- $\bar{B}^0 \to D^0\pi^+\ell^-\bar{\nu}_\ell$
- $\bar{B}^0 \to D^+\pi^0\ell^-\bar{\nu}_\ell$

In this analysis we concentrate only on the first mode since D^{*0} is below the $D^+\pi^-$ threshold. This is quite interesting since resonant $D\pi$ contributions, which could otherwise swamp the nonresonant signal, will not contribute and neither will interference between them and the nonresonant contribution. The reconstruction method makes use of the hermicity of our detector to reconstruct the missing neutrino [7,8]. At present only electrons are used due to their lower momentum threshold for identification.

For this analysis we have 3 different Monte-Carlo samples to study our signal and backgrounds. The nonresonant contributions are studied using the Goity & Roberts model [6]. For the resonant semileptonic background studies we use the ISGW 2 model [4]. We also have a sample of generic $B\bar{B}$ Monte Carlo events for $b \to c \to Xl\nu$ studies, also called secondary leptons. Every Monte-Carlo sample represents approximately 6 times the luminosity of our data sample.

A few words about the Goity & Roberts Model [6]. This model is based around the B_{l4} decays. These decays are semileptonic $B \to D$ with emission of a single pion. The B_{l4} decays can be described by an effective Lagrangian which implements chiral and heavy-quark symmetries. An area of interest of the B_{l4} decays is their contribution to ρ^2, the slope of the Isgur-Wise function [4], which impacts on the extraction of $|V_{cb}|$ from data. A rough theoretical estimate from this model for the total B_{l4} branching ratio in the D mode is about 1% .

For the event selection we use hadronic events with tracks that originate from the vicinity of e^+e^- interaction region. For the neutrino reconstruction we require only one lepton candidate per event, since any extra lepton would likely be accompanied by a neutrino. This extra neutrino would then complicate the use of the energy-momentum constraints and make it more difficult to reconstruct the event accurately. The D^+ meson is reconstructed in the $D^+ \to K^-\pi^+\pi^+$ mode. We select all $K\pi\pi$ combinations within 15 MeV from the PDG [3] value of the mass of the D^+. This corresponds to a 2.5σ cut. The D candidates are then combined with the electron found in the event plus an additional pion. Implicit charge correlation is assumed in this process.

We then reconstruct the neutrino using the rest of the event. We assign the most likely particle mass hypothesis to each track and then define the missing energy and missing momentum variables:

$$E_{\text{miss}} = 2E_{\text{beam}} - E_{D\pi e} - E_{\text{other}} \qquad (6)$$

$$\vec{P}_{\text{miss}} = -\vec{P}_{D\pi e} - \vec{P}_{\text{other}} \qquad (7)$$

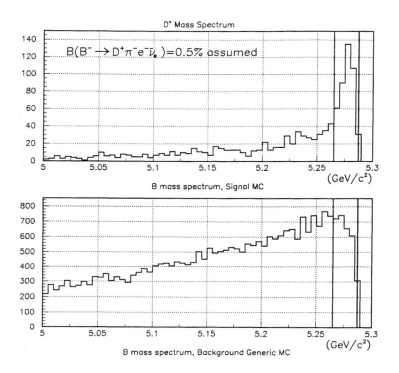

FIGURE 3. Monte-Carlo plots for the decay channel $B^- \to D^+\pi^-e^-\bar{\nu}_e$ for signal and background. Both plots are scaled such that $\mathcal{B}(B^- \to D^+\pi^-e^-\bar{\nu}_e) = 0.5\%$.

Since the resolution on E_{miss} is much worse than the resolution on the $\vec{P}_{\text{miss}}$ ($\sigma_{E_{\text{miss}}} \approx 260$ MeV and $\sigma_{\vec{P}_{\text{miss}}} \approx 110$ MeV) we shall define our neutrino using only the missing momentum variables, defining $E_\nu = |\vec{P}_{\text{miss}}|$ and $\vec{P}_\nu = \vec{P}_{\text{miss}}$. We then make use of neutrino quality cuts that will select signal events and reject background events. The beam constrained mass is then obtained from the expression:

$$M^2(B) = E_{\text{beam}}^2 - |\vec{p}_{D^+} + \vec{p}_{\pi^-} + \vec{p}_{e^-} + \vec{p}_{\bar{\nu}_e}|^2 \qquad (8)$$

Figure 3 shows the scaled signal and background B mass distribution. The signal is scaled to a hypothetical branching fraction of 0.5% for our decay channel. The reconstruction efficiency is expected to be 4.2%. The main difficulty one has to deal with is the relative abundance of remaining background to our signal. This arises when the final pion in our decay is combined with the D meson and the electron. Further cuts are presently under investigation to deal with this problem.

We expect approximately 60 events using this technique, assuming as above that $\mathcal{B}(B^- \to D^+\pi^- e^- \bar{\nu}_e) = 0.5\ \%$.

IV CONCLUSIONS AND ACKNOWLEDGMENTS

We have demonstrated that there is a significant deficit in the experimental measurements of semileptonic B decays. We measured the D_1^0 contribution and set an upper limit on the D_2^{*0} contribution. We are presently working on the nonresonant $D\pi$ contributions which can be expected to have a significant impact towards solving this deficit.

We gratefully acknowledge the effort of the CESR staff in providing us with excellent luminosity and running conditions. This work was supported by the National Science Foundation, the U.S. Department of Energy, the Heisenberg Foundation, the Alexander von Humboldt Stiftung, the Natural Sciences and Engineering Research Council of Canada, le Fonds Québécois pour la Formation de Chercheurs et l'Aide à la Recherche, and the A.P. Sloan Foundation.

REFERENCES

1. Y. Kubota et al., *Nucl.Instrum. Methods Phys. Res. A* **320**, 66 (1992).
2. CLEO Collaboration, A. Anastassov et al., *Phys. Rev. Lett.* **80**, 4127 (1998).
3. Particle Data Group, R.M. Barnett et al., *Phys. Rev. D* **54**, 1 (1996).
4. D. Scora and N. Isgur, *Phys. Rev. D* **52**, 2783 (1995).
5. CLEO Collaboration, P. Avery et al., *Phys. Lett. B* **331**, 236 (1994).
6. J.L. Goity and W. Roberts, *Phys. Rev. D* **51**, 3459 (1995).
7. J.P. Alexander et al., *Phys. Rev. Lett.* **77**, 5000 (1996).
8. M. Athanas et al., *Phys. Rev. Lett.* **79**, 2208 (1997).

Nuclear Flow Inversion in the Boltzmann-Uehling-Uhlenbeck model

D. PERSRAM and C. GALE

*Department of Physics, McGill University, 3600 University Street
Montréal, Québec H3A 2T8, Canada*

Abstract. We follow a numerical approach to model the collision of two heavy nuclei from which information about the nuclear mean field can be extracted. The model used in this work is the so-called *BUU* model named after Boltzmann, Uehling and Uhlenbeck. It involves a solution of the Vlasov transport equation coupled with a quantum collision integral. Solution of the equation is obtained through the self-consistent Lattice Hamiltonian method. Some results which show a transition from attractive to repulsive mean field dynamics are discussed and are in qualitative agreement with experimental data.

INTRODUCTION

For the past two decades the field of heavy ion collisions has seen many advances on both the experimental and theoretical fronts. Traditionally, the theoretical input has come from the solution of transport theories such as the BUU [1-3] and QMD [4,5] models. Both have been quite successful in explaining many observables extracted from experimental studies at energies ranging from a few hundred MeV per nucleon up to a few GeV per nucleon. Recently, advances in experimental techniques has opened up the lower energy regime ($\sim \epsilon_f$) and the need for more precise numerical models has followed. Previous BUU solutions have employed the so called "test particle" method [3]. However, it has been shown that at low energies this method may lead to solutions that violate energy conservation [6]. This aspect is expected to worsen as the bombarding energy is decreased. The energy conservation problem inherent in the test particle method has largely been circumvented through the "Lattice Hamiltonian" [7] solution.

At high incident beam energy (several times the fermi energy ϵ_f), many two body nucleon-nucleon collisions are predicted and observed to occur. Thus the nucleus-nucleus dynamics is governed by both mean field effects as well as hard scattering. As the beam energy is lowered however, the corresponding increase in phase space density does not permit many nucleon-nucleon collisions as the Pauli exclusion principle forbids scattering into occupied states. Thus, we expect the mean field to play an increasingly important role in the nucleus-nucleus dynamics. The nuclear mean field can contribute to an attractive potential (indeed, a nucleus at equilibrium must be bound by the mean field) as well as a repulsive potential. This can be seen in figure 1. At low density, an attractive mean field is present and at high density, the mean field becomes repulsive. Note also that the energy per nucleon is a minimum at equilibrium nuclear matter density as expected. From these arguments, it was proposed that performing a study of the nuclear flow (defined later) excitation function could perhaps reveal the presence of a transition from repulsive to attractive nuclear mean field dynamics. This transition is referred to as the "nuclear flow inversion". An example of this "inferred" flow inversion can be found in reference [8,9].

THE VLASOV/BUU EQUATION

Starting from quantum mechanics, we wish to derive a transport equation which can be used to describe the time evolution of a system consisting of a large number of nucleons. As this system is purely fermionic, our starting wave function(Ψ) must be totally asymmetric. The Schrödinger equation gives the time evolution as:

$$i\hbar \frac{\partial}{\partial t}|\Psi> = H|\Psi>$$

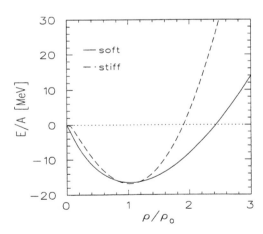

FIGURE 1. Total energy per nucleon for two compressibilities of the nuclear EOS. The potential used is the generalized Skyrme potential that depends on density only. Parameterizations are taken from [10].

The kinetic part of the Hamiltonian H in a general basis can be written as:

$$\hat{T} = \sum_{\alpha\beta} <\alpha|\hat{t}|\beta> a_\alpha^\dagger a_\beta$$

$$\begin{aligned}\Longrightarrow \quad <\hat{T}> &= <\Psi|\hat{T}|\Psi> \\ &= \sum_{\alpha\beta} <\alpha|\hat{t}|\beta><\Psi|a_\alpha^\dagger a_\beta|\Psi> \\ &= \sum_{\alpha\beta} t_{\alpha\beta}\rho_{\beta\alpha}\end{aligned}$$

Here, $\rho_{\beta\alpha} \equiv <\Psi|a_\alpha^\dagger a_\beta|\Psi>$ is the *one body density matrix* (OBDM). And the single particle kinetic energy operator respects: $\hat{t}|p> = p^2/2m|p>$.

We can obtain a similar expression in terms of the OBDM for the potential energy operator through a similar but more involved calculation. Here however, we must assume that the total wave function Ψ is written as a Slater Determinant. This assumption allows us to adopt a mean field picture as the potential term speaks only of the OBDM. At this point, the two, three ... particle interaction terms are lumped into the mean field. In interest of brevity we show below only the result. (Note that $\hat{v}$ below only operates on a distinct set of states.)

$$<\hat{V}> = \frac{1}{2}\sum_{\alpha\beta\gamma\delta} <\alpha\beta|\hat{v}|\gamma\delta> (\rho_{\gamma\alpha}\rho_{\delta\beta} - \rho_{\delta\alpha}\rho_{\gamma\beta})$$

We now have the ingredients to calculate the expectation values for the kinetic and potential energy of our many body wave function (systems of nucleons). However, a little foresight can carry us further. Since the observables we are interested in are both written in terms of the OBDM, we might be able to exploit this to our advantage. If we move from a general basis to the configuration space basis for the OBDM and perform a Wigner integral transform on this new OBDM, we arrive at

$$f(\vec{r},\vec{p}) = \frac{1}{(2\pi\hbar)^3}\int d\vec{s}\, e^{\frac{-i}{\hbar}\vec{p}\cdot\vec{s}} \rho_{\vec{r}+\frac{\vec{s}}{2},\vec{r}-\frac{\vec{s}}{2}},$$

where $\rho_{\vec{r}\vec{r}'}$ is the coordinate space representation of the OBDM. Integration of $f(\vec{r},\vec{p})$ over momentum and configuration space gives us quantities that behave like the ordinary configuration and momentum space densities. The Wigner transform above makes quantum mechanics look like ordinary classical mechanics. We can now treat $f(\vec{r},\vec{p})$ as a classical space space density and examine its time dependence. What we have arrived at is often referred to as the "Vlasov equation". It is shown below for a potential which may depend on *momentum* [10] as well as the local nuclear density (as seen in the Skyrme interaction from figure 1). The Vlasov equation reads as follows where H is now the Hamiltonian:

$$\frac{\partial}{\partial t}f(\vec{r},\vec{p}) + \nabla_{\vec{p}}H \cdot \nabla_{\vec{r}}f(\vec{r},\vec{p}) - \nabla_{\vec{r}}H \cdot \nabla_{\vec{p}}f(\vec{r},\vec{p}) = 0.$$

Up to now, since the adoption of the mean field picture, we have lost all two-particle correlations. Also note that the Vlasov equation as written above is devoid of any collision terms (just a restatement of the former). However, we do expect hard nucleon-nucleon collisions to be present in a medium containing many nucleons. Thus, we require a method to restore by hand some of these terms. The "Boltzmann quantum collision integral" can rescue us here. Essentially, we replace the zero on the right hand side of the Vlasov equation by a collision source term which respects the fermionic nature of the nucleons. Namely, the Pauli exclusion principle. In a numerical solution of this new equation (the "BUU" equation), the analytical form of this collision integral integral is not as important as how it is implemented. The interested reader is referred to [2,11] for details.

Lattice Hamiltonian Solution of the BUU equation

Solution of the BUU equation requires that we know the phase space density $f(\vec{r},\vec{p})$ for all particles (nucleons). First we assume that f is separable and then assign N particles positions and momenta on a grid $(\delta x, \delta p)$ at site (α, β) as follows:

$$f_{\alpha\beta}(\vec{r},\vec{p}) = \sum_{i=1}^{N} R(\vec{r}_\alpha - \vec{r}_i) P(\vec{p}_\beta - \vec{p}_i).$$

In the above equation, R and P represent configuration and momentum space form factors which specify the "range" or width of the particles. We follow reference [7] for R and adopt a delta function distribution for P [3]. These are shown below in one dimension.

$$R(x_\alpha - x_i) = \frac{1}{(N\delta x)^2}(\delta x - |x_\alpha - x_i|)\Theta(\delta x - |x_\alpha - x_i|)$$
$$P(p_\beta^x - p_i^x) = \delta(p_\beta^x - p_i^x)$$

From the above, we now see that all the particles are distributed on a finite grid in configuration space $(\delta x \neq 0)$ and an infinitesimal grid (continuum) in momentum space $(\delta p = 0)$. Insertion of the above form factor into the BUU equation leads to the following set of equations:

$$\frac{\partial \vec{r}}{\partial t} = \nabla_{\vec{p}} H \qquad \frac{\partial \vec{p}}{\partial t} = -\nabla_{\vec{r}} H.$$

So we see that if the particles obey Hamiltonian's equations of motion, then the above space space density represents a solution of the BUU equation minus collisions (or the Vlasov equation). The next step is to assign to the particles some starting position and momentum and in discreet time steps calculate the force on each particle self-consistently. The positions and momenta at the end of the time step are iterated from the aforementioned force. Collisions are treated by checking at each time step the relative separation of each particle with all others and comparing to some assigned cross section (taken from well known experimental nucleon-nucleon cross section measurements). A Monte Carlo procedure is then performed to ascertain whether or not the particles are allowed to scatter. This test is not a sufficient condition however. Recall that the quantum collision integral must obey the Pauli exclusion principle. Implementation of this is achieved by sampling phase space close to the new (scattered) momenta and positions. If the local population is too high, then the collision is forbidden on the grounds of their fermionic nature. So, given some initial conditions (two ground state, non-interacting nuclei Lorentz boosted towards each other), the time evolution of the system is calculated numerically and all relevant information can be extracted at the conclusion of the interaction. See reference [3] for details.

NUCLEAR FLOW

Implementation of the BUU equation allows one to study the time evolution of a collision of two heavy ions. In fact, in this method one has access to the entire history of all particles. For example, the local density and number of binary collisions as a function of time for every particle is at all times accessible. However, in experimental studies, one can only observe the final state as typical time durations of such a collision are quite small ($\sim 100 fm/c$ or about 10^{-22} seconds!). In addition, there are typically ~million experimentally detected events. Spectra and multiplicities etc. are averaged over many or all such events. This averaging is also possible in BUU studies of heavy ion collisions. Here, one can simulate hundreds (typically done in parallel) of events and extract average quantities from these. One such observable is "nuclear flow". This observable gives us insight into large scale (larger than a nucleon) collective motion. For example, one can determine in the course of a reaction whether nucleons on average are deflected away from, or attracted to, the interaction or nuclei overlap region. For a head-on collision, the later is maximal and is minimal for a peripherial collision. Figure 2 shows an example where the target and projectile spectator caps are slightly deflected *away* from the interaction region. In this picture, the *nuclear flow* is positive. For attractive scattering, the nuclear flow is negative.

Quantitative Flow Analysis

The previously mentioned figure demonstrates a case in which repulsive forces dominate any attractive forces. This is illustrated by the positive(repulsive) scattering of the spectators. Some positive pressure has been generated in the overlap region which in turn pushes the "caps" away from the interaction region. If we define the beam axis as the $+\hat{z}$ axis and the $\hat{x}$ axis parallel to the impact parameter "b", then we should find that on average, nucleons with both $< p_x >$ and $< p_z >$ positive ($\hat{x}$ points to the top of the page) have suffered a repulsive interaction. On the other hand, nucleons with $< p_x >$ negative and $< p_z >$ positive have partially orbited the participant region and have been scattered to negative angles, or attracted. This gives us a means to test whether or not a nucleus-nucleus interaction is on average attractive or repulsive.

^{14}N+^{154}SM FLOW INVERSION

BUU Results

Monte Carlo BUU simulations have been run at several energies and impact parameters for ^{14}N projectiles on ^{154}Sm targets. The energies range from $\sim \epsilon_f \rightarrow 150 MeV/A$. Both semi-central ($b/b_{max} \sim 0.45$) and peripherial ($b/b_{max} \sim 0.90$) impact parameters were examined. The average transverse momenta per nucleon ($< p_x > /A$) in the forward hemisphere (a polar lab angle cut of $25° < \Theta_{lab} < 35°$ was used for purposes of comparison with experimental measurements) was examined for four energies and two impact parameters. The results are displayed in figure 3. The figure demonstrates that at low bombarding energy, the nuclear potential is dominantly attractive resulting in a negative average transverse momenta per nucleon in the forward hemisphere (negative flow). As the energy is increased, the mean field becomes dominantly repulsive, thus making $< p_x >$ positive (positive flow). For peripherial collisions, we see that the nuclei experience a larger attractive force. This can be attributed to the smaller overlap region (participant region). A smaller overlap region implies less collisions and thus less repulsion. The "balance energy", defined as the energy at which the mean field switches from attractive to repulsive is seen to increase with impact parameter.

Experimental Results

Recent experimental results from NSCL-MSU on flow inversion measurements have also shown a direct flow inversion signal in the ^{14}N+^{154}Sm system [13]. The

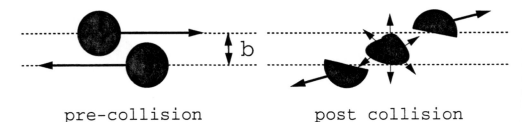

pre-collision   post collision

FIGURE 2. Initially, before the collision on the left hand side, the projectile (entering from the left) approaches the target with a non-zero impact parameter "b". After the collision we are left with a mixed and expanding participant region and two separated spectator regions. In this case, the spectators are slightly deflected away from the interaction region.

measurement made was of the γ polarization from the decay of the excited residual target. The sign of the polarization of the γ tells us the nature of the nuclear interaction. For $P_\gamma > 0$ the interaction is dominantly attractive and for $P_\gamma < 0$ the interaction is dominantly repulsive. Details of this measurement can be found in [14]. We just quote the results. Figure 4 displays the P_γ detected in coincidence with α-particle fragments. The results are in qualitative agreement with the BUU calculation. The flow is dominantly attractive at low bombarding energy and switches signs at high bombarding energy. Also note that the balance energy increases with impact parameter. Although the two quantities (experimental and theoretical) are different, they both provide the same information in regards to the flow inversion.

CONCLUSION

We have implemented the Lattice Hamiltonian solution of the Vlasov/Boltzmann equation for the case of two colliding heavy ions. The change in sign in the average transverse momenta per nucleon is indicative of a nuclear flow inversion. The balance energy is found to be an increasing function of impact parameter. Direct measurements of the γ polarization emitted from the residual target nucleus also directly confirm the flow inversion and show that the balance energy is an increasing function of impact parameter. We see qualitative agreement between the theoretical

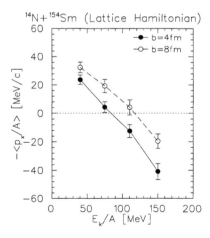

FIGURE 3. Negative average transverse momenta per nucleon in the angular range $25° < \Theta_{lab} < 35°$ for collisions of ^{14}N on ^{154}Sm as a function of lab bombarding energy. Results are from the Lattice Hamiltonian solution of the BUU equation with the "MDYI" momentum dependent nuclear potential [12].

predictions and the experimental observables. Further studies will be to investigate in the lattice hamiltonian method with our momentum dependent potential the role of the in-medium nucleon-nucleon cross section in the balance energy as the later is expected to be quite sensitive to this cross section as previous calculations suggest [8,15].

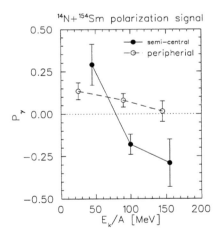

FIGURE 4. Measured γ polarization in coincidence with ejected α particles in the angular range $25° < \Theta_{lab} < 35°$ for collisions of ^{14}N on ^{154}Sm as a function of lab bombarding energy. The data is taken from [13].

REFERENCES

1. E. A. Uehling and G. E. Uhlenbeck, *Phys. Rev.* **43**, 552 (1933).
2. G. F. Bertsch, H. Kruse and S. Das Gupta, *Phys. Rev.* **C29**, 673 (1984).
3. G. F. Bertsch and S. Das Gupta, *Phys. Rep.* **160**, 189 (1988).
4. J. Aichelin and H. Stöcker, *Phys. Lett.* **B176**, 14 (1986).
5. J. Aichelin *Phys. Rep.* **202**, 233 (1991).
6. C. Gale and S. Das Gupta, *Phys. Rev.* **C42**, 1577 (1990).
7. R. J. Lenk and V. R. Pandharipande, *Phys. Rev.* **C39**, 2242 (1989).
8. D. Krofcheck et. al. *Phys. Rev.* **C46**, 1416 (1992).
9. G. D. Westfall et. al. *Phys. Rev. Lett.* **71**, 1986 (1993).
10. C. Gale, G. F. Bertsch and S. Das Gupta, *Phys. Rev.* **C35**, 1666 (1987).
11. J. Aichelin and G. F. Bertsch, *Phys. Rev.* **C31**, 1730 (1985).
12. C. Gale et. al. *Phys. Rev.* **C41**, 1545 (1990).
13. R. C. Lemmon et. al. *MSU Preprint*, **MSUCL-1084** (1998).
14. M. B. Tsang et. al. *Phys. Rev. Lett.* **57**, 559 (1986).
15. G. F. Bertsch, W. G. Lynch and M. B. Tsang *Phys. Lett.* **B189**, 384 (1987).

$SUSY$ Threshold Corrections and Up-Down Unification in the superpotential

C. Hamzaoui[a] and M. Pospelov[a,b]

[a] *Département de Physique, Université du Québec à Montréal*
C.P. 8888, Succ. Centre-Ville, Montréal, Québec, Canada, H3C 3P8
[b] *Budker Institute of Nuclear Physics, Novosibirsk, 630090, Russia*

Abstract. We study the possibility of the model with identical Up and Down Yukawa matrices in the superpotential. Large $\tan\beta$ and the additive renormalization of Yukawa matrices at the supersymmetric threshold in this case can be responsible for the realistic mass spectrum and the origin of the Kobayashi-Maskawa mixing. We show that Up-Down unification is possible in a moderate quark-squark alignment scenario with an average squark mass of the order 1 TeV and with $\tan\beta > 15$.

INTRODUCTION

The observed hierarchical patterns in the masses of quarks and leptons and in the Kobayashi-Maskawa mixing pose a serious problem for particle theory. Many theories have been put forward in an attempt to guess a symmetry which can explain the observed mass spectrum and mixings. Some of them exploit horizontal symmetries [1,2], gauged or nongauged, anomalous symmetries [3], possible compositness [4], etc.

Another theoretical goal, closely related to the mass problem, is the possibility to reduce the number of the free parameters in the fundamental theory in comparison with that of the standard model. The supersymmetric GUT theories offer this possibility citeR. The scale of the unification for Yukawa couplings in this case is believed to coincide with the scale of the gauge unification and therefore is very high, 10^{16}GeV or so.

In this talk, we attempt to construct a model with a minimal number of free parameters in the superpotential. This is a model with identical Up and Down quark Yukawa matrices in the superpotential and general form of the soft-breaking parameters, not constrained by usual assumption of universality and proportionality. The Kobayashi-Maskawa mixing itself in this situation originates from the supersymmetric threshold corrections. We show that this possibility is not excluded.

There are several possible physical motivations related to the U-D unification. For example, it can be a consequence of a horizontal symmetry responsible for the generation of the Yukawa couplings which "does not feel hypercharge", i.e. which does not distinguish between H_1 and H_2, U and D superfields and therefore can only generate identical Y_u and Y_d. Another example is the case of the supersymmetric left-right theory based on the $SU(3) \times SU(2)_L \times SU(2)_R \times U(1)_{B-L}$ group [6] where the unification of U and D right-handed superfields is protected by an extra gauge symmetry.

U-D UNIFICATION AT LOW ENERGY

The standard superpotential of the minimal supersymmetric standard model (MSSM),

$$W = \epsilon_{ij}[-Q^i H_2^j \mathbf{Y}_u U + Q^i H_1^j \mathbf{Y}_d D + L^i H_1^j \mathbf{Y}_e E + \mu H_1^i H_2^j], \tag{1}$$

contains the same number of free dimensionless parameters as the Yukawa sector of the standard model.

In addition, in the soft-breaking sector there are other couplings which have a potential influence on flavor physics. Among different scalar masses, the soft-breaking sector has the squark mass terms

$$\tilde{U}^\dagger \mathbf{M}_U^2 \tilde{U} + \tilde{D}^\dagger \mathbf{M}_D^2 \tilde{D} + \tilde{Q}^\dagger \mathbf{M}_Q^2 \tilde{Q}; \tag{2}$$

and the trilinear terms

$$\epsilon_{ij}\left(-\tilde{Q}^i H_2^j \mathbf{A}_u \tilde{U} + \tilde{Q}^i H_1^j \mathbf{A}_d \tilde{D}\right) + H.c.; \tag{3}$$

as the possible sources of flavor transitions.

Counting all free parameters in the model, one comes to a huge number 105 [7]. This enormous number of free parameters originates mainly from the soft-breaking sector and cannot be reduced *a priori*, without knowledge of the ways the supersymmetry breaking occurs. It is customary to assume, at the scale of the breaking, the following, very restrictive conditions are fulfilled:

$$\mathbf{M}_Q^2 = m_Q^2 \mathbf{1}; \quad \mathbf{M}_D^2 = m_D^2 \mathbf{1}; \quad \mathbf{M}_U^2 = m_U^2 \mathbf{1} \quad \text{"degeneracy"} \tag{4}$$
$$\mathbf{A}_u = A_u \mathbf{Y}_u; \quad \mathbf{A}_d = A_d \mathbf{Y}_d \quad \text{"proportionality"}, \tag{5}$$

and similarly for leptons. These conditions, if held, ensure that the physics of flavor comes entirely from the superpotential. But it might not necessarily be the case. For example, these conditions are not held in superstring inspired models (See Ref. [8] and references therein). Neither are they in the simplest flavor models operating with horizontal symmetries [9].

To this end, it is interesting to abandon strict conditions (4)-(5) in the soft-breaking sector and to explore the possibility of having a smaller number of free parameters in the superpotential. As an ultimate example of this, let us analyze the theory with the low-energy unification of Up and Down Yukawa matrices. Instead of working with the superpotential of the form (1) with two independent matrices $\mathbf{Y}_u$ and $\mathbf{Y}_d$, we consider the theory with $\mathbf{Y}_u \equiv \mathbf{Y}_d$ so that the superpotenial can be written in the following compact form:

$$W = \epsilon^{ij}\epsilon^{kl}[Q^i\Phi^{jk}\mathbf{Y}Q_R^l + \frac{1}{2}\mu\Phi^{jk}\Phi^{il}] \tag{6}$$

where $Q_R = (U, D)^T$ and $\Phi = (H_1; H_2)$. From here and below we consistently omit the leptonic part. At the same time we assume similar U-D unification in the soft-breaking sector as well:

$$... + \tilde{Q}^\dagger \mathbf{M}_L^2 \tilde{Q} + \tilde{Q}_R^\dagger \mathbf{M}_R^2 \tilde{Q}_R + \epsilon^{ij}\epsilon^{kl}\tilde{Q}^i\Phi^{jk}\mathbf{A}\tilde{Q}_R^l + H.c... \tag{7}$$

At the scale of the breaking, matrix $\mathbf{Y}$ can be chosen diagonal and it will stay the same when it is run down to the supersymmetric threshold one. It means that, at the tree level plus logarithmic renormalization, $M_u \sim M_d$ and all Kobayashi-Maskawa mixing angles are zero. In this case the observed mixing angles and masses come from the supersymmetric threshold corrections. These corrections induce additional terms in the Yukawa interaction containing the conjugated Higgs fields, H_1^*, H_2^*, and in our case Φ^*. As a result, below the threshold, the effective interaction of fermions with Higgs doublets can be written in the form

$$\mathcal{L}_{eff} = \bar{Q}\mathbf{Y}_1\tau_2\Phi\tau_2 Q_R + \bar{Q}\mathbf{Y}_2\Phi^*Q_R + ... \tag{8}$$

The ellipses stands here for the possible terms of bigger dimension which also may influence the fermion spectrum (for example, $M_{sq}^{-2}\bar{Q}_R\mathbf{Y}_1\tau_2\Phi\tau_2 Q\mathrm{Tr}(\Phi^\dagger\Phi)$). In Fig. 1 we list the diagrams which contribute to the matrices $\mathbf{Y}_1$ and $\mathbf{Y}_2$.

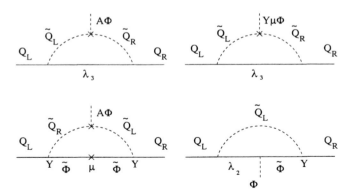

FIGURE 1. The diagrams generating threshold corrections for $\mathbf{M}_u$ and $\mathbf{M}_d$.

To some extent, one can view this unification as being inspired by supersymmetric left-right models where the relevant part of the superpotential is usually written in the following form:

$$W = Q\mathbf{Y}_1\tau_2\Phi_1\tau_2 Q_R + Q\mathbf{Y}_2\tau_2\Phi_2\tau_2 Q_R + \sum_{i,j=1,2} \mu_{ij}\text{Tr}\tau_2\Phi_i\tau_2\Phi_j^T \qquad (9)$$

If the analysis of the threshold correction can generate realistic $\mathbf{M}_u$ and $\mathbf{M}_d$ then one can eliminate one of the bidoublets, reducing (9) to the more economic form (6).

In a number of papers [11], the conditions (4)-(5) were imposed on the soft-breaking sector. However, in some specific variants of the unified models [12] the departure from these conditions can reproduce masses of first generation and the Cabbibo angle.

The typical size of the corrections to the mass of b-quark when the mass of the gluino is equal to the masses of squarks is of the order of m_b itself, $\Delta m_b/m_b \sim 0.4\mu/m_{squark}$. The ratio μ/m_{squark} is presumably of the order 1 but can be larger. For $m_{squark} \sim 1\text{TeV}$, μ/m_{sq} can be as large as $\sqrt{2}m_{sq}/v \sim 5$. In this case m_b, as well as the other quark masses and mixing angles, can be completely of radiative origin. In the following we consider the possibility of the low-energy unification of Yukawa couplings, not confining our analysis to the case of $\mu \simeq m_{squark}$ and not specifying any particular model. In particular, we have to check if the following three conditions are satisfied:

1. The matrix $\mathbf{A}$ is consistent with scale independent constraints from the absence of color breaking minima.

2. Radiatively generated masses and mixing angles correspond to the observed hierarchy.

3. The predictions for FCNC are acceptable.

PHENOMENOLOGICAL CONSEQUENCES OF U-D UNIFICATION

It is convenient to choose the basis in which matrix $\mathbf{Y}$ is diagonal;

$$\mathbf{Y} = \text{diag}(y_1; y_2; y_3) \qquad (10)$$

and y_3 basically coincides with the top quark Yukawa coupling, $y_3 \simeq m_t\sqrt{2}/v \simeq 1$.

The value of $\tan\beta \equiv v_u/v_d$ has to be large but it is not fixed to m_t/m_b due to the substantial renormalization of m_b. Therefore, to $O(\tan^{-1}\beta)$ accuracy we can adopt the following approximation for the squark mass matrices:

$$M_{\tilde{u}}^2 = \begin{pmatrix} \mathbf{M}_L^2 + \frac{v^2}{2}\mathbf{Y}^\dagger\mathbf{Y} & \frac{v}{\sqrt{2}}\mathbf{A}^\dagger \\ \frac{v}{\sqrt{2}}\mathbf{A} & \mathbf{M}_R^2 + \frac{v^2}{2}\mathbf{Y}\mathbf{Y}^\dagger \end{pmatrix} \qquad (11)$$

$$M_{\tilde{d}}^2 = \begin{pmatrix} \mathbf{M}_L^2 + \frac{v^2}{2}\mathbf{Y}^\dagger\mathbf{Y} & \frac{v}{\sqrt{2}}\mu^*\mathbf{Y}^\dagger \\ \frac{v}{\sqrt{2}}\mu\mathbf{Y} & \mathbf{M}_R^2 + \frac{v^2}{2}\mathbf{Y}\mathbf{Y}^\dagger \end{pmatrix} \quad (12)$$

The general formulae for the mass matrices, tree level plus radiative corrections, can be presented in the following form:

$$\frac{\sqrt{2}}{v}\mathbf{M}_{uij} = \mathbf{Y}_{ij} + \frac{2\alpha_s m_\lambda}{3\pi}\int \frac{d^4p}{(2\pi)^4(p^2-m_\lambda^2)}\left[\frac{1}{p^2-\mathbf{M}_L^2}\mathbf{A}\frac{1}{p^2-\mathbf{M}_R^2}\right]_{ij} + ...$$

$$\frac{\sqrt{2}}{v}\mathbf{M}_{dij} = \frac{\mathbf{Y}_{ij}}{\tan\beta} + \frac{2\alpha_s\mu m_\lambda}{3\pi}\int \frac{d^4p}{(2\pi)^4(p^2-m_\lambda^2)}\left[\frac{1}{p^2-\mathbf{M}_L^2}\mathbf{Y}\frac{1}{p^2-\mathbf{M}_R^2}\right]_{ij} + ... \quad (13)$$

Here, ellipses stands for the chargino and neutralino corrections and next order corrections in v^2/m^2 which we neglect at the moment.

The form of the matrix A suggested by the absence of dangerous directions in the field space along which color breaking minima can emerge is the following [13]:

$$\mathbf{A} = \begin{pmatrix} 0 & 0 & A_{13} \\ 0 & 0 & A_{23} \\ A_{31} & A_{32} & A_{33} \end{pmatrix} \quad (14)$$

The same form is suggested by the analysis of the FCNC processes [14]. Restricting our analysis only to this form of $\mathbf{A}$ matrix, we satisfy the first condition mentioned in the previous section automatically.

Let us consider the corrections to the mass of the b-quark. Neglecting for a moment all flavor changing effects in the squark sector, we can identify m_b with the $(\mathbf{M_d})_{33}$ matrix element of the $\mathbf{M_d}$ and write the relation among observable m_b taken at the scale of the squark mass, $\tan\beta$ and dimensionless ratios μ/m_{sq}, $x = m_\lambda/m_{sq}$. For simplicity we assume also the approximate equality of the left- and right-handed squark masses.

$$\frac{2m_b^2}{v^2} = |y_3|^2\left|\frac{1}{\tan\beta} + \frac{2\alpha_s}{3\pi}e^{i\phi}\frac{|\mu|}{m_{sq}}F(m_\lambda/m_{sq})\right|^2 \quad (15)$$

$$F(x) = \frac{x}{1-x^2} + \frac{x^3\ln x^2}{(1-x^2)^2}$$

Taking $y_3 \sim 1$, $m_b(1TeV) \sim 2.8 GeV$ and $F = F_{max} = F(2.1) = 0.57$, we plot the allowed values of $\tan\beta$ and $|\mu|$ in Fig. 2. The phase ϕ between tree-level contribution and radiative corrections to m_b is unknown [1] which leads to a certain allowed area on $\tan\beta$ - $|\mu|$ plane. The case of relatively low $\tan\beta$ corresponds to the destructive interference between the tree level contribution and the loop correction. With $\tan\beta \sim 15$, it corresponds to the 80% mutual cancellation between tree-level

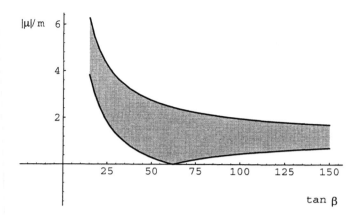

FIGURE 2. Allowed area on the $|\mu| - \tan\beta$ plane in the case of maximal threshold corrections.

value and radiative correction which we take as maximally allowed degree of the fine tuning.

The necessity for the off-diagonal entries in the squark mass-matrices to be nonzero comes from two reasons. First, they can be the only source for the off-diagonal mass matrices leading to nonvanishing Kobayashi-Maskawa mixing angles. Second, their existence leads to the substantial renormalization of u, d, s, c quark masses which is needed to account for the relations $m_s/m_c > m_b/m_t$; $m_d/m_u \gg m_b/m_t$. The 100% renormalization of the charm mass, for example, may result from the combination of the flavor changing entries in $\mathbf{A}$ and $\mathbf{M}_R^2$, and so on. At the same time, it is preferable to keep the off-diagonal elements of the squark mass matrices at the lowest possible level to avoid large FCNC contributions from the box diagrams. Treating these flavor transitions as the mass insertions, i.e. assuming that they are small in some sense, we arrive at the following set of the order-of-magnitude relations connecting observable masses and mixing angles to flavor structure of the soft-breaking sector:

$$\frac{\sqrt{2}}{v}\Delta m_c \sim \eta_c \frac{A_{23}(\mathbf{M}_R^2)_{23}}{m_{sq}^3}$$

$$\frac{\sqrt{2}}{v}\Delta m_s \sim \eta_s \frac{(\mathbf{M}_L^2)_{23}(\mathbf{M}_R^2)_{23}}{m_{sq}^4}$$

---

[1] Possible constraints on ϕ from the limits on the neutron EDM ar relaxed when squark mass ~ 1 TeV.

$$\frac{\sqrt{2}}{v}\Delta m_d \sim \eta_d \frac{(\mathbf{M}_L^2)_{13}(\mathbf{M}_R^2)_{13}}{m_{sq}^4} \tag{16}$$

$$\theta_{23} \sim \frac{(\mathbf{M}_L^2)_{23}}{m_{sq}^2}$$

$$\theta_{13} \sim \frac{(\mathbf{M}_L^2)_{13}}{m_{sq}^2}$$

Here m_{sq} is the average squark mass and numerical coefficients η_i represent loop factors. For our estimates we take η_i to be of the order of $\eta_b \sim \sqrt{2}m_b/v$. As to the mixing angle between first and second generation, it can be generated either in the down sector [12], or in the up sector through the correction to the matrix element $(\mathbf{M}_u)_{12} \sim \frac{\sqrt{2}}{v}\eta A_{13}(\mathbf{M}_R^2)_{23}/m_{sq}^3$.

If the squark masses were diagonal, one would observe an approximate relation $m_b/m_t \simeq m_s/m_c$ which is violated in reality. There are several possible ways to avoid this problem depending on which part of the allowed values of μ - $\tan\beta$ plane we choose. If $\tan\beta$ is in the neighbourhood of m_t/m_b, it is preferable to have $y_2 v/\sqrt{2} \simeq m_c$ and the strange quark mass being completely of radiative origin. The latter condition requires $(\mathbf{M}_L^2)_{23}/m_{sq}^2 \sim \lambda^2$ and $(\mathbf{M}_R^2)_{23}/m_{sq}^2 \sim 1$. For lower values of $\tan\beta$, when the tree-level contribution to the mass of bottom quark is compensated by loop corrections to give observable m_b, the element $(\mathbf{M}_R^2)_{23}$ can be made smaller and $y_2 v/\sqrt{2} \simeq m_s$.

Combining together relations (16), we write the phenomenologically acceptable form of the squark mass matrices which can produce correct flavor physics through the loop mechanism:

$$\mathbf{M}_L^2 \sim m^2 \begin{pmatrix} 1 & \lambda^4 & \lambda^2 \\ \lambda^4 & 1 & \lambda^2 \\ \lambda^2 & \lambda^2 & 1 \end{pmatrix}; \quad \mathbf{M}_R^2 \sim m^2 \begin{pmatrix} 1 & \lambda & \lambda \\ \lambda & 1 & 1 \\ \lambda & 1 & 1 \end{pmatrix}. \tag{17}$$

Every entry in (17) denotes an order of magnitude of corresponding matrix element in terms of the power of Wolfenstein parameter λ. The squark masses chosen in the form (17), plugged in the general formula (13), reproduce correctly the hierarchy among quark masses and mixing angles and therefore satisfies the second condition formulated at the end of the previous section. $(\mathbf{M}_{L(R)}^2)_{12}$ is not fixed by (16) and for its value we take $(\mathbf{M}_{L(R)}^2)_{12} \sim (\mathbf{M}_{L(R)}^2)_{13}(\mathbf{M}_{L(R)}^2)_{23}/m_{sq}^2$.

Similar forms of the squark mass matrices can appear in supersymmetric theories with horizontal symmetries responsible for mass hierarchy [9] (quark-squark alignement). We can invert the set of argument and conclude that in the quark-squark alignment picture with large $\tan\beta$ regime and $\mu m_\lambda \sim m_{sq}^2$ the radiative corrections to the mass matrices and mixings are important. At the same time, the bare superpotential of the model can be of the form (6), i.e. simpler than that of the conventional MSSM.

The FCNC processes in the neutral Kaon sector arises both at the tree level and at one-loop level. When the FCNC contribution of the box diagrams generated by the mass matrices (17) is considered, the constraints [15] implies that the lowest possible value for the squark mass has to be of the order 1 TeV. Similar constraints arise from the analysis of $b \to s\gamma$ process [16]. This can be still viewed as an acceptable value for the scalar quark mass scale.

Perhaps the most serious consequence of the low-energy unification of Up and Down Yukawa matrices and radiative mechanism for Kobayashi-Maskawa mixing is the appearence of the FCNC transitions mediated by H_d field. Since the resulting Yukawa interaction resembles that of a generic left-right model with the right-hadned mixing angles not smaller than left-handed ones, $|V_{Rij}| > |V_{Lij}|$, we can use the limits on the FCNC Higgs masses obtained in [17], namely $M_A > 10$ TeV.

CONCLUSIONS

The supersymmetric mass problem has one interesting aspect, not always properly emphasized. With the general flavor-changing soft-breaking terms, it is hard to interpret the dimensionless coefficient in the superpotential in terms of the observed fermionic masses and mixing angles. This opens up a possibility of a supersymmetric theory with much less number of the free parameters in the superpotential than it is assumed in the conventional MSSM. We have shown the phenomenological possibility of the low-energy U-D unification in the supersymmetric models. The Kobayashi-Maskawa mixing in this case is the result of the supersymmetric threshold corrections. The analysis of these corrections shows that $\tan\beta$ is not fixed by the requirement of the unification. U-D unification allows it to be in a rather large range $15 \leq \tan\beta \leq \infty$.

It was shown in Refs. [2] that the scale of the physics responsible for the flavor hierarchy can be as low as few TeV scale. Our conclusion is that the condition of the unification of all Yukawa couplings does not require to raise this scale. The scale of the unification, i.e. supersymmetric threshold in the case considered, can be as low as 1TeV without causing unacceptable mixing in the neutral kaon sector.

The possibility of the U-D unification at low energies with the radiative mechanism for Kobayashi-Maskawa mixing has an interesting application to the left-right supersymmetric models. It allows one to reduce the number of Higgs bidoublets, making it more similar to the conventional (nonsupersymmetric) left-right models. Unfortunately, in the case of the manifest left-right symmetry, where left-handed and right-handed squark masses are equal, the radiatively induced fermion masses and mixings do not correspond to the observables.

Acknowledgements
This work is supported in part by N.S.E.R.C. of Canada.

REFERENCES

1. C.D. Froggatt and H.B. Nielsen, Nucl. Phys. **B147** (1979) 277.

2. M. Leurer, Y. Nir and N. Seiberg, Nucl. Phys. **B398** (1993) 319.
3. P. Binétruy, S. Lavignac and P. Ramond, Nucl. Phys. **B477** (1996) 353; A.N. Nelson and D. Wright, Phys. Rev. **D56** (1997) 1598.
4. M.L. Luty and R.N. Mohapatra, Phys. Lett. **B396** (1997) 161; D.B. Kaplan, F. Lepeintre and M. Shmaltz, Phys. Rev. **D56** (1997) 7193.
5. S. Dimopoulos, L.J. Hall and S. Raby, Phys. Rev. Lett. **68** (1992) 1984, Phys. Rev. **D45** (1992) 4192; P. Ramond, R.G. Roberts and G.G. Ross, Nucl. Phys. **B406** (1993) 19.
6. R. Mohapatra and A. Razin, Phys. Rev. **D54** (1996) 5835.
7. M. Dine, hep-ph/9612389.
8. M. Dine, A.Kagan, R. Leigh, Phys. Rev. **D48** (1993) 2214.
9. Y. Nir and N. Seiberg, Phys. Lett. **B309** (1993) 337; M. Leurer, Y. Nir and N. Seiberg, Nucl. Phys. **B420** (1994) 468.
10. A.G. Cohen, D.B. Kaplan and A.E. Nelson Phys. Lett. **B388** (1996) 588; A.G. Cohen, D.B. Kaplan, F. Lepeintre, and A.E. Nelson, Phys. Rev. Lett. **78** (1997) 2300.
11. L. Hall, R. Rattazzi and U. Sarid, Phys. Rev. **D50** (1994) 7048; M. Carena et al. Nucl. Phys. **B426** (1995) 269; T. Blazek, S. Raby and S. Pokorski, Phys. Rev. **D52** (1995) 4151.
12. R. Hemplfing, Phys. Rev. **D49** (1994) 6168.
13. J.A. Casas and S. Dimopoulos, Phys. Lett. **B387** (1996) 107.
14. D. Choudhury et. al., Phys. Lett. **B342** (1995) 180.
15. F. Gabbiani et. al., Nucl. Phys. **B447** (1996) 321; P.H. Chankowski and S. Pokorski, hep-ph/9703442.
16. T. Blazek and S. Raby, hep-ph 9712257; H. Baer et al., hep-ph 9712305.
17. R.N. Mohapatra, G. Senjanovich and M. Tran, Phys. Rev. **D28** (1983) 546; G. Ecker, W. Grimus and H. Neufeld, Phys. Lett. **B127** (1983) 365; F. G. Gilman and M.H. Reno, Phys. Lett. **B127** (1983) 426; F.G. Gilman and M.H. Reno Phys. Rev. **D29** (1983) 937: J.F. Gunion, J. Grifols, A. Mendez, B. Kayser and F. Olness, Phys. Rev. **D40** (1989) 1546; M. E. Pospelov, Phys. Rev. **D56** (1997) 259.

From Super QCD to QCD

Francesco Sannino[1]

Department of Physics, Yale University, New Haven, CT 06520-8120, USA.

Abstract. We present a "toy" model for breaking supersymmetric gauge theories at the effective Lagrangian level. We show that it is possible to achieve the decoupling of gluinos and squarks, below a given supersymmetry breaking scale m, in the fundamental theory for super QCD once a suitable choice of supersymmetry breaking terms is made. A key feature of the model is the description of the ordinary QCD degrees of freedom via the auxiliary fields of the supersymmetric effective Lagrangian. Once the anomaly induced effective QCD meson potential is deduced we also suggest a decoupling procedure, when a flavored quark becomes massive, which mimics the one employed by Seiberg for supersymmetric theories. It is seen that, after quark decoupling, the QCD potential naturally converts to the one with one less flavor. Finally we investigate the N_c and N_f dependence of the η' mass.

GENERAL STRATEGY

In the last few years there has been an enormous progress in understanding supersymmetric gauge theories via effective Lagrangians. Such a progress is partially due to some papers of Seiberg [1] and Seiberg and Witten [2] in which a number of "exact results" were obtained. There are already several review articles [3–5].

It is natural to expect that information obtained from the more highly constrained supersymmetric gauge theories can be used to learn more about ordinary gauge theories. Here we illustrate the general strategy behind a "toy" model presented in [6] for breaking super symmetric gauge theories at the effective Lagrangian level.

Let us consider an "exact" effective super potential W which can be constructed for a given, confining, supersymmetric gauge theory. The superpotential, by construction, correctly saturates all the supersymmetric quantum anomalies. The contribution to the bosonic part of the potential, contained in the superpotential, before imposing the equation of motion for the auxiliary fields, is:

$$-V_0(\mathcal{F}, \phi) = \int d^2\theta \ W(\mathcal{S}, \mathcal{T}) + \text{H.c.} , \qquad (1)$$

---

[1] Electronic address: sannino@apocalypse.physics.yale.edu

where the chiral superfields $\mathcal{S}$ and $\mathcal{T}$ schematically describe gauge invariant supersymmetric composite operators whose bosonic components (ϕ) respectively contain gluino-ball and squark-antisquark mesons. $\mathcal{F}$ are the set of auxiliary fields associated with the chiral superfields $\mathcal{S}$ and $\mathcal{T}$. We note that the previous potential term, due to supersymmetry, is holomorphic in the fields, i.e. $V_0(\mathcal{F},\phi) = \chi(\mathcal{F},\phi) + \chi^\dagger(\mathcal{F}^*,\phi^*)$ and χ is a function of the complex fields $\mathcal{F}$ and ϕ. The composite operators $\mathcal{F}$ are seen to describe the ordinary glue-ball and mesonic objects (see for example Eq. (8)).

Let us imagine to add SUSY breaking terms in the fundamental Lagrangian whose order parameters (i.e. gluino mass and squark mass terms) can be *schematically* represented by $\mathcal{M}$. This will induce in the low energy effective theory a SUSY breaking potential $V_B(\mathcal{M},\phi)$ which should be added to the one in Eq. (1). The full potential for finite $\mathcal{M}$ can then be written as

$$V(\mathcal{F},\phi,\mathcal{M}) = V_0(\mathcal{F},\phi) + V_B(\mathcal{M},\phi) \ . \tag{2}$$

If $\mathcal{M} \ll \Lambda_S$, where Λ_S is the SUSY invariant scale of the theory we are close to the supersymmetric limit and $\mathcal{F}$ must be eliminated via its equation of motion

$$\frac{\partial V}{\partial \mathcal{F}} = 0 \quad \text{for} \quad \mathcal{M} \ll \Lambda_S \ . \tag{3}$$

In the absence of the Kähler term, the previous equation simply reproduces the supersymmetric vacuum solution for the bosonic fields ϕ. To recover the "soft" SUSY breaking effects [8], beside modeling V_B via supersymmetric spurions, one must consider a model for the Kähler term which, in turn, should be invariant under the anomalous transformations. It is amusing to note that Kähler terms in the effective theory can be regarded as higher order in a derivative expansion with respect to the invariant scale of the theory and that $\mathcal{M}/\Lambda_S$ corrections arise when also Kähler terms are present.

If $\mathcal{M} \gg \Lambda_S$ the light degrees of freedoms are now the ordinary fields (quarks, gluons, etc.). These seems to be contained in the auxiliary fields of the effective Lagrangian description. Hence we expect $\mathcal{F}$ to remain, while ϕ, which describes the colorless objects made of gluinos and squarks, to decouple. This can be obtained by assuming, as proposed in [6], the following equation of motion

$$\frac{\partial V}{\partial \phi} = 0 \quad \text{for} \quad \mathcal{M} \gg \Lambda_S \ . \tag{4}$$

This provides the relation

$$\phi = \phi(\mathcal{F},\mathcal{M},\Lambda_S) \ , \tag{5}$$

By substituting the previous expression in Eq. (2) we obtain the potential:

$$V = V(\mathcal{F},\mathcal{M},\Lambda_S) \ . \tag{6}$$

A smooth decoupling is achieved if, at least at one loop in the underlying theories, the scales $\mathcal{M}$ and Λ_S combine in the unique scale Λ associated with the ordinary gauge theory.

Of course the knowledge of V_B is essential. We partially fix the breaking potential [6] by requiring the anomalies (trace anomaly as well as global anomalies) to match at the one loop level and assuming holomorphy for the breaking potential. The latter assumption is partially supported by the holomorphic behavior of the one-loop QCD coupling constant.

FROM SUPER YANG MILLS TO YANG MILLS

In this section we illustrate, in some detail, how the previous strategy works in the SUSY Yang Mills case [6]. The effective Lagrangian for Super Yang Mills was given [7] by Veneziano and Yankielowicz (VY) and is described by the Lagrangian

$$\mathcal{L} = \frac{9}{\alpha} \int d^2\theta d^2\bar{\theta} \left(SS^\dagger\right)^{\frac{1}{3}} + \left\{ \int d^2\theta \, S \left[\ln\left(\frac{S}{\Lambda^3_{SYM}}\right)^{N_c} - N_c\right] + \text{H.c.} \right\}, \qquad (7)$$

where Λ_{SYM} is the super $SU(N_c)$ Yang Mills invariant scale and the chiral superfield S stands for the composite object $S = \frac{g^2}{32\pi^2} W^\alpha_a W_{\alpha a}$. Here g is the gauge coupling constant and W^α_a is the supersymmetric field strength. At the component level with $S(y) = \phi(y) + \sqrt{2}\theta\psi(y) + \theta^2 F(y)$ we have

$$\phi \approx \lambda^2, \quad \psi \approx \sigma^{mn}\lambda_a F_{mn,a} \quad \text{and} \quad F \approx -\frac{1}{2} F_a^{mn} F_{mn,a} - \frac{i}{4}\epsilon_{mnrs} F_a^{mn} F_a^{rs}. \qquad (8)$$

λ^α_a is the gluino field, $F_{mn,a}$ the gauge field strength.

We interpret the complex field ϕ as representing scalar and pseudoscalar gluino balls while ψ is their fermionic partner. The auxiliary field F is seen to contain scalar and pseudoscalar glueball type objects.

Equation (7) describes the vacuum of the theory and saturates the anomalous Ward identities at tree level. These anomalies arise in the axial current of the gluino field, the trace of the energy momentum tensor and in the special superconformal current. In supersymmetry these three anomalies belong to the same supermultiplet and hence are not independent. For example

$$\theta^m_m = 3 N_c (F + F^*) = -\frac{3 N_c g^2}{32\pi^2} F_a^{mn} F_{mn,a}, \qquad (9)$$

$$\partial^m J^5_m = 2i N_c (F - F^*) = \frac{N_c g^2}{32\pi^2} \epsilon_{mnrs} F_a^{mn} F_a^{rs}. \qquad (10)$$

where $J^5_m = \bar{\lambda}_a \bar{\sigma}_m \lambda_a$ is the axial current. The effective Lagrangian yields [7] the gluino condensation of the form $\langle\phi\rangle = -\frac{g^2}{32\pi^2}\langle\lambda^2\rangle = \Lambda^3 e^{\frac{2\pi i k}{N_c}}$ where $k = 0, 1, 2, \cdots, (N_c - 1)$.

Masiero and Veneziano investigated the "soft" supersymmetry breaking regime [8] by introducing a "gluino mass term" in the Lagrangian

$$\mathcal{L} = \cdots + m\left(\phi + \phi^*\right), \tag{11}$$

with the softness restriction $m \ll \Lambda_{SYM}$. The results of this model [8] indicate that the theory is "trying" to approach the ordinary Yang Mills case.

It seems very desirable to extend this model to the case of large $m (\gg \Lambda_{SYM})$ in which the superparticles decouple from the theory and the theory gets reexpressed in terms of ordinary glueball fields. In Ref. [6] we proposed a toy model which accomplishes these goals. Our approach is based on the general strategy presented in the previous paragraph which we specialize, here, for the super Yang Mills case:

i) We shall concentrate completely on the superpotential. This contains all the information on the anomaly structure and seems to be the least model dependent part of the effective Lagrangian.

ii) We will show that the generalization of the supersymmetry breaking term Eq. (11) to

$$V_B = -m^\delta \phi^\gamma + \text{H.c.}, \tag{12}$$

where $\delta = 4 - 3\gamma$ and $\gamma = \frac{12}{11}$ automatically accomplishes the decoupling of the underlying gluino degree of freedom at the scale m. The deviation of the exponent γ from unity is being thought of as an effective description of the evolution of the symmetry breaker Eq. (11) for large m.

iii) Since the Yang Mills fields of interest are contained in F the heavy gluino ball field ϕ is eliminated by its equation of motion $\frac{\partial V}{\partial \phi} = 0$ (see Eq. (4)).

The potential of our model

$$V(F, \phi) = -F \ln \left(\frac{\phi}{\Lambda_{SYM}^3}\right)^{N_c} - m^\delta \phi^\gamma + \text{H.c.}, \tag{13}$$

provides the equation of motion, $\frac{\partial V}{\partial \phi} = 0$ for eliminating ϕ: $\phi^\gamma = -\frac{N_c F}{\gamma m^\delta}$. Our physical requirement is that the presence of the symmetry breaker Eq. (12) should convert the anomalous quantity θ_m^m into the appropriate one for the ordinary Yang Mills theory. This is in the same spirit as the well known [9] criterion for decoupling a heavy flavor (at the one loop level) in QCD. We compute θ_m^m at tree level obtaining

$$\theta_m^m = \frac{4N_c}{\gamma} (F + F^*) = -\frac{4N_c}{\gamma} \left(\frac{g^2}{32\pi^2} F_a^{mn} F_{mn,a}\right). \tag{14}$$

Now the 1-loop anomaly in the underlying theory is given by

$$\theta_m^m = -b \frac{g^2}{32\pi^2} F_a^{mn} F_{mn,a}, \tag{15}$$

where $b = 3N_c$ for supersymmetric Yang Mills and $b = \frac{11}{3}N_c$ for ordinary Yang Mills. In order that Eq. (14) match Eq. (15) for ordinary Yang Mills we evidently require $\gamma = \frac{12}{11}$ as mentioned above. With ϕ eliminated in terms of F the potential becomes

$$V(F) = -\frac{11N_c}{12} F \left[\ln\left(\frac{-11N_c F}{12 m^{\frac{8}{11}} \Lambda_{SYM}^{\frac{36}{11}}} \right) - 1 \right] + \text{H.c.} \; . \tag{16}$$

We now check that this is consistent with a physical picture in which the gauge coupling constant evolves according to the super Yang Mills beta–function above scale m and according to the Yang Mills beta–function below scale m. Since the coupling constant at scale μ is given by $\left(\frac{\Lambda_{SYM}}{\mu}\right)^b = \exp\left(\frac{-8\pi^2}{g^2(\mu)}\right)$, the matching at $\mu = m$ requires $\left(\frac{\Lambda_{SYM}}{m}\right)^b = \left(\frac{\Lambda_{YM}}{m}\right)^{b_{YM}}$, which yields $\Lambda_{YM}^4 = m^{\frac{8}{11}} \Lambda_{SYM}^{\frac{36}{11}}$, in agreement with the combination appearing in Eq. (16). The Lagrangian in Eq. (16) manifestly depends only on quantities associated with the Yang Mills theory, the gluino degree of freedom having been consistently decoupled. Equation (16) thus seems to be a reasonable candidate for the potential term of a model describing the trace anomaly in Yang Mills theory.

The model is seen to contain a scalar glueball field $\text{Re}F$ and a pseudoscalar glueball field $\text{Im}F$. In order to relate our present results to previous investigations [2] we also eliminate $\text{Im}F$ by its equation of motion which yields $\text{Im}F = 0$. Substituting this back into Eq. (16) and using the notation $H = \frac{11N_c}{3} \frac{g^2}{32\pi^2} F_a^{mn} F_{mn,a}$, leads to the potential function

$$V(H) = \frac{H}{4} \ln\left(\frac{H}{8e\Lambda_{YM}^4}\right) . \tag{17}$$

This may be considered as a zeroth order model [10–12] for Yang Mills theory in which the only field present is a scalar glueball. $V(H)$ has a minimum at $\langle H \rangle = 8\Lambda_{YM}^4$, at which point $\langle V \rangle = -2\Lambda_{YM}^4$. This corresponds to a magnetic–type condensation of the glueball field H. A number of phenomenological questions have been discussed using toy models based on Eq. (17) [12–14].

It can also be shown that $m_\psi \to \infty$ in the case $m \to \infty$; thus, as expected, ψ decouples [6].

FROM SUPER QCD TO QCD AND QUARK DECOUPLING

The more complicated case of adding matter fields with number of flavors N_f less than number of colors N_c with $N_c \neq 2$ has been analyzed in [6,15]. Here we summarize some of the relevant results. The needed "mesonic" composite superfield

---

[2]) See discussion in Ref. [6]

is the complex $N_f \times N_f$ matrix $T_{ij} = Q_i \tilde{Q}_j = t_{ij} + \sqrt{2}\theta\psi_{Tij} + \theta^2 M_{ij}$, where i and j are flavor indices. Q and $\tilde{Q}$ are the quark anti-quark chiral superfields. It can be seen that the mesonic auxiliary field $M \approx -\psi_Q \psi_{\tilde{Q}}$ contains the ordinary quark-antiquark meson field while $t = \phi_Q \phi_{\tilde{Q}}$ describes the squark anti-squark composite operator. In Ref. [16] a straightforward generalization of the supersymmetric potential presented in Eq. (7) for $N_f < N_c$ was derived. By a suitable decoupling of the squark as well as gluino degrees of freedom (see Ref. [6] for more details) the following potential for ordinary QCD can be deduced

$$V(M) = -C(N_c, N_f) \left[\frac{\Lambda_{QCD}^{\frac{11}{3}N_c - \frac{2}{3}N_f}}{\det M} \right]^{\frac{12}{11(N_c - N_f)}}, \qquad (18)$$

where Λ_{QCD} is the invariant QCD scale and $C(N_c, N_f)$ is a definite positive quantity which cannot be fixed by requiring the potential to satisfy the anomalies. We also eliminated the glue-ball degrees of freedom via their equation of motion (see Ref. [15]).

The potential for the meson variables in Eq. (18) is similar to the effective Affleck-Dine-Seiberg (ADS) [17] superpotential for massless super QCD theory with $N_f < N_c$

$$W_{ADS}(T) = -(N_c - N_f) \left[\frac{\Lambda_S^{3N_c - N_f}}{\det T} \right]^{\frac{1}{N_c - N_f}}, \qquad (19)$$

where Λ_S is the invariant scale of SQCD.

An intriguing feature of the potential in Eq. (18) is that it presents a fall to the origin rather than a run-away vacuum associated with the ADS superpotential. The fall to the origin can be fixed by adding an anomalous free non holomorphic term in the manner outlined in [6], which in turn requires spontaneous chiral symmetry breaking. It is worth noticing [15] that one can directly derive Eq. (18) from QCD, if one assumes, besides the correct anomalous transformations, also one-loop holomorphicity in the QCD coupling constant [15,19].

As for the SUSY case [1] we can partially deduce the N_f and N_c dependence of C by defining a decoupling procedure for quarks. In Ref. [15] it has been shown that by adding the following, generalized quark mass operator,

$$V_m = -m^\Delta M_{N_f}^{N_f \Gamma} + \text{H.c.} \,, \qquad (20)$$

to the potential in Eq. (18), with $\Delta = 4 - 3\Gamma$, is possible to obtain a complete decoupling when a flavored quark becomes massive. This procedures mimics the one employed by Seiberg for supersymmetric gauge theories. It is seen that, after decoupling, the QCD potential naturally converts to the one with one less flavor provided that $\Gamma = 12/11$ and the coefficient C has the following functional form

$$C\left(N_c, N_f\right) = \left(N_c - N_f\right) D\left(N_c\right)^{\frac{1}{N_c - N_f}} . \qquad (21)$$

$D\left(N_c\right)$ is an unknown N_c dependent function. It is interesting to contrast the coefficient of the "holomorphic" part of the QCD potential Eq. (18) with Seiberg's result [1] $C\left(N_c, N_f\right) = N_c - N_f$ for the coefficient of the ADS superpotential (Eq. (19)). Clearly, in the SUSY case the analog of $D\left(N_c\right)$ is just a constant. This feature arises in the SUSY case because of the existence of squark fields which can break the gauge and flavor symmetries by the Higgs mechanism. The possibility of a non-constant $D\left(N_c\right)$ factor can thus be taken as an indication that there is no Higgs mechanism present in QCD-like theories.

Finally the knowledge about $C(N_c, N_f)$ can be used to suggest that the well known [18] large N_c behavior of the η' (pseudoscalar singlet) meson mass should also include an N_f dependence of the form:

$$M_{\eta'}^2 \propto \frac{N_f}{N_c - N_f} \Lambda^2 \qquad (N_f < N_c) . \qquad (22)$$

It is amusing to observe that when N_f is close to N_c the resulting pole in Eq. (22) suggests a possible enhancement mechanism for the η' mass. This would explain the unusually large value of this quantity in the realistic three flavor case.

In future we would like to understand the very important and yet elusive case $N_f = N_c$, where as argued in Ref. [15] we expect non holomorphic terms to be relevant, since the coefficient $C\left(N_c, N_f\right)$, in analogy with the SUSY case, of the anomalous potential vanishes for $N_f = N_c$.

In Ref. [20], we computed the one–loop effective action in the specific case of $N_f = N_c + 1$ and $N_c = 2$ while keeping only the auxiliary fields on the external legs, and in the presence of supersymmetry breaking terms. This procedure amounts to integrate out order by order, in a loop expansion, the scalar fields. It was shown how a non-trivial kinetic term for the auxiliary field naturally emerges, reinforcing our assumption that the latter can be associated with a physical field once the supersymmetric particles decouple.

It is also very interesting to explore the $N_f > N_c$ case which might shed some light on the zero temperature chiral restoration and a possible relation with the conformal window [21].

ACKNOWLEDGMENTS

I am happy to thank Joseph Schechter for very helpful discussions. I also thank CERN Theoretical Division, where part of this work was completed. The present work has been partially supported by the US DOE under contract DE-FG-02-92ER-40704.

REFERENCES

1. N. Seiberg, Phys. Rev. **D49**, 6857 (1994); Nucl. Phys. **B435**, 129 (1995).
2. N. Seiberg and E. Witten, Nucl. Phys. **B426**, 19 (1994); erratum **B430**, 485 (1994); **B431**, 484 (1994).
3. K. Intriligator and N. Seiberg, Nucl. Phys. Proc. Suppl. **45BC**, 1, 1996.
4. M.E. Peskin, *Duality in Supersymmetric Yang-Mills Theory*, (TASI 96): Fields, Strings, and Duality, Boulder, CO, 2-28 June 1996, hep-th/9702094.
5. P. Di Vecchia, *Duality in Supersymmetric Gauge Theories*, Surveys High Energy Phys. **10**, 119 (1997), hep-th/9608090.
6. F. Sannino and J. Schechter, Phys. Rev. **D57**, 170, 1998. See also references therein.
7. G. Veneziano and S. Yankielowicz, Phys. Lett. **113B**, 321 (1982).
8. A. Masiero and G. Veneziano, Nucl. Phys. **B249**, 593 (1985). N. Evans, S.D.H. Hsu and M. Schwetz, Phys. Lett. **B404**, 77 (1997). L. Alvarez-Gaume and M. Marino, Int. J. Mod. Phys. **A12**, 975 (1997). L. Alvarez-Gaume, J. Distler, C. Kounnas, M. Marino, Int. J. Mod. Phys. **A11**, 4745 (1996). L. Alvarez-Gaume, M. Marino, F. Zamora, hep-th/9703072 and hep-th/9707017. M. Marino and F. Zamora, hep-th/9804038.
9. E. Witten, Nucl. Phys. **B104**, 445 (1976); **B122**, 109 (1977). M. Shifman, A. Vainshtein and V. Zakharov, Nucl. Phys. **B147**, 385 (1979); **B147**, 448 (1979).
10. J. Schechter, Phys. Rev. **D21**, 3393 (1980). C. Rosenzweig, J. Schechter and G. Trahern, Phys. Rev. **D21**, 3388 (1980). P. Di Vecchia and G. Veneziano, Nucl. Phys. **B171**, 253 (1980). E. Witten, Ann. of Phys. **128**, 363 (1980). P. Nath and A. Arnowitt, Phys. Rev. **D23**, 473 (1981). A. Aurilia, Y. Takahashi and D. Townsend, Phys. Lett. **95B**, 265 (1980). K. Kawarabayashi and N. Ohta, Nucl. Phys. **B175**, 477 (1980).
11. A.A. Migdal and M.A. Shifman, Phys. Lett. **114B**, 445 (1982). J.M. Cornwall and A. Soni, Phys. Rev. **D29**, 1424 (1984); **D32**, 764 (1985).
12. A. Salomone, J. Schechter and T. Tudron, Phys. Rev. **D23**, 1143 (1981). J. Ellis and J. Lanik, Phys. Lett. **150B**, 289 (1985). H. Gomm and J. Schechter, Phys. Lett. **158B**, 449 (1985).
13. H. Gomm, P. Jain, R. Johnson and J. Schechter, Phys. Rev. **D33**, 801 (1986).
14. H. Gomm, P. Jain, R. Johnson and J. Schechter, Phys. Rev. **D33**, 3476 (1986).
15. D.H. Hsu, F. Sannino and J. Schechter, YCTP-P28-97, hep-th/9801097. To appear in Phys. Lett. B.
16. T.R. Taylor, G. Veneziano and S. Yankielowicz, Nucl. Phys. **B218**, 493 (1983).
17. I. Affleck, M. Dine and N. Seiberg, Nucl. Phys. **B241**, 493 (1984); **B256**, 557 (1985).
18. E. Witten, Nucl. Phys. **B156**, 269 (1979). G. Veneziano, Nucl. Phys. **B159**, 213 (1979).
19. A QCD Lagrangian with a similar "holomorphic" piece plus a needed non holomorphic piece was discussed in sect. VII of Ref. [13]. This was partly motivated by a complex form for the two QCD coupling parameters which had been speculated in J. Schechter, *Proceedings of the third annual MRST meeting*, University of Rochester, April 30 – May 1 (1981).
20. N. Kitazawa and F. Sannino, YCTP-1-98, hep-th/9802017; YCTP-P4-98, hep-th/9803225.

21. T. Appelquist and F. Sannino, YTCP-P12-98, hep-ph/9806409. R.D. Mawhinney, CU-TP-839, hep-lat/9705031; CU-TP-802, hep-lat/9705030; D. Chen, R.D. Mawhinney, Nucl. Phys. Proc. Suppl. **53**, 216, 1997.

ℜ-Parity "ℜℓ" Us ℜochester

Mike Bisset[1], Otto C. W. Kong, Cosmin Macesanu, and Lynne H. Orr

*Department of Physics and Astronomy,
University of ℜochester, ℜochester NY 14627-0171*

Abstract. ℜ-parity violating terms in the minimal supersymmetric extension of the Standard Model are studied within a parametrization characterized by a single vacuum expectation value for the scalar components of the $Y = -1/2$ superfields. The benefits of this efficient parametrization are that it enables phenomenological analyses of ℜ-parity violation without *a priori* assumptions (such as setting most ℜ-parity violating terms to zero in an *ad hoc* way). The tree-level fermionic mass eigenstates are derived and shown to be particularly simple in this scheme, involving only the bilinear μ_i ℜ-parity violating terms. Suprisingly large values of these μ_i are consistent with the experimantal constraints. This is outlined.

In honor of representing the ℜ in $\mathcal{MRST}$, it seemed appropriate to choose this title rather than one that is perhaps more revealing. I will report on a very efficient parametrization for studying ℜ-parity violating effects in supersymmetric models. ℜ-parity, which might be described as 'conservation of SUSY-ness', is a discrete symmetry tacked on rather arbitrarily to the supersymmetric framework to distinguish particles from superparticles. ℜ-parity conservation wipes out in one fell swoop a whole host of B- or L-number violating couplings, including ones that could lead to catastrophic scalar superparticle mediated proton decays. However, the complete annihilation of all such couplings is not necessary to yield consistency with proton decay and other experimental constraints. For instance, prohibiting only the B-number violating couplings would be sufficient to prevent the fatal proton decays. Allowing L-number and ℜ-parity violating couplings can lead to distinctive predictions for the results of future experiments. And, as we shall see, suprisingly large ℜ-parity violating couplings are viable.

---

[1] Talk presented by M.B.

The most general Lagrangian compatible with supersymmetry may be written in 4-component language as:

$$\mathcal{L}_{SUSY} = \sum_i (D_\mu S_i)^\dagger (D^\mu S_i) + \frac{i}{2} \sum_i \overline{\psi}_i \not{D} \psi_i$$

$$-\frac{1}{4} \sum_A F_{\mu\nu A} F^{\mu\nu}{}_A + \frac{i}{2} \sum_A \overline{\lambda}_A \not{D} \lambda_A$$

$$-\sqrt{2} \sum_i \left[S_i^\dagger (t_A) \overline{\psi}_i \frac{1-\gamma_5}{2} \lambda_A + \text{h.c.} \right]$$

$$-\frac{1}{2} \sum_A \left[\sum_i S_i^\dagger t_A S_i + \xi_A \right]^2 - \sum_i \left| \frac{\partial W}{\partial \hat{S}_i} \right|^2_{\hat{S}=S}$$

$$-\frac{1}{2} \sum_{i,j} \left\{ \overline{\psi}_i \left[\frac{1-\gamma_5}{2} \right] \left[\frac{\partial^2 W}{\partial \hat{S}_i \, \partial \hat{S}_j} \right]_{\hat{S}=S} \psi_j + \text{h.c.} \right\}, \quad (1)$$

where the scalar and Majorana fermionic components of a chiral superfield, $\hat{S}$, are denoted by S and ψ, respectively; the λ_A are the gauginos and W is the superpotential. The most general renormalizable superpotential for the supersymmetric standard model without $\Re$-parity can be written as:

$$W = \varepsilon_{ab} \left[\mu_\alpha \hat{L}_\alpha^a \hat{H}_u^b + h_{ik}^u \hat{Q}_i^a \hat{H}_u^b \hat{U}_k^c + \lambda'_{i\alpha k} \hat{Q}_i^a \hat{L}_\alpha^b \hat{D}_k^c + \lambda_{\alpha\beta k} \hat{L}_\alpha^a \hat{L}_\beta^b \hat{E}_k^c \right] + \lambda''_{ijk} \hat{D}_i^c \hat{D}_j^c \hat{U}_k^c ,$$

where (a, b) are $SU(2)$ indices, (i, j, k) are family (flavor) indices, and (α, β) are (extended) flavor indices from 0 to 3 with $\hat{L}_\alpha$'s denoting the four doublet superfields with $Y = -1/2$. λ and λ'' are antisymmetric in the first two indices as required by $SU(2)$ and $SU(3)$ product rules respectively. The bilinear μ_i couplings and the trilinear λ_{ijk}, λ'_{ijk}, and λ''_{ijk} couplings give $\Re$-parity violating ($\Re$PV) interactions. If these all vanish, one recovers the expression for the $\Re$-parity conserving ($\Re$PC) minimal supersymmetric standard model (MSSM), with $\hat{L}_0$ identified as $\hat{H}_d$. In that case, the two Higgses acquire vacuum expectation values (VEV's) and the lepton and quark Yukawa couplings are given by $h_{ik}^e \equiv 2\lambda_{i0k} (= -2\lambda_{0ik})$, $h_{ik}^d \equiv \lambda'_{i0k}$, and $-h_{ik}^u$, respectively.

The full Lagrangian for the model also includes a soft SUSY-violating part,

$$\mathcal{L} = \mathcal{L}_{SUSY} + \mathcal{L}_{soft} \quad (2)$$

where $\mathcal{L}_{soft}$ contains both $\Re$PC and $\Re$PV terms, introducing a disgusting number of new independent parameters into the theory (if no assumptions about relationships among the parameters trickling down from higher energy scales are introduced). Fortunately, none of these $\Re$PV terms enter into the tree-level fermion mass matrices on which this report concentrates, and the only $\Re$PC terms that enter are the soft gaugino masses, M_1 and M_2, which are assumed to be related by gaugino unification, $M_1 = \frac{5}{3}\tan^2\theta_w M_2 \equiv xM_2$.

The other sources of $\Re$PV interactions in the model are vacuum expectation values (VEV's) for the sneutrinos ($\tilde{\nu}_i$), denoted by $\mathbf{v}_i \equiv \sqrt{2}\langle \tilde{\nu}_i \rangle = \sqrt{2}\langle \hat{L}_i \rangle$ It is assumed that neither a right-handed neutrino nor the sneutrino that would be its scalar superpartner is present. In this sense the model contains the absolute minimum number of new particles needed to supersymmetrize the SM. If ν_R's and their associated sneutrinos are added, then these sneutrinos could also acquire VEV's and complicate the general analysis presented here. Even without right-handed neutrinos we will see that a Majorana neutrino mass is generated at tree-level via the neutrinos mixing with gauginos and higgsinos.

The tree-level neutral fermion mass mixing matrix has contributions from lines 3 and 5 of Eq.(1) as well as M_1 and M_2 from $\mathcal{L}_{soft}$. In an arbitrary flavor basis it is given by

$$\mathcal{M}_\mathcal{N}^{(o)} = \begin{pmatrix} M_1 & 0 & \frac{g_1 v_u}{2} & -\frac{g_1 v_d}{2} & -\frac{g_1 v_1}{2} & -\frac{g_1 v_2}{2} & -\frac{g_1 v_3}{2} \\ 0 & M_2 & -\frac{g_2 v_u}{2} & \frac{g_2 v_d}{2} & \frac{g_2 v_1}{2} & \frac{g_2 v_2}{2} & \frac{g_2 v_3}{2} \\ \frac{g_1 v_u}{2} & -\frac{g_2 v_u}{2} & 0 & -\mu_0 & -\mu_1 & -\mu_2 & -\mu_3 \\ -\frac{g_1 v_d}{2} & \frac{g_2 v_d}{2} & -\mu_0 & 0 & 0 & 0 & 0 \\ -\frac{g_1 v_1}{2} & \frac{g_2 v_1}{2} & -\mu_1 & 0 & 0 & 0 & 0 \\ -\frac{g_1 v_2}{2} & \frac{g_2 v_2}{2} & -\mu_2 & 0 & 0 & 0 & 0 \\ -\frac{g_1 v_3}{2} & \frac{g_2 v_3}{2} & -\mu_3 & 0 & 0 & 0 & 0 \end{pmatrix}. \quad (3)$$

It is not necessary, however, to retain all five VEV's in parametrizing the model, because the freedom associated with the choice of flavor basis creates redundancy amongst the parameters. A $U(4)$ rotation can be used to set all the v_i to zero, leaving a single VEV for $\langle \hat{L}_0 \rangle$. This is analogous to the $U(3)$ rotational freedom used to diagonalize the up-quark sector, leaving the six quark masses and the 4 parameters of the CKM-matrix as the inputs to be determined. In our basis, which we call the single-VEV parametrization, the down-type Higgs $\hat{H}_1 (\equiv \hat{L}_0)$ is identified as the $Y = -1/2$ doublet that bears the full VEV among the $\hat{L}_\alpha$'s. In this basis the tree-level neutral fermion mass mixing matrix becomes

$$\mathcal{M}_\mathcal{N} = \begin{pmatrix} M_1 & 0 & \frac{g_1 v_u}{2} & -\frac{g_1 v_d}{2} & 0 & 0 & 0 \\ 0 & M_2 & -\frac{g_2 v_u}{2} & \frac{g_2 v_d}{2} & 0 & 0 & 0 \\ \frac{g_1 v_u}{2} & -\frac{g_2 v_u}{2} & 0 & -\mu_0 & -\mu_1 & -\mu_2 & -\mu_3 \\ -\frac{g_1 v_d}{2} & \frac{g_2 v_d}{2} & -\mu_0 & 0 & 0 & 0 & 0 \\ 0 & 0 & -\mu_1 & 0 & 0 & 0 & 0 \\ 0 & 0 & -\mu_2 & 0 & 0 & 0 & 0 \\ 0 & 0 & -\mu_3 & 0 & 0 & 0 & 0 \end{pmatrix}. \quad (4)$$

Diagonalizing this matrx yields the neutralino and neutrino mass eigenvalues. Two neutrino eigenstates are left massless, while the third one gains a mass through the $\Re$PV-couplings (μ_i's) to the higgsino. The massive eigenstate is in general a mixture of all three neutrino states; however, a simple rotation decouples the two massless states, leaving a 5×5 mass matrix given by

$$\mathcal{M}_\mathcal{N}^{(5)} = \begin{pmatrix} M_1 & 0 & \frac{g_1 v_u}{2} & -\frac{g_1 v_d}{2} & 0 \\ 0 & M_2 & -\frac{g_2 v_u}{2} & \frac{g_2 v_d}{2} & 0 \\ \frac{g_1 v_u}{2} & -\frac{g_2 v_u}{2} & 0 & -\mu_0 & -\mu_5 \\ -\frac{g_1 v_d}{2} & \frac{g_2 v_d}{2} & -\mu_0 & 0 & 0 \\ 0 & 0 & -\mu_5 & 0 & 0 \end{pmatrix}, \quad (5)$$

where
$$\mu_5 = \sqrt{\mu_1^2 + \mu_2^2 + \mu_3^2} \; ; \tag{6}$$
and the corresponding massive neutrino state is given by
$$|\nu_5\rangle = \frac{\mu_1}{\mu_5}|\nu_1\rangle + \frac{\mu_2}{\mu_5}|\nu_2\rangle + \frac{\mu_3}{\mu_5}|\nu_3\rangle \; . \tag{7}$$

Eigenvalues can be found numerically, or various approximation strategies can be employed. If μ_5 is small, then a "seesaw" approximation yields
$$m_{\nu_5} = \frac{\det \mathcal{M}_N^{(5)}}{\det \mathcal{M}_{4\times 4}} = -\frac{1}{2}\frac{\mu_5^2 v^2 \cos^2\beta \, (xg_2^2 + g_1^2)}{\mu_0 \, [2xM_2\mu_0 - (xg_2^2 + g_1^2)v^2 \sin\beta \cos\beta]} \tag{8}$$
where $v_d = v\cos\beta$ and $v_u = v\sin\beta$. Alternatively, the EW-symmetry breaking terms proportional to the VEV's can be treated as perturbations, leading to the formula
$$m_{\nu_5} = -\frac{1}{4}\frac{\mu_5^2 v^2 \cos^2\beta \, (xg_2^2 + g_1^2)}{(\mu_0^2 + \mu_5^2)xM_2} \Rightarrow \mu_5^2 < \frac{4x\mu_0^2 M_2 m_{\nu_5(\text{bound})}}{v^2 \cos^2\beta\,(xg_2^2 + g_1^2) - 4xM_2 m_{\nu_5(\text{bound})}} \; . \tag{9}$$

As M_2 increases, the denominator above drops to zero, beyond which there is *no* bound on μ_5. Using the τ-neutrino machine bound of 18.2 MeV [1] this occurs at $M_2 \sim 275$ GeV for $\tan\beta = 45$. And μ_5 values comparable to those of M_2 and μ_0 are allowed throughout the interesting region of the parameter space, as shown in the exact numerical results in Figure 1. Comparing the results for $\tan\beta = 2$ and $\tan\beta = 45$ we see the weakening of the restriction on μ_5 arising from the $\cos\beta$ suppression factor seen in the approximate mass expressions above. (This gives factor of $\sim 10^{-3}$ for $\tan\beta = 45$.)

The single-VEV parametrization is not the only possible choice to eliminate the spurious degrees of freedom. Another popular parametrization exploiting the flavor rotations has all μ_α's except μ_0 as well as two of the v_i's set to zero

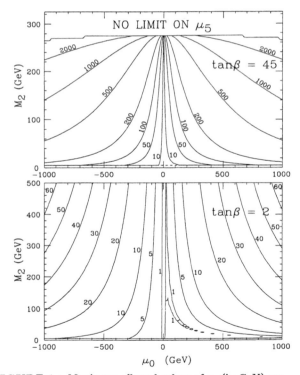

FIGURE 1.: Maximum allowed values of μ_5 (in GeV) consistent with $m_{\nu_\tau} < 18.2$ MeV ($\mu_1 : \mu_2 : \mu_3 = 0 : 0 : 1$). The region above or outside of a given contour is excluded for μ_5's above the indicated value.

(rotated away) [2]. This we term the single-μ parametrization. (In this context sneutrinos refer to the scalar components of the three $\hat{L}_i$ superfields as defined in the basis where the μ_i's are zero.) To compare the relative merits of these two choices, we look next at the charged fermion sector.

The tree-level mixing mass matrix for the charged fermion sector in an arbitrary flavor basis has the form

$$\mathcal{M}_C^{(o)} = \begin{pmatrix} M_2 & \frac{g_2 v_u}{\sqrt{2}} & 0 & 0 & 0 \\ \frac{g_2 v_d}{\sqrt{2}} & \mu_0 & -h_{i1}^e \frac{v_i}{\sqrt{2}} & -h_{i1}^e \frac{v_i}{\sqrt{2}} & -h_{i1}^e \frac{v_i}{\sqrt{2}} \\ \frac{g_2 v_1}{\sqrt{2}} & \mu_1 & h_{11}^e \frac{v_d}{\sqrt{2}} + 2\lambda_{i11} \frac{v_i}{\sqrt{2}} & h_{12}^e \frac{v_d}{\sqrt{2}} + 2\lambda_{i12} \frac{v_i}{\sqrt{2}} & h_{13}^e \frac{v_d}{\sqrt{2}} + 2\lambda_{i13} \frac{v_i}{\sqrt{2}} \\ \frac{g_2 v_2}{\sqrt{2}} & \mu_2 & h_{21}^e \frac{v_d}{\sqrt{2}} + 2\lambda_{i21} \frac{v_i}{\sqrt{2}} & h_{22}^e \frac{v_d}{\sqrt{2}} + 2\lambda_{i22} \frac{v_i}{\sqrt{2}} & h_{23}^e \frac{v_d}{\sqrt{2}} + 2\lambda_{i23} \frac{v_i}{\sqrt{2}} \\ \frac{g_2 v_3}{\sqrt{2}} & \mu_3 & h_{31}^e \frac{v_d}{\sqrt{2}} + 2\lambda_{i31} \frac{v_i}{\sqrt{2}} & h_{32}^e \frac{v_d}{\sqrt{2}} + 2\lambda_{i32} \frac{v_i}{\sqrt{2}} & h_{33}^e \frac{v_d}{\sqrt{2}} + 2\lambda_{i33} \frac{v_i}{\sqrt{2}} \end{pmatrix},$$

where the i's are summed over. Going to the single-VEV parametrization, this becomes

$$\mathcal{M}_C = \begin{pmatrix} M_2 & \frac{g_2 v_u}{\sqrt{2}} & 0 & 0 & 0 \\ \frac{g_2 v_d}{\sqrt{2}} & \mu_0 & 0 & 0 & 0 \\ 0 & \mu_1 & m_1 & 0 & 0 \\ 0 & \mu_2 & 0 & m_2 & 0 \\ 0 & \mu_3 & 0 & 0 & m_3 \end{pmatrix} \qquad (10)$$

after the remaining available leptonic flavor rotations are used to diagonalize the 3 × 3 charged lepton submatrix. (Here $m_i = h_{ii}^e \frac{v_d}{\sqrt{2}}$ are the Yukawa mass terms for the leptons.) In contrast, in the single-μ parametrization the lepton Yukawa matrix h_{ik}^e *cannot* then be taken as diagonalized, since a set of λ-couplings associated with the nonzero v still remain. In the single-VEV parametrization, the (tree-level) mass matrices for *all* the fermions *do not* involve any trilinear $\Re$PV couplings. These very simple mass matrices are obtained *without any a priori assumptions*; we have simply chosen to parametrize the model in a specific flavor basis [3]. The single-μ parametrization was introduced originally in studies of $\Re$PV effects on the leptons under the assumption that the trilinear $\Re$PV couplings were zero [2]. Extending its usage to the most general $\Re$PV scenario is clearly difficult.

Unitary matrices, U_L and U_R, can be found such that

$$U_L^\dagger \mathcal{M}_C U_R = \mathrm{diag}\{\bar{M}_{c1}, \bar{M}_{c2}, \bar{m}_1, \bar{m}_2, \bar{m}_3\}, \qquad (11)$$

yielding the chargino and charged lepton masses. Note that so long as the μ_i's are not all zero, the L_i's are not exactly identifiable as the physical charged leptons (though the L_i's align fairly well with the charged-lepton mass eigenstates for small μ_i's), nor are the charginos and neutralinos, for instance, the same states as in the $\Re$-parity conserving MSSM. This means that the Yukawa mass terms in Eq.(10) are not the physical lepton masses. Perturbatively [1], the m_i's can be approximated by the $\bar{m}_i$'s, the physical charged lepton masses.

---

[1] see [4] and [5] for approximate expressions for the chargino masses and the U_L and U_R matrix elements.

For the exact computations, one can numerically integrate from $\mu_i = 0$ (for which the m_i are exactly the $\bar{m}_i$) to the final μ_i values. This is necessary to find an acceptable set of m_i's that yield the correct physical charged lepton masses for a given set of μ_i's. The charginos are as of yet undiscovered so that their masses are not fixed. These also depend on the μ_i's, or, to a very good approximation, on μ_5 since the charged lepton masses are relatively tiny. For example, the minimum μ_5 values required to give both chargino masses above 90 GeV for $\tan\beta = 45$ are shown in Figure 2.

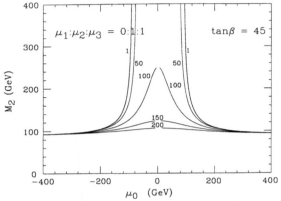

FIGURE 2.: Minimum values of μ_s (in GeV) required to give both chargino masses above 90 GeV The area above or outside of a given contour has both chargino masses > 90 GeV for μ_s's above the indicated value.

The couplings of the charged leptons and charginos to the gauge bosons come from the covariant derivatives in the second term in line 1 (or line 2) of Eq.(1). These will depend on the matrix elements of U_L and U_R. For instance, the Z^0-boson coupling to the mass eigenstates is given by

$$\mathcal{L}_{int}^{Z\bar{\chi}^-\chi^-} \equiv \frac{g_2}{2\cos\theta_w} Z^\mu \bar{\chi}_i^- \gamma_\mu \left(\tilde{A}_{ij}^L \frac{1-\gamma_5}{2} + \tilde{A}_{ij}^R \frac{1+\gamma_5}{2} \right) \chi_j^- . \quad (12)$$

where

$$\tilde{A}_{ij}^L = U_L^{i1} U_L^{j1} + \delta_{ij}(1 - 2\sin^2\theta_w)$$
$$\tilde{A}_{ij}^R = 2U_R^{i1} U_R^{j1} + U_R^{i2} U_R^{j2} - 2\delta_{ij}\sin^2\theta_w \quad (13)$$

There are a large number of well-measured (and not-so-well measured) quantities which are determined by these couplings, a partial laundry list of which is presented in Table 1. In Ref. [6], many of these constraints are studied assuming the trilinear $\Re$PV couplings vanish. We expand on their work, but need not make the foregoing assumption. Also, the single-VEV parametrization allows the constraints to be cast explicitly in terms of the magnitudes of the μ_i's. The relative sizes of μ_1, μ_2, and μ_3 will be very important here. See [4] and [5] for a detailed treatment. Here I will just illustrate our results with one of the most stringent constraints, that from B.R.$(\mu^- \longrightarrow e^- e^+ e^-) < 1.0^{-12}$ [7]. Figure 3 shows contours of μ_5 (to parallel Figure 1 on neutrino masses) assuming a ratio $\mu_1 : \mu_2 : \mu_3 = 1 : 1 : 0$. Again, we see a substantial weakening of the constraint for large $\tan\beta$, which is again borne out by perturbative formulæ. These formulæ as well as the exact numerical results

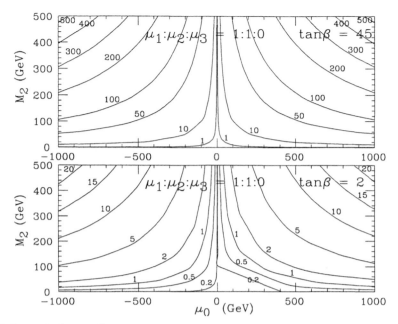

FIGURE 3. Maximum allowed values of μ_s (in GeV) ($\mu_1 : \mu_2 : \mu_s = 1 : 1 : 0$) consistent with B.R.$(\mu^- \longrightarrow e^- e^+ e^-) < 1.0^{-12}$ The region above or outside of a given contour is excluded for μ_s's above the indicated value.

show that the rate for $\mu^- \longrightarrow e^- e^+ e^-$ depends on the product $|\mu_1 \mu_2|$ and thus can be eluded if either μ_1 or μ_2 is tiny or zero. Restrictions from many of the other Z^0-mediated processes are much weaker since their respective experimental bounds are not as strong. Again, see [4] and [5] for a detailed treatment.

Several caveats to the results presented here need to be addressed. First, all parameters have been assumed to be real — potentially rich CP violating features of the model have not been considered yet. Second, if ν_5 is not aligned with the τ-lepton, then the experimental bound on m_{ν_5} is weakened considerably since the τ can then dribble off into one of the massless (at tree-level) neutrinos to which it also couples. Estimates of the bound in this case are around 149 MeV [8]. There are potentially much stronger bounds on neutrino masses from cosmological considerations which however depend on the decay modes and other assumptions so that a neutrino mass above an MeV is not definitely ruled out [9]. There are also other experimental constraints on neutrino masses and mixings not addressed yet.

Though the single-VEV parametrization, with explicit bilinear $\Re$PV-terms, does not manifestly exhibit ʋ's it is nevertheless possible to use this framework to describe a spontaneously broken $\Re$-parity scenario, which is defined by the existence of ʋ's in some basis where *all* explicit $\Re$PV terms vanish. A rotation of the $\hat{L}_\alpha$ would connect such a basis to that of the single-VEV parametrization. Note that

TABLE 1. Summary of Constraints on $\Re$-Parity Violation

Z^0-coupling:	$W^\pm$-coupling:
• B.R.$(\mu \to eee)$ • B.R.$(\tau \to eee)$ • B.R.$(\tau^\pm \to \mu^\pm ee)$ • B.R.$(\tau^\pm \to \mu^\mp ee)$ • B.R.$(\tau^\pm \to e^\pm \mu\mu)$	• π decays $R^{\pi e}_{\pi\mu}$ • τ decays $R^{\tau e}_{\tau\mu}$ • decays to e's $R^{\mu e}_{\tau e}$ • ν-less double-β dacay
• B.R.$(\tau^\pm \to e^\mp \mu\mu)$ • B.R.$(\tau \to \mu\mu\mu)$ • B.R.$(Z^0 \to e^\pm \mu^\mp)$ • B.R.$(Z^0 \to e^\pm \tau^\mp)$ • B.R.$(Z^0 \to \mu^\pm \tau^\mp)$ • B.R.$(Z^0 \to \chi^\pm \ell^\mp)$ • B.R.$(Z^0 \to \chi^\pm \chi^\mp)$ • e-μ universality • e-τ universality • μ-τ universality • e-μ L-R asymmetry • τ-e L-R asymmetry • τ-μ L-R asymmetry • $\Delta\Gamma_Z$ • $\Delta\Gamma_Z^{invisible}$ • B.R.$(Z^0 \to \chi_i^0 \chi_j^0, \chi_j^0 \nu)$; $j \neq 1$	**mass constraints:** • ν_5 mass • $\chi_1^\pm$ mass See [5] for explicit values for these quantites and a detailed analysis.

such a rotation to the single-VEV basis in general will introduce not only bilinear (μ_i) terms but also trilinear couplings and $\Re$PV terms from $\mathcal{L}_{soft}$ as well.

There are numerous discussions in the literature which consider the MSSM with only a few $\Re$PV terms added (often one or two trilinear $\Re$PV terms are added) while the v_i's and μ_i's are *both* assumed to be zero. Imposing such assumptions at the Lagrangian level is not a (flavor) basis or parametrization independent procedure. Additional assumptions needed to impose these conditions on top of the allowed freedom from flavor rotations are not clearly specified. This confusion is eliminated by beginning with the Lagrangian in the single-VEV parametrization and then specifying the values of the $\Re$PV terms. In the single-VEV parametrization, setting the μ_i's to zero would be an additional *ad hoc* assumption. All in all, while specific $\Re$PV models may be more naturally formulated in another parametrization (low-energy phenomenology does not distinguish between different flavor bases, but assumed conditions from higher energy scales could select the single-μ parametrization or a mixed basis with both μ_i's and v_i's), the single-VEV parametrization does provides a particularly efficient framework for general phenomenological studies.

In summary, in the single-VEV parametrization, the sneutrino VEV's are set to zero; however, the bilinear $\Re$PV couplings cannot then also be set to zero for free. In this basis trilinear $\Re$PV couplings play no role in the tree-level fermion masses.

At large tanβ, there are strong suppression factors which allow very large μ_i values to not conflict with various experimental bounds. The clean way in which the single-VEV parametrization deals with the fermion masses will hopefully enable a more comprehensive phenomenological analysis of both scalar-mediated fermion decays and scalar (Higgs boson and slepton) decays. Here the parameter space will expand considerably since trilinear $\mathcal{R}$PV couplings and many terms from $\mathcal{L}_{soft}$ now contribute significantly. Nonetheless, eliminating 3 of the 5 VEV's does simplify the analysis of the scalar sector, as some preliminary calculations presented in [5] clearly show, and a more comprehensive analysis of the full model now seems feasible using the single-VEV parametrization.

This work was supported in part by the U.S. Department of Energy, under grant DE-FG02-91ER40685 and by the U.S. National Science Foundation, under grant PHY-9600155.

REFERENCES

1. R. Barate et al. (ALEPH Collaboration), CERN-PPE-97-138, (1997).
2. R. Barbieri, D.E. Brahm, L.J. Hall, and S.D.H. Hsu, Phys. Lett. **B238** (1990) 86; L.J. Hall and M. Suzuki, Nucl. Phys. **B231** (1984) 419.
3. E. Nardi, Phys. Rev. **D55** (1997) 5772.
4. M. Bisset, O.C.W. Kong, C. Macesanu, and L.H. Orr, Phys. Lett. **B430** (1998) 274.
5. M. Bisset, O.C.W. Kong, C. Macesanu, and L.H. Orr, U$\mathcal{R}$-1524.
6. A. Joshipura and M. Nowakowski, Phys. Rev. **D51** (1995) 5271; M. Nowakowski and A. Pilaftsis, Nucl. Phys. **B461** (1996) 19.
7. R.M. Barnett et al. (Particle Data Group), Phys. Rev. **D54** (1996) 1.
8. A. Bottini, N. Fornengo, C.W. Kim, and G. Mignola, Phys. Rev. **D53** (1996) 6361.
9. K.S. Babu, T.M. Gould, and I.Z. Rothstein, Phys. Lett. **B321** (1994) 140; A.D. Dolgov, S. Pastor, and J.W.F. Valle, Phys. Lett. **B383** (1996) 193; K. Kainulainen, hep-ph/9608215. M. Kawasaki, P. Kernan, H.-S. Kang R.J. Scherrer, G. Steigman, and T.P. Walker, Nucl. Phys. **B419** (1994) 105.

Low Energy Z_R^0 Based on an $SO(10)$ SUSY-GUT

Calvin S. Kalman

Physics Department, Concordia University, Montréal, Québec, H3G 1M8, Canada

Abstract. We find agreement with the experimental data for a mass of Z_R^0 in the range of 800 GeV – 2000 GeV.

I INTRODUCTION

Amaldi et al. [1] point out using data from DELPHI at LEP that in the minimal non-supersymmetric standard model with one Higgs doublet, a single unification point is excluded by more than 7 standard deviations. In contrast the minimum supersymmetric standard model leads to a good agreement with a single unification scale. There is some further experimental evidence for SUSY. Connors et al. [2] show SUSY leads to the temperature fluctuations found in the recent COBE measurement. Gordon Kane points out that there are at least nine phenomenological indications that nature is supersymmetric [3].

Exact supersymmetry would require the mass of the scalar partner of the electron (selectron) to be degenerate in mass with the electron. Such a light particle should have been observed by now. Moreover, as shown by Fayet [4], exact supersymmetry would also mean that the anomalous magnetic moment of all the leptons would be identically zero. This occurs because every loop in the calculation of the magnetic moment has a counterpart loop composed of opposite-type particles to cancel them. We must then consider a supersymmetric theory broken globally or locally.

As shown by Dimopoulos and Georgi [5], global breaking of supersymmetry is unacceptable for three reasons:

(1) Such a model still contains light scalar particles, which should have been observed by now, in particular, a scalar partner of one of the quarks (squark) must always be lighter than the lightest up or down quark.

(2) In global supersymmetry the vacuum energy is an order parameter. Breaking occurs if $E_{\text{vac}} \neq 0$, implying a nonzero cosmological constant.

(3) As shown by Hawking et al. [6], there remain quadratic divergences in the calculation of the Higgs-boson mass because of the couplings of the Higgs-boson to gravitons.

We must then break supersymmetry locally, which automatically incorporates gravity into any theory [7,8]. In broken local supersymmetry, usually referred to as supergravity, a spin 3/2 partner (gravitino) to the graviton naturally occurs. The contribution to the Higgs-boson mass arising from the coupling of the graviton to the Higgs-boson is canceled by the contribution of the gravitino.[1]

Problems (1) and (2) are also cured in supergravity.

Consider then a supersymmetric extension of the standard model. The most general superpotential has the form:

$$W = h_u Q H_u U^c + h_d Q H_d D^c + h_e L H_d E^c + \mu_1 H_u H_d + \mu_2 H_u L \\ + f_{pqr} Q_r L D_r^c + \tilde{h}_{[pq]r} L_p L_q + \lambda_1 E^c H_d H_d + \lambda_{[pq]r} D_q D_r^c. \quad (1)$$

The last term violates baryon number conservation and leads to rapid proton decay. The next to last four terms violate lepton number conservation. If the coefficients μ_2, f_{pqr}, $\tilde{h}_{[pq]r}$, λ_1 and $\lambda_{[pq]r}$ are set to zero, the terms cannot be regenerated at the tree level due to the nonrenormalization theorem of supersymmetric field theory. Setting the coefficients to zero corresponds to the "standard" supersymmetric model [11]. No baryon or lepton number violating terms are present a priori in left-right models. The original motivation to introduce such a model based upon $SU(2)_L \times SU(2)_R \times U(1)_{B-L}$ was to provide a possible motivation for parity violation in weak interactions. In this framework the weak interaction respects all space-time symmetries, as do the strong, electromagnetic, and gravitational interactions. The asymmetry observed in nature at low energies is then attributed to the noninvariance of the vacuum under parity symmetry [12]. A bonus of this approach is that it reproduces all the features of $SU(2)_L \times U(1)_Y$ at low energies.

Another compelling reason to consider LR models is found in CP violation. In the Cabibbo-Kobayashi-Maskawa parameterization of generation mixing, for three generations, all CP violations are dependent on only one parameter, δ_{CKM} (the CKM phase), and there is no hint as to why the observed CP-violation has milliweak strength. The LR model can give rise to CP violation for only two generations and can account for its strength by relating it to the suppression of $V + A$ currents [13].

Furthermore if the neutrino has a mass [14], then this class of model becomes the most natural framework in which to work. In, addition if it turns out that quarks and leptons are themselves the results of a more fundamental substructure [15], and that the forces operating at the substructure level are similar to QCD [16], then there are strong arguments which point to $SU(2)_L \times SU(2)_R \times U(1)_{B-L}$ as

---

[1] In exact supergravity there is an exact cancelation of the quadratic divergence arising from a coupling of the Higgs boson to gravitons. In broken supergravity the coupling still does not pose a problem and moreover, as seen, for example in broken supergravity models based upon grand unified schemes involving the standard model [9], the masses of the particles can be related to the gravitino mass. This occurs because the breaking of supergravity (super Higgs effect) [10] drives the parameter $-m^2$ in the Higgs potential $V(f) = \lambda \phi^4 - m^2 \phi^2$ negative, producing the spontaneous symmetry breaking occurring in the ordinary Higgs effect.

the weak-interaction symmetry rather than $SU(2)_L \times U(1)_Y$. The $B-L$ quantum number [17] (baryon number minus lepton number) is the only anomaly-free quantum number left ungauged in the standard model, a fact which seems to suggest a deeper symmetry structure. By replacing the gauge generator $U(1)_Y$, which has no physical significance, with $U(1)_{B-L}$, all the generators of the theory acquire a physical meaning.

II LOW-ENERGY Z_R^0 WITHIN THE $SO(10)$ SUSY GUT

The simplest SUSY GUT containing $SU(2)_L \times SU(2)_R \times U(1)_{B-L}$ is $SO(10)$. Not only does supersymmetric $SO(10)$ resolve many problems experienced in other theories but also there is the beginning of experimental evidence for these theories [18]. There is however, no point in pursuing left-right symmetric theories unless there is a reasonable chance of observing the right-handed Z in experiments in the near future. For non-supersymmetric $SO(10)$, there are a number of models which feature low-energy parity restoration [19]. For example Deshpande et. al. [20] obtain a low value for M_R the $SU(2)_R$ breaking scale but at the cost of introducing a complicated Higgs structure. Their model also does not have a simple see-saw mechanism for the neutrino masses and it is all a little less elegant than the usual LR model. Besides with this solution to neutrino masses, one will probably break R-parity too [21]. There have been some attempts to do this for supersymmetric $SO(10)$ as well. Among these, Datta, Pakvasa and Sarkar [22] use an approach that allows LR symmetry to survive down to ~ 1 TeV at the cost of introducing higher dimensional operators in the $SO(10)$ Lagrangian.

In this paper we redo the non-supersymmetric $SO(10)$ calculations of Chang, Mohapatra and Parida [23], Chang, Mohapatra, Gipson, Marshak and Parida [24], and Gipson and Marshak [25] for supersymmetric $SO(10)$ using the latest LEP data. Basically we use the latest LEP data and the renormalization group to find out what parameters are consistent with low energy right-handed Z and possibly W bosons and a common unification point for the coupling constants. (In this we follow the pattern described by Barger and Phillips [26] and Lopez, Nanopoulos and Zichichi [27] for the minimal supersymmetric model.)

III THE CALCULATION

The notation is as similar to Mohapatra [11]:

$$SO(10) \xrightarrow[210]{M_U} 2213P \xrightarrow[210]{M_P} 2213 \xrightarrow[210]{M_R} 2113 \xrightarrow{M_S} \xrightarrow[126]{M_{R^0}} 213$$

We employ $2213P$ as a shorthand for $SU(3)_c \times SU(2)_L \times U(1)_{I_{3R}} \times U(1)_{B-L} \times P$; similarly the other entries are corresponding subgroups. The numbers across the bottom line specify the dimensionality of a representation of $SO(10)$. As used above it means breaking the symmetry group G at a mass M_X using Higgs that are elements of this particular representation.

The particle content at each scale is as follows (See He and Meljanac [28], Lee [29] and Lee and Mohapatra [30] for more information and for details on the notation.).

At all scales the following particles will be present:

$$Q(3,2,1,1/(2\sqrt{6})), \quad Q^c(\overline{3},1,2,-1/(2\sqrt{6})),$$
$$L(1,1,2,-1/(2\sqrt{2/3})), \quad L^c(1,2,1,1/(2\sqrt{2/3})).$$

Additionally at other scales the following particles will contribute.

For $M_P < \mu < M_U$:

$$w_L^3(3,3,1,\sqrt{2/3}), \quad w_L^{\overline{3}}(,3,1,-\sqrt{2/3}),$$
$$w_R^3(3,1,3,\sqrt{2/3}), \quad w_R^{\overline{3}}(,1.3,-w_L^{\overline{3}}),$$
$$\Delta_L(1,3,1,\sqrt{3/2}), \quad \Delta_R(1,3,1,\sqrt{3/2}),$$
$$\delta_L(1,3,1,-\sqrt{3/2}), \quad \delta_R(1,3,1,-\sqrt{3/2}),$$
$$d(1,2,2,0), \quad B_{5^-}(1,1,1,0).$$

For $M_{R^+} < \mu < M_P$, d will no longer contribute. The other particles remain.

For $M_{R^0} < \mu < M_{R^+}$, the extra particles present are:

$$\begin{pmatrix} H_u^+ \\ H_u^0 \end{pmatrix}(1,2,1/2,0), \quad \begin{pmatrix} H_d^0 \\ H_d^- \end{pmatrix}(1,2,-1/2,0), \quad \Delta_{R^0}(1,1,1,\sqrt{3/2}).$$

Additionally, $\delta_{R^0}(1,1,1,-\sqrt{3/2})$ will remain until M_S is reached.

For $M_Z < \mu < M_{R^0}$, the only extra particles are:

$$\begin{pmatrix} H_u^+ \\ H_u^0 \end{pmatrix}(1,2,0), \quad \begin{pmatrix} H_d^0 \\ H_d^- \end{pmatrix}(1,2,0).$$

We now need the two loop renormalization group equations for the coupling constants:

$$\frac{d\alpha_i}{dt} = \frac{\alpha_i^2}{2\pi}\left(b_{0i} + \sum_j b_{ij}\frac{\alpha_j}{4\pi}\right) \tag{2}$$

where $t = ln(\mu)$, μ is the energy of the scale, $\alpha = g_i^2/4\pi$, and g_i is the coupling constant of the group i. We use the following shorthand notations:

$$\alpha_1 = \frac{5}{3\cos^2\theta_W}\alpha_{em}, \quad \alpha_2 = \alpha_L = \frac{\alpha_{em}}{\sin^2\theta_W} \quad \text{and} \quad \alpha_3 = \alpha_S. \tag{3}$$

The β functions,

$$\beta_i = \frac{\alpha_i}{2\pi}\left(b_{0i} + \sum_j b_{ij}\frac{\alpha_j}{4\pi}\right), \tag{4}$$

for the groups used in this paper are derived using the formalism developed by Del Aguila, Coughlan and Quiros [31] based on the work on one loop β functions by Gross and Wilczek, and Politzer [32] and on two loop β functions by Jones and by Caswell [33].

Write $y_i = 1/\alpha_i$. Then Eq. (2) takes the form

$$\frac{dy_i}{dt} = -\left(\frac{b_{0i}}{2\pi} + \sum_j b_{ij}\frac{y_i^{-1}}{8\pi^2}\right). \tag{5}$$

We then have to numerically integrate the following equations:
$M_Z < \mu < M_{R^0}$:

$$\begin{aligned}
y_1' &= -23/10\pi - 217/(400\pi^2 y_1) - 22.5/(40\pi^2 y_2) - 44/(40\pi^2 y_3) \\
y_2' &= 7/6\pi - 1.5/(8\pi^2 y_1) - 97/(48\pi^2 y_2) - 1.5/(\pi^2 y_3) \\
y_3' &= 7/2\pi - 1.1/(8\pi^2 y_1) - 4.5/(8\pi^2 y_2) + 26/(8\pi^2 y_3).
\end{aligned} \tag{6}$$

$M_{R^0} < \mu < M_S$ (for scenario I, where supersymmetry is broken above the Z_R^0 mass). Here $\alpha_1 = \alpha_{B-L}$, $\alpha_2 = \alpha_{I_{3R}}$, $\alpha_3 = \alpha_L$ and $\alpha_4 = \alpha_S$. Following Deshpande et. al. [20], note that $y_{I_{3R}}(M_{R_0}) = \frac{5}{3}y_Y(M_{R_0}) - \frac{2}{3}y_{B-L}(M_{R_0})$.

$$\begin{aligned}
y_1' &= -7/2\pi - 78/(16\pi^2 y_1) + 27/(4\pi^2) - 21/(16\pi^2 y_2) \\
&\quad -33/(16\pi^2 y_3) - 0.5/(\pi^2 y_4) \\
y_2' &= -2.5/\pi - 21/(16\pi^2 y_1) + 9/(4\pi^2) - 9/(8\pi^2 y_2) - 1.5/(\pi^2 y_4) \\
y_3' &= 7/6\pi - 3/(16\pi^2 y_1) - 233/(48\pi^2 y_3) - 2/(\pi^2 y_4) \\
y_4' &= 3.5/\pi - 1/(16\pi^2 y_1) - 3/(32\pi^2 y_2) - 9/(16\pi^2 y_3) + 56/(8\pi^2 y_4)
\end{aligned} \tag{7}$$

$M_S < \mu < M_{R^+}$:

$$\begin{aligned}
y_1' &= -6/\pi - 43/(8\pi^2 y_1) + 9/(\pi^2) - 15/(8\pi^2 y_2) \\
&\quad -33/(8\pi^2 y_3) - 1/(\pi^2 y_4) \\
y_2' &= -4/\pi - 15/(8\pi^2 y_1) + 6/(\pi^2) - 914/(8\pi^2 y_2) - 3/(\pi^2 y_4) \\
y_3' &= -3/(8\pi^2 y_1) - 1/(8\pi^2 y_2) - 25/(8\pi^2 y_3) - 3/(\pi^2 y_4) \\
y_4' &= 1.5/\pi - 1/(8\pi^2 y_1) - 1.5/(8\pi^2 y_2) - 9/(8\pi^2 y_3) - 7/(4\pi^2 y_4)
\end{aligned} \tag{8}$$

$M_{R^+} < \mu < M_P$: Here, $\alpha_1 = \alpha_{B-L}$, $\alpha_2 = \alpha_R$ (Note that following Deshpande et. al. [20], $\alpha_R(M_{R^+}) = \alpha_{I_{3R}}(M_{R^+}) + 1/6\pi$), $\alpha_3 = \alpha_L$ and $\alpha_4 = \alpha_S$.

$$\begin{aligned}
y_1' &= -24/\pi - 179/(8\pi^2 y_1) - 177/(8\pi^2 y_2) - 177/(8\pi^2 y_3) - 17/(\pi^2 y_4) \\
y_2' &= -9/4\pi - 17/(8\pi^2 y_1) - 59/(4\pi^2 y_2) - 9/(4\pi^2 y_3) - 8/(\pi^2 y_4)
\end{aligned}$$

$$y'_3 = -9/4\pi - 17/(8\pi^2 y_1) - 9/(4\pi^2 y_2) - 59/(4\pi^2 y_3) - 8/(\pi^2 y_4)$$
$$y'_4 = -1.5/\pi - 17/(8\pi^2 y_1) - 33/(8\pi^2 y_2) - 33/(8\pi^2 y_3) - 41/(4\pi^2 y_4) \quad (9)$$

$M_P < \mu < M_U$:

$$y'_1 = -24/\pi - 179/(8\pi^2 y_1) - 177/(8\pi^2 y_2) - 177/(8\pi^2 y_3) - 17/(\pi^2 y_4)$$
$$y'_2 = -2.5/\pi - 17/(8\pi^2 y_1) - 73/(8\pi^2 y_2) - 9/(4\pi^2 y_3) - 8/(\pi^2 y_4)$$
$$y'_3 = -2.5/\pi - 17/(8\pi^2 y_1) - 9/(4\pi^2 y_2) - 73/(8\pi^2 y_3) - 8/(\pi^2 y_4)$$
$$y'_4 = -1.5/\pi - 17/(8\pi^2 y_1) - 33/(8\pi^2 y_2) - 33/(8\pi^2 y_3) - 41/(4\pi^2 y_4) \quad (10)$$

Following Deshpande et. al. [20], note that $y_1 = y_U$, $y_2 = y_U + 1/6\pi$, $y_3 = y_U + 1/6\pi$, $y_4 = y_U + 1/4\pi$.

IV RESULTS AND CONCLUSIONS

In doing the numerical integration, our procedure is to begin with a value of M_U and y_U, introduce values of M_P, M_{R^+}, M_{R^0} and M_S as required to numerically integrate the renormalization group equations to obtain values of y_1, y_2 and y_3 at the Z (Z_L^0) mass. These are compared with the values of y_1, y_2 and y_3 at the Z (Z_L^0) mass derived from the values of $\sin^2 \theta_W$, y_{em} and α_S given in Langacker and Polonsky [34]: $y_1 \approx 59$, $y_2 \approx 28\text{-}30$, and $y_3 \approx 7.5\text{-}8.4$. The values of M_{R^+} and M_{R^0} are presumably roughly the masses of W_R and Z_R^0 respectively.

Table I. Mass scales

$y_U =$ α_U^{-1}	M_U (GeV)	M_P (GeV)	M_{R^+} (GeV)	M_S (GeV)	M_{R^0} (GeV)
36	3.2×10^7	3.2×10^7	3.2×10^7	1.9×10^7	800
36	3.2×10^7	3.2×10^7	3.2×10^7	1.9×10^7	2000

y_1		y_2		y_3	
calc	exp't	calc	exp't	calc	exp't
60	59	28	28–30	7.5	7.5–8.4
59	59	28	28–30	7.6	7.5–8.4

We find agreement with the experimental data for a value of M_{R^0} in the range of 800–2000 GeV (see Table I). The fact that the unification mass is inconsistent with known limits on proton decay can be viewed as a puzzle to be solved rather than a difficulty with the theory. There are several possible ways out of this difficulty which need to be explored. One possibility is the existence of other intermediate scales [35]. Another is supergravity. As discussed earlier, it is most likely that any SUSY GUT will ultimately be a part of a supergravity theory and then other effects will occur delaying unification at least until the supergravity scale M_G and possibly until the Planck scale M_{Pl} [36].

REFERENCES

1. U. Amaldi et al., Phys. Lett. **B260**, 447 (1991).
2. Lawrence Connors, Ashley J. Deans, and John S. Hagelin, Maharishi International Univ. Report No. MIU–THP–62/92.
3. G.L. Kane "Is the World Supersymmetric? Do we already know?" in *Moriond 1993*, proceedings of the 28th Rencontres de Moriond: Electroweak Interactions and Unified Theories, Les Arcs, France, March, 1993; University of Michigan Report No. UM–TH–93–10.
4. P. Fayet, in *Unification of the Fundamental Particle Interactions,* proceedings of the Europhysics Study Conference, Erice, Italy, edited by S. Ferrara, J. Ellis, and P. van Nieuwenhauzen, (Plenum, New York,1980).
5. S. Dimopoulos and H. Georgi, Nucl. Phys. **B193**, 150 (1981).
6. S. Hawking et al., Phys. Lett. **B86**, 175 (1979); Nucl. Phys. **B170**, 283 (1980).
7. P. Nath and R. Arnowitt, Phys. Lett. **56B**, 177 (1975); R. Arnowitt, P. Nath and B. Zumino, Phys. Lett. **56B**, 81 (1975).
8. D. Freedman, P. van Nieuwenhuizen, and S. Ferrara, Phys. Rev. **D13**, 3214 (1976).
9. P. Nath, R. Arnowitt and A.H. Chamsedine, Harvard University Report No. HUTP–83/A077, 1983 (unpublished).
10. P. Fayet, Phys. Lett. **B70**, 461 (1977); H.P. Niles, Phys. Lett. **B115**, 193 (1982); A.H. Chamsedine, R. Arnowitt and P. Nath, Phys. Rev. Lett. **49**, 970 (1982); R. Barbieri, S. Ferrara and C. A. Savoy, Phys. Lett. **B119**, 343 (1982); L. Hall, J. Lykken and S. Weinberg, Phys. Rev. **D27**, 2359 (1983); L. Alverez-Gaumé, J. Polchinski and M.B. Wise, Nucl. Phys. **B221**, 495 (1983).
11. R.N. Mohapatra, *Unification and Supersymmetry. The Frontiers of Quark–Lepton Physics,* 2nd Ed. (Springer–Verlag, New York, 1992).
12. G. Senjanovic and R.N. Mohapatra, Phys. Rev. **D12**, 1502 (1975).
13. R.N. Mohapatra and J.C.Pati, Phys. Rev. **D11**, 566 (1975); **D11** 2558 (1975).
14. R.N. Mohapatra and G. Senjanovic, Phys. Rev. **D23**, 165 (1981).
15. I.A. D'Souza and C.S. Kalman, *Preons: Models of Leptons, Quarks and Gauge Bosons as Composite Objects* (World Scientific, Singapore, 1992).
16. R. Barbieri, R.N. Mohapatra, and A. Maisiero, Phys. Lett. **105B**, 363 (1981).
17. A. Davidson, Phys. Rev. **D20**, 776 (1979).
18. Howard Haber lists ten reasons why supersymmetry is the leading candidate for physics beyond the standard model: Santa Cruz Institute for Particle Physics Report No. SCIPP 93/22, 1993 (unpublished).
19. See, for example, Thomas G. Rizzo and G. Senjanovic, Phys. Rev. **D24**, 704 (1981).
20. N.G. Deshpande, E. Keith and T.G. Rizzo, Phys. Rev. Lett. **70**, 3189 (1993).
21. R.N. Mohapatra (private communication).
22. Alakabha Datta, Sandip Pakvasa and Utpal Sarkar, Phys. Lett. **B313**, 83 (1993); Phys. Rev **D50**, 2192 (1994).
23. D. Chang, R.N. Mohapatra and M.K. Parida, Phys. Rev. **D30**, 1052 (1984).
24. D. Chang, R.N. Mohapatra, J.M. Gipson, R.E. Marshak and M.K. Parida, Phys. Rev. **D31**, 1718 (1985).
25. John M. Gipson and R. E. Marshak, Phys. Rev. **D31**, 1705 (1985).

26. V. Barger and R. J. N. Phillips, "Closing in on Supersymmetry," presented at A Symposium in Honor of Tetsuro Kobayashi's 63rd Birthday, Tokyo, Japan, Mar. 1993. University of Wisconsin–Madison Report No. MAD/PH/762, May, 1993 (unpublished).
27. Jorge L. Lopez, D. V. Nanopoulos and A. Zichichi, Prog. Part. Nucl. Phys. **33**, 303 (1994).
28. Ziao–Gang He and S. Meljanac, Phys. Rev. **D41**, 1620 (1990).
29. Dae–Gyu Lee, Phys. Rev. **D49**, 1417, (1994).
30. Dae–Gyu Lee and R. N. Mohapatra, Phys. Lett. **B324**, 376 (1994); Phys. Rev. **D51**, 1353 (1995).
31. F. Del Aguila, G.D. Coughlan and M. Quiros, Nucl. Phys. **B307**, 633 (1988).
32. D.J. Gross and F. Wilczek, Phys. Rev. Lett. **30**, 1343 (1973); Phys. Rev. **D8**, 3633 (1973); H. D. Politzer, Phys. Rev. Lett. **30**, 1346 (1973).
33. D.R.T. Jones, Nucl. Phys. **B75**, 531 (1974); **B87**, 127 (1975); Phys. Rev. **D25**, 581 (1982); W. Caswell, Phys. Rev. Lett. **33**, 244 (1974).
34. Paul Langacker and Nir Polonsky, Phys. Rev. **D52**, 3081 (1995).
35. Stephen P. Martin and Pierre Ramond, Phys. Rev. **D51**, 6515 (1995); B. Brahmachari and R.N. Mohapatra, Phys. Lett. **B357**, 566 (1995).
36. Nir Polonsky and Alex Pomarol, Phys. Rev. **D51**, 6532 (1995).

Epilogue

The elusive goal of physicists everywhere is to unravel the mysteries of the universe. Alas, some mysteries remained unsolved at the end of the conference, in particular the apparent malfunction of a certain speaker's laser pointer following his talk. The conspiratorially minded have suggested the presence of a "second gunman," hidden behind a nearby grassy knoll. There are also unconfirmed eyewitness accounts of this person adding extra water to the orange juice that was served on the first day of the conference. Perhaps these matters must remain unsolved until MRST '99.

MRST '98: Toward the Theory of Everything

Reception & Registration – Tuesday 19:00–21:00
Thompson House, 3650 McTavish

Registration & Welcome – Wednesday 9:00–10:15
Strathcona Anatomy and Dentistry Building, room 1/12

Gravity I – Wednesday 10:15–11:30
Chair: Simon Catterall

Robert Mann (Waterloo)	Topological Black Holes
Martin Kamela (McGill)	Constant Curvature Effective Actions
Ariel Edery (Montréal)	Deflection of Light in Conformal (Weyl) Gravity

Physics at Lepton Colliders – Wednesday 11:30–12:20
Chair: Ken Ragan

Stephen Godfrey (Carleton)	Leptoquark Production & Identification at High Energy Lepton Colliders
Cosmin Macesanu (Rochester)	Gluon Radiation in Top Quark Production

Heavy Quarks I – Wednesday 14:15–15:30
Chair: Patrick O'Donnell

Salam Tawfiq (Toronto)	Strong Decay of Charmed Baryons
David London (Montréal)	CP-Violation Issues in B Physics and Parity Doubles
Alakabha Datta (Toronto)	CP-Violation in Quasi-inclusive B Decays

Coffee 15:30–16:00

Model Building – Wednesday 16:00–17:40
Chair: Guy Moore

Thomas Gregoire (Montréal)	Dileptons and the 3-3-1 Model
Otto Kong (Rochester)	Two-Higgs-Doublet-Models and Radiative CP Violation
Sachindeo Vaidya (Syracuse)	Skyrmions, Spectral Flow, and Parity Doubles
Marc Baillargeon (UQÀM)	General Texture Zeroes Mass Matrices

Heavy Quarks II – Thursday 9:00–10:40
Chair: Greg Mahlon

Michael Luke (Toronto)	Power Counting in Nonrelativistic Effective Field Theories
Sean Fleming (Toronto)	Photoproduction of h_c
Christian Bauer (Toronto)	Extracting Heavy Quark Matrix Elements from Inclusive B Decays
Patrick O'Donnell (Toronto)	Hyperfine Interactions in Charm and Bottom

Coffee 10:40–11:10

2D Field Theory – Thursday 11:10–12:25
Chair: C.R. Hagen

Rashmi Ray (Montréal)	Spin Effects in the Integral Quantum Hall System
Alex Travesset (Syracuse)	Quantum Hall Skyrmions and Anomalies
Bogomil Gerganov (Cornell)	Integrability, Boson-Fermion Duality, and Bukhvostov-Lipatov Model

Strongly Interacting Theories – Thursday 14:20-16:00
Chair: Michael Luke

Deirdre Black (Syracuse)	Evidence for a Scalar $\kappa(900)$ Resonance in πK Scattering
Stef Roux (Toronto)	Possible Exceptions to the Most Attractive Channel Hypothesis
Kevin Sprague (Western)	The Condensate Component of the Anomalous Magnetic Moment of a Condensing Fermion
Tibor Torma (Toronto)	Negative Contributions to S and T in Strongly Interacting Theories

Coffee 16:00–16:30

Conformal Field Theory, Duality & Strings – Thursday 16:30–17:45
Chair: Marcia Knutt

Alexander Buchel (Cornell)	Deriving $N = 2$ S-dualities from Scaling for Product Gauge Groups
Chi–Wei Lee (Rochester)	Symmetries of Large N_c Matrix Models for Closed Strings
Marco Ameduri (Cornell)	Perturbed Conformal Field Theory: A Tool for Investigating Integrable Models

Banquet – Gargantua & Pantagruel, 3873 St. Denis – 19:00

Gravity II – Friday 9:30–10:20
Chair: Robb Mann

Eric Gregory (Syracuse)	Simplicial Quantum Gravity and the Renormalization Group
Simeon Warner (Syracuse)	Phase Structure of 3-D Triangulations with a Boundary

Coffee 10:20–10:50

Experiment – Friday 10:50–12:05
Chair: Doug Stairs

Claude Theoret (McGill)	High Energy Gamma Ray Astronomy with STACEE Astroparticle Physics
Rene Janicek (McGill)	The Experimental Status of the Deficit in Inclusive Charm Semileptonic B Decays
Declan Persram (McGill)	Dynamics of Heavy Ion (N+Sm) Collisions

Supersymmetry – Friday P.M. 14:00–15:40
Chair: Jim Cline

Maxim Pospelov (UQÀM)	Up-Down Unification Just Above the SUSY Threshold
Francesco Sannino (Yale)	From Super QCD to QCD
Mike Bisset (Rochester)	Efficient Parameterization for Supersymmetric Models without R-Parity
Calvin Kalman (Concordia)	Low Energy Right-Handed Z_R^0 Based on an $SO(10)$ SUSY-GUT

End of Conference–enjoy Montréal!

Conference Participants

Abdel-Rehim, Abdou M.	Syracuse	abdou@suhep.phy.syr.edu
Ameduri, Marco	Cornell	marco@mail.lns.cornell.edu
Azuelos, George	U. Montréal	azuelos@lpshpd.lps.umontreal.ca
Baillargeon, Marc	UQÀM	baillarg@mercure.phy.uqam.ca
Bauer, Christian	Toronto	bauer@medb.physics.utoronto.ca
Bisset, Mike	Rochester	bisset@urhepf.pas.rochester.edu
Black, Deirdre	Syracuse	black@physics.syr.edu
Buchel, Alexander	Cornell	asb10@cornell.edu
Burrell, Craig	Toronto	craig@medb.physics.utoronto.ca
Catterall, Simon	Syracuse	smc@npac.syr.edu
Cline, Jim	McGill	jcline@physics.mcgill.ca
Datta, Alakabha	Toronto	datta@lugaid.physics.utoronto.ca
Davignon, Didier	U. Montréal	davignod@lps.umontreal.ca
Depommier, Pierre	U. Montréal	pom@pshpd.lps.umontreal.ca
Edery, Ariel	U. Montréal	edery@lps.umontreal.ca
Fleming, Sean	Toronto	fleming@medb.physics.utoronto.ca
Frank, Mariana	Concordia	mfrank@vax2.concordia.ca
Gerganov, Bogomil	Cornell	beg@ruph.cornell.edu
Girard, Patrick	McGill	girard@hep.physics.mcgill.ca
Godfrey, Stephen	Carleton	godfrey@physics.carleton.ca
Gregoire, Thomas	U. Montréal	gregoirt@jsp.umontreal.ca
Gregory, Eric B.	Syracuse	gregory@npac.syr.edu
Hagen, C. R.	Rochester	Hagen@URHEP.pas.rochester.edu
Hamzaoui, Cherif	UQÀM	hamzaoui@mercure.phy.uqam.ca
Janicek, Rene	McGill	rene@lns.cornell.edu
Kalman, Calvin	Concordia	Kalman@vax2.concordia.ca
Kalyniak, Patricia	Carleton	kalyniak@physics.carleton.ca
Kamela, Martin	McGill	kamela@physics.mcgill.ca
Karl, Gabriel	Guelph	kgabriel@uoguelph.ca
Kiriushcheva, Natalia	Western	nkiriusc@julian.uwo.ca
Kitazawa, Noriaki	Yale	kitazawa@zen.physics.yale.edu
Knutt, Marcia	McGill	knutt@hep.physics.mcgill.ca
Kong, Otto C.W.	Rochester	kong@pas.rochester.edu
Kordas, Kostas	McGill	kordas@physics.mcgill.ca
Kuzmin, Sergiy	Western	nkiriusc@julian.uwo.ca
Lee, Chi-Wei (Herbert)	Rochester	cwhlee@pas.rochester.edu
London, David	U. Montréal	london@lps.umontreal.ca
Luke, Michael	Toronto	luke@medb.physics.utoronto.ca

Conference Participants (continued)

Macesanu, Cosmin	Rochester	mcos@pas.rochester.edu
Mahlon, Greg	McGill	mahlon@hep.physics.mcgill.ca
Majumder, Abhijit	McGill	majumder@physics.mcgill.ca
Mann, Robert	Waterloo	mann@avatar.uwaterloo.ca
Mazini, Rachid	U. Montréal	mazini@lps.umontreal.ca
Milek, Marko	McGill	milek@hep.physics.mcgill.ca
Moore, Guy	McGill	guymoore@hep.physics.mcgill.ca
Mouline, Saad	UQÀM	saad@mercure.phy.uqam.ca
Newhouse, Jim	U. St. Thomas	jnewhouse@atcorp.com
O'Donnell, Patrick J.	Toronto	pat@medb.physics.utoronto.ca
Patel, Popat	McGill	patel@hep.physics.mcgill.ca
Persram, Declan	McGill	declan@physics.mcgill.ca
Pospelov, Maxim	UQÀM	pospelov@mercure.phy.uqam.ca
Ragan, Ken	McGill	ragan@hep.physics.mcgill.ca
Rahman, Tanvir	McGill	tanvir@physics.mcgill.ca
Ray, Rashmi	U. Montréal	rray@lps.umontreal.ca
Roux, F. Stef	Toronto	stef@medb.physics.utoronto.ca
Sannino, Francesco	Yale	sannino@genesis1.physics.yale.edu
Schechter, Joe	Syracuse	schechte@suhep.phy.syr.edu
Sprague, Kevin B.	Western	kbs@pineapple.apmaths.uwo.ca
Stairs, Doug	McGill	stairs@physics.mcgill.ca
Tawfiq, Salam	Toronto	tawfiq@medb.physics.utoronto.ca
Theoret, Claude G.	McGill	theoret@physics.mcgill.ca
Toharia, Manuel	UQÀM	toharia@mercure.phy.uqam.ca
Torma, Tibor	Toronto	kakukk@physics.utoronto.ca
Travesset, Alex	Syracuse	alex@suhep.phy.syr.edu
Vaidya, Sachindeo	Syracuse	sachin@suhep.syr.edu
Warner, Simeon	Syracuse	simeon@physics.syr.edu

AUTHOR INDEX

A

Ameduri, M., 195

B

Baillargeon, M., 80
Bauer, C., 108
Bisset, M., 254
Black, D., 153
Boudjema, F., 80
Browder, T. E., 54
Buchel, A., 178
Burgess, C., 1

C

Catterall, S. M., 202, 212

D

Datta, A., 54
Doncheski, M. A., 19

E

Edery, A., 10
Elias, V., 169

F

Fleming, S., 101

G

Gale, C., 227
Gerganov, B., 144
Godfrey, S., 19
Grégoire, T., 63
Gregory, E. B., 202

H

Hamzaoui, C., 80, 236
He, X.-G., 54

J

Janicek, R., 220

K

Kalman, C. S., 263
Kamela, M., 1
Kong, O. C. W., 72, 254
Körner, J. G., 37

L

Lee, C.-W. H., 187
Lin, F.-L., 72
Lindig, J., 80
London, D., 46
Luke, M., 91

M

Macesanu, C., 28, 254
Mehen, T., 101

O

O'Donnell, P. J., 37, 117
Orr, L. H., 28, 254

P

Pakvasa, S., 54
Paranjape, M. B., 10
Persram, D., 227
Pospelov, M., 236

R

Rajeev, S. G., 187
Ray, R., 125
Renken, R., 212
Roux, F. S., 163

S

Sannino, F., 245
Sprague, K. B., 169

T

Tawfiq, S., 37
Thorleifsson, G., 202
Travesset, A., 134

W

Warner, S., 212